T0269139

Lecture Notes in Physics

Volume 920

The Lecture Notes in Physics

The series Lecture Notes in Physics (LNP), founded in 1969, reports new developments in physics research and teaching-quickly and informally, but with a high quality and the explicit aim to summarize and communicate current knowledge in an accessible way. Books published in this series are conceived as bridging material between advanced graduate textbooks and the forefront of research and to serve three purposes:

- to be a compact and modern up-to-date source of reference on a well-defined topic
- to serve as an accessible introduction to the field to postgraduate students and nonspecialist researchers from related areas
- to be a source of advanced teaching material for specialized seminars, courses and schools

Both monographs and multi-author volumes will be considered for publication. Edited volumes should, however, consist of a very limited number of contributions only. Proceedings will not be considered for LNP.

Volumes published in LNP are disseminated both in print and in electronic formats, the electronic archive being available at springerlink.com. The series content is indexed, abstracted and referenced by many abstracting and information services, bibliographic networks, subscription agencies, library networks, and consortia.

Proposals should be sent to a member of the Editorial Board, or directly to the managing editor at Springer:

Christian Caron
Springer Heidelberg
Physics Editorial Department I
Tiergartenstrasse 17
69121 Heidelberg/Germany
christian.caron@springer.com

More information about this series at http://www.springer.com/series/5304

Franz Wegner

Supermathematics and its Applications in Statistical Physics

Grassmann Variables and the Method of Supersymmetry

 Springer

Franz Wegner
Institut für Theoretische Physik
Universität Heidelberg
Heidelberg, Germany

ISSN 0075-8450   ISSN 1616-6361   (electronic)
Lecture Notes in Physics
ISBN 978-3-662-49168-3   ISBN 978-3-662-49170-6   (eBook)
DOI 10.1007/978-3-662-49170-6

Library of Congress Control Number: 2016931278

Springer Heidelberg New York Dordrecht London

Printed on acid-free paper

Springer International Publishing AG Switzerland is part of Springer Science+Business Media
(www.springer.com)

To Anne-Gret,
Annette, and Christian

Preface

This book arose from my interest in disordered systems. It was known, for some time, that disorder in a one-particle Hamiltonian usually leads to localized states in one-dimensional chains. Anderson had argued that in higher-dimensional systems, there may be regions of localized and extended states, separated by a mobility edge. In 1979 and 1980, it became clear that this Anderson transition could be described in terms of a nonlinear sigma model. Lothar Schäfer and myself reduced the model to one described by interacting matrices by means of the replica trick. Efetov, Larkin, and Khmel'nitskii performed a similar calculation. They, however, started from a description by means of anticommuting components. In 1982 Efetov showed that a formulation without the replica trick was possible using supervectors and supermatrices with equal number of commuting and anticommuting components.

I had the pleasure of giving many lectures and seminars on disordered systems and critical systems, and also on fermionic systems, where Grassmann variables play an essential role. Among them were seminars in the Sonderforschungsbereich (collaborative research center) on *stochastic mathematical models* with mathematicians and physicists and in the Graduiertenkolleg (research training group) on *physical systems with many degrees of freedom* and seminars with Heinz Horner and Christof Wetterich. In particular, I remember a seminar with Günther Dosch on Grassmann variables in statistical mechanics and field theory.

Some of the applications of Grassmann variables are presented in this volume. The book is intended for physicists, who have a basic knowledge of linear algebra and the analysis of commuting variables and of quantum mechanics. It is an introductory book into the field of Grassmann variables and its applications in statistical physics.

The algebra and analysis of Grassmann variables is presented in Part I. The mathematics of these variables is applied to a random matrix model, to path integrals for fermions (in comparison to the path integrals for bosons) and to dimer models and the Ising model in two dimensions.

Supermathematics, that is, the use of commuting and anticommuting variables on an equal footing, is the subject of Part II. Supervectors and supermatrices, which contain both commuting and Grassmann components, are introduced.

In Chaps. 10–14, the basic formulae for such matrices and the generalization of symmetric, real, unitary, and orthogonal matrices to supermatrices are introduced. Chapters 15–17 contain a number of integral theorems and some additional information on supermatrices. In many cases, the invariance of functions under certain groups allows the reduction of the integrals to those where the same number of commuting and anticommuting components is canceled.

In Part III, supersymmetric physical models are considered. Supersymmetry appeared first in particle physics. If this symmetry exists, then bosons and fermions exist with equal masses. So far, they have not been discovered. Thus, either this symmetry does not exist or it is broken. The formal introduction of anticommuting space-time components, however, can also be used in problems of statistical physics and yields certain relations or allows the reduction of a disordered system in d dimensions to a pure system in $d - 2$ dimensions. Since supersymmetry connects states with equal energies, it has also found its way into quantum mechanics, where pairs of Hamiltonians, $Q^\dagger Q$ and $QQ^\dagger$, yield the same excitation spectrum. Such models are considered in Chaps. 18–20.

In Chap. 21, the representation of the random matrix model by the nonlinear sigma model and the determination of the density of states and of the level correlation are given. The diffusive model, that is, the tight-binding model with random on-site and hopping matrix elements, is considered in Chap. 22. These models show collective excitations called diffusions and if time-reversal holds, also cooperons. Chapter 23 discusses the mobility edge behavior and gives a short account of the ten symmetry classes of disorder, of two-dimensional disordered models, and of superbosonization.

I acknowledge useful comments by Alexander Mirlin, Manfred Salmhofer, Michael Schmidt, Dieter Vollhardt, Hans-Arwed Weidenmüller, Kay Wiese, and Martin Zirnbauer. Viraf Mehta kindly made some improvements to the wording.

Heidelberg, Germany   Franz Wegner
September 2015

Contents

Acronyms

$\mathscr{A}_{0,1}$ Even, odd elements, Sect. 2.2, p. 8

$\nu(.), \nu$ Z_2-degree, Sect. 2.2, p. 8, and Sect. 10.1, p. 103

$\mathscr{M}$ Class of matrices, Sect. 10.1, p. 103

$\mathscr{P}$ Parity operator, Sect. 2.2, p. 8

$*$ Complex conjugate 1st kind, Sect. 6.1, p. 45

$\times$ Complex conjugate 2nd kind, Sect. 6.1, p. 45

t Conventional transposed, Sect. 4.4, p. 32, and Sect. 10.1, p. 103

T Super transposed, Sect. 10.1, p. 103

π Parity transposed, Sect. 10.5, p. 110

$\sigma^{(.,.)}$ Sect. 10.1, p. 103

C Sect. 12.3, p. 120

ϵ Sect. 12.3, p. 120

$\dagger$ Hermitian adjoint first kind, Sect. 3.3, p. 16, and Sect. 13.1.1, p. 123

$\ddagger$ Hermitian adjoint second kind, Sect. 13.1.2, p. 124

ord   Ordinary part, body, Sect. 2.3, p. 10

nil   Nilpotent part, soul, Sect. 2.3, p. 10

det   Determinant, Sect. 3.3, p. 16

sdet   Superdeterminant, Sect. 10.3, p. 106

str   Supertrace, Sect. 10.4, p. 109

pf   Pfaffian, Sect. 5.1, p. 37

spf   Superpfaffian, Sect. 12.2, p. 118

$(.,.)_0$ Scalar product of 0th kind, Sect. 12.3, p. 120

OSp   Orthosymplectic group, Sect. 12.3, p. 120

$(.,.)_1$ Scalar product of 1st kind, Sect. 13.2.1, p. 125

UPL_1 (Pseudo)unitary group of 1st kind, Sect. 13.2.1, p. 125

$(.,.)_2$ Scalar product of 2nd kind, Sect. 13.2.2, p. 126

UPL_2 (Pseudo)unitary group of 2nd kind, Sect. 13.2.2, p. 126

UOSp   (Pseudo)unitary-orthosymplectic group, Sect. 14.1, p. 131

g.c.   grand canonical, Sect. 7.2, p. 49

Part I
Grassmann Variables and Applications

The mathematics of Grassmann variables is introduced in this first Part. After an introduction (Chap. 1) Grassmann algebra (Chap. 2), Grassmann analysis (Chaps. 3 and 5) and conjugation (Chap. 6) are developed. An introduction to exterior algebra is given in Sects. 2.2 and 3.4.

Products of Grassmann variables anticommute in contrast to c-numbers, whose products commute. In many cases we compare the properties of these Grassmann variables with those of c-numbers, but we will not combine them, in this part, into what is called supermathematics. This will be done in Parts two and three.

One of the most important applications of Grassmann variables are path integrals for fermions (Chap. 7), which are considered together with path integrals for bosons. Another application is the determination of the number of dimer configurations on two-dimensional lattices (Chap. 8), and the solution of the two-dimensional Ising model (Chap. 9). This part also includes a first application to the random matrix problem (Chap. 4). The various applications can be read independently.

Chapter 1
Introduction

Abstract A short introduction into the history and the present use of supermathematics is given.

1.1 History

A short historic introduction is given to Grassmann, the inventor of anticommuting variables, and to Berezin, who introduced the analysis of Grassmann variables and applications in physics.

Hermann Günther Grassmann (Stettin 1809–Stettin 1877), a high school teacher in Stettin, presented in his book [99], in 1844 *Lineare Ausdehnungslehre (Theory of Linear Extension)*, an algebraic theory of extended quantities, where he introduced what is now known as the wedge or exterior product. Saint-Venant published similar ideas of exterior calculus, in 1845, for which he claimed priority over Grassmann. Grassmann's work contained important developments in linear algebra, in particular the notion of linear independence. He advocated that once geometry is put into algebraic form, the number three has no privileged role as the number of spatial dimensions; the number of possible dimensions is in fact unbounded. It was, however, a revolutionary text, too far ahead of its time to be appreciated and too difficult to read. Grassmann submitted it as a Ph.D. thesis, but Möbius said he was unable to evaluate it and sent it to Ernst Kummer, who rejected it without giving it a careful reading. Thus, Grassmann never obtained a university position. Appreciation for his work started only with William Rowan Hamilton (1853) and Hermann Hankel (1866). In addition to his mathematical works, Grassmann was also a linguist. Among his works his dictionary and his translation of the Rigveda (still in print) were most revered among linguists. He devised a sound law of Indo-European languages, named Grassmann's aspiration law (Also, there is another law by Grassmann concerning the mixing of color). These philological accomplishments were honored during his lifetime; in 1876 he received an honorary doctorate from the University of Tübingen. The *laudatio* mentions both his mathematical and

© Springer-Verlag Berlin Heidelberg 2016

F. Wegner, *Supermathematics and its Applications in Statistical Physics*,
Lecture Notes in Physics 920, DOI 10.1007/978-3-662-49170-6_1

linguistic excellence.[1] Grassmann's biography can be found in the books by Petsche [209, 210].

Just as we know Grassmann as the creator of what we call today Grassmann variables, Felix Alexandrowich Berezin (Moscow 1931–Kolyma region 1980) is the founder of supermathematics. His most important accomplishments were what we now call the Berezin integral over anticommuting Grassmann variables and the closely related construction of the Berezinian: the generalization of the Jacobian. In fact the integral over a Grassmann variable is actually its derivative. This might have, at first glance looked useless, but it turned out to be very fruitful. His main works besides many published articles, are his books on the application of Grassmann variables to fermionic fields 'The Method of Second Quantization' [24] 1966 and the book 'Introduction to Superanalysis' [25], which due to his untimely death collects his papers on this subject and supplementary papers by colleagues. Perhaps the first authors to use Grassmann variables for fermions were Candlin [44], who introduced coherent fermionic states as early as 1956, and J.L. Martin [174, 175], who considered the fermionic oscillator in 1959.

1.2   Applications

A short account of the use of Grassmann variables and of supersymmetric ideas is given.

So what can be done with Grassmann variables? First, they can be used to describe fermionic systems: The classical limit of bosonic fields are complex fields. Bosonic creation operators commute with each other as do bosonic annihilation operators. In contrast fermionic creation and annihilation operators anticommute among themselves. Thus, the classical analogue are numbers, which anticommute. Indeed, path integrals for fermions can be expressed in terms of Grassmann variables just as path integrals for bosons are expressed in terms of complex variables.

Another interesting application is in two-dimensional lattice models. Both, dimer problems and the two-dimensional Ising model can be elegantly formulated by integrals over Grassmann variables.

Unlike for integrals over Gauss functions of real and complex numbers typically yielding the inverse of the determinant of the coefficient matrix of the quadratic form in the exponent or its square root, it is precisely the opposite for integrals over

---

[1]The original Latin text can be found in [209, 210, 223] "qui raro hac aetate exemplo mathematicae peritiam coniunxit cum scientia rerum philologicarum et in utroque studiorum genere scriptor extitit clarissimus maxime vero acutissima vedicorum carminum interpretatione nomen suum reddidit illustrissimum", a tentative English translation would be "who, what is rare in this time, brings together exemplary knowledge in mathematics and in linguistic and excels in both sciences as brilliant author, made himself the most famous name by his perceptive translation of the Rig-Veda hymns."

Gauss functions of Grassmann variables: here one obtains the determinant of the coefficient matrix or its square root. Thus, if one introduces both types of variables, the determinants cancel, which is very useful, if the integral over the Gaussians constitutes something similar to a partition function. An example of this for random matrices is given in Chap. 4. Specifically we will study the random-matrix problem and particles in random potentials in greater detail in Part III.

Grassmann variables have properties quite opposite to real and complex numbers. They are a kind of antipodes to complex numbers; unfortunately these antipodes are rather degenerate, since their variety of functions is much less than that of commuting variables: a function of one Grassmann variable can only be linear. Therefore they cannot be used for all problems of disorder. Nevertheless they have a large field of applications.

References

[24] F.A. Berezin, *The Method of Second Quantization* (Academic Press, New York, 1966)

[25] F.A. Berezin, *Introduction to Superanalysis* (Reidel, Dordrecht, 1987)

[44] D.J. Candlin, On sums over trajectories for systems with Fermi statistics. Nuovo Cim. **4**, 231 (1956)

[99] H. Grassmann, *Lineare Ausdehnungslehre* (Wigand, Leipzig, 1844)

[174] J.L. Martin, General classical dynamics, and the 'classical analogue' of a Fermi Oscillator. Proc. Roy. Soc. **A251**, 536 (1959)

[175] J.L. Martin, The Feynman principle for a Fermi system. Proc. Roy. Soc. **A251**, 543 (1959)

[209] H.-J. Petsche, *Graßmann* (German). Vita Mathematica, vol. 13 (Springer, Birkhäusser, Basel, 2006)

[210] H.-J. Petsche, M. Minnes, L. Kannenberg, *Hermann Grassmann: Biography* (English) (Birkhäusser, Basel, 2009)

[223] K. Reich, Über die Ehrenpromotion Hermann Grassmanns an der Universität Tübingen im Jahre 1876, in *Hermann Grassmanns Werk und Wirkung*, ed. by P. Schreiber (Ernst-Moritz-Arndt-Universität Greifswald, Fachrichtungen Mathematik/Informatik, Greifswald, 1995), S. 59

Chapter 2
Grassmann Algebra

Abstract The elements of the Grassmann algebra, and the operations addition and multiplication are defined. Distinction is made between even and odd elements. A few remarks on exterior algebra then follow.

2.1 Elements of the Algebra

The elements of the algebra, addition and multiplication are defined.

The use of Grassmann variables in the context of physical problems and an introduction to these variables can be found for example in the books by Zinn-Justin [297] and by Efetov [65]. From the more mathematical side the books by Berezin [25] and by de Witt [55] are recommended.

To begin with, we have a basis of r vectors ζ_i, $i = 1, \ldots r$. This basis is then enlarged by the introduction of products of the vectors ζ_i. This product obeys the associative law and the law of anticommutativity,

$$\zeta_i \zeta_j = -\zeta_j \zeta_i, \tag{2.1}$$

which implies $\zeta_i^2 = 0$. Including the empty product, one obtains 2^r basis elements

$$
\begin{aligned}
&1 \\
&\zeta_i && i = 1 \ldots r \\
&\zeta_i \zeta_j = -\zeta_j \zeta_i && i < j \\
&\ldots \\
&\zeta_1 \zeta_2 \ldots \zeta_r.
\end{aligned}
\tag{2.2}
$$

The elements of the algebra are then the linear combinations of these 2^r basis vectors

$$a = a^{(0)} + \sum_i a_i^{(1)} \zeta_i + \sum_{i<j} a_{ij}^{(2)} \zeta_i \zeta_j + \ldots, \tag{2.3}$$

where the coefficients $a_{i_1 \ldots i_m}^{(m)}$ are complex numbers. We denote the set of elements a given in (2.3) by $\mathscr{A}$.

© Springer-Verlag Berlin Heidelberg 2016
F. Wegner, *Supermathematics and its Applications in Statistical Physics*,
Lecture Notes in Physics 920, DOI 10.1007/978-3-662-49170-6_2

Addition is defined as is usual in vector spaces: the coefficients with equal indices $m, i_1, \ldots i_m$ are added. Thus, addition is commutative and associative,

$$a + b = b + a, \quad (a + b) + c = a + (b + c) = a + b + c. \tag{2.4}$$

Multiplication of the monomials

$$a = a^{(k)} \zeta_{i_1} \zeta_{i_2} \ldots \zeta_{i_k}, \tag{2.5}$$

$$b = b^{(l)} \zeta_{j_1} \zeta_{j_2} \ldots \zeta_{j_l} \tag{2.6}$$

yields

$$ab = a^{(k)} b^{(l)} \zeta_{i_1} \zeta_{i_2} \ldots \zeta_{i_k} \zeta_{j_1} \zeta_{j_2} \ldots \zeta_{j_l}. \tag{2.7}$$

If at least one factor ζ_i agrees with one factor ζ_j, then ab vanishes. Multiplication of polynomials, like (2.3), follows from the requirement that the law of distributivity holds,

$$(a + b)c = ac + bc, \quad a(b + c) = ab + ac. \tag{2.8}$$

Then it is easy to show that multiplication is associative,

$$(ab)c = a(bc) = abc. \tag{2.9}$$

Example 2.1.1 $a = 5\zeta_2, b = 3\zeta_1\zeta_3$ yield $ab = 15\zeta_2\zeta_1\zeta_3 = -15\zeta_1\zeta_2\zeta_3$.

Example 2.1.2 $a = 3\zeta_1\zeta_3 + 5\zeta_2$ yields $a^2 = (3\zeta_1\zeta_3 + 5\zeta_2)(3\zeta_1\zeta_3 + 5\zeta_2) = 9\zeta_1\zeta_3\zeta_1\zeta_3 + 15\zeta_1\zeta_3\zeta_2 + 15\zeta_2\zeta_1\zeta_3 + 25\zeta_2\zeta_2 = 9 \cdot 0 - 15\zeta_1\zeta_2\zeta_3 - 15\zeta_1\zeta_2\zeta_3 + 25 \cdot 0 = -30\zeta_1\zeta_2\zeta_3$.

2.2 Even and Odd Elements, Graded Algebra

Even and odd elements and their algebraic properties are defined.
Let us introduce the linear parity operator $\mathscr{P}$,

$$\mathscr{P}(\zeta) = -\zeta. \tag{2.10}$$

This operator multiplies a monomial of order k in the Grassmann variables by $(-)^k$. Thus a in (2.5) and b in (2.6) obey

$$\mathscr{P}(a) = (-)^k a, \quad \mathscr{P}(b) = (-)^l b. \tag{2.11}$$

For the monomials a, b in (2.5), (2.6) one obtains

$$ab = (-)^{kl}ba. \tag{2.12}$$

A product of an even (odd) number of factors of ζ and their linear combinations are called even (odd) elements of the algebra. Each element, $a \in \mathscr{A}$, can be decomposed uniquely in a sum of an even and an odd element, $\mathscr{A} = \mathscr{A}_0 \oplus \mathscr{A}_1$,

$$a = a_0 + a_1, \quad a_i \in \mathscr{A}_i, \tag{2.13}$$

$$a_0 = a^{(0)} + \sum_{i<j} a_{ij}^{(2)} \zeta_i \zeta_j + \sum_{i<j<k<l} a_{ijkl}^{(4)} \zeta_i \zeta_j \zeta_k \zeta_l + \dots, \tag{2.14}$$

$$a_1 = \sum_i a_i^{(1)} \zeta_i + \sum_{i<j<k} a_{ijk}^{(3)} \zeta_i \zeta_j \zeta_k + \dots \tag{2.15}$$

This decomposition into even and odd elements is the reason that this algebra is called a *graded algebra*. Generally, a graded algebra has the property that all elements can be decomposed into elements of *degree* ν,

$$a = \sum_\nu a_\nu, \quad a_\nu \in \mathscr{A}_\nu. \tag{2.16}$$

Sums of elements in $\mathscr{A}_\nu$ belong to $\mathscr{A}_\nu$. Products of elements in $\mathscr{A}_\nu$ and $\mathscr{A}_{\nu'}$ belong to $\mathscr{A}_{\nu+\nu'}$.

The following expressions, from (2.14), (2.15), belong to $\mathscr{A}_0$ and $\mathscr{A}_1$

$$a_0 + b_0, \quad a_0 b_0 = b_0 a_0, \quad a_1 b_1 = -b_1 a_1 \in \mathscr{A}_0, \tag{2.17}$$

$$a_1 + b_1, \quad a_0 b_1 = b_1 a_0, \quad a_1 b_0 = b_0 a_1 \in \mathscr{A}_1. \tag{2.18}$$

This grading is a Z_2-grading, since the degree ν assumes only the values 0 and 1, and calculation is modulo 2.

In the following, we will mainly deal with elements a, which are either even ($a \in \mathscr{A}_0$) or odd ($a \in \mathscr{A}_1$) (Problem 2.3 is an exception). The degree of an element a will often be denoted by $\nu(a)$. Thus (2.12) may be rewritten as

$$ba = (-)^{\nu(a)\nu(b)} ab. \tag{2.19}$$

One observes that $a^2 = 0$ for all $a \in \mathscr{A}_1$.

2.3 Body and Soul, Functions

Body (ordinary part) and soul (nilpotent part) are introduced.
  The contribution $a^{(0)}$ from a in (2.3) is called the *body (ordinary part)* of a

$$a^{(0)} = \operatorname{ord} a, \tag{2.20}$$

with everything else being the *soul (nilpotent part)* of a

$$\operatorname{nil} a = a - a^{(0)}, \tag{2.21}$$

since, as one sees immediately

$$(\operatorname{nil} a)^{r+1} = 0. \tag{2.22}$$

Actually, one has already $(\operatorname{nil} a)^p = 0$ for $2p > r + 1$.
  If $f : \mathscr{A} \to \mathscr{A}$ is a function, which can be differentiated sufficiently often, then $f(a)$ can be defined by its Taylor expansion

$$f(a) = f(a^{(0)}) + f'(a^{(0)})\operatorname{nil} a + \frac{1}{2} f''(a^{(0)})(\operatorname{nil} a)^2 + \ldots \tag{2.23}$$

Due to (2.22) the expansion only has a finite number of terms.
  The generalization to functions of several variables constitutes no problem if the variables are even, since these variables and their nilpotent parts commute. Functions of even and odd variables can always be represented as polynomials in the odd variables, where the coefficients are functions of the even variables.

2.4 Exterior Algebra I

A short excursion (first part) to exterior algebra is given.
  From

$$a(\zeta) = a_1 \zeta_1 + a_2 \zeta_2 + a_3 \zeta_3, \quad b(\zeta) = b_1 \zeta_1 + b_2 \zeta_2 + b_3 \zeta_3, \tag{2.24}$$

and similarly for c, one obtains the exterior products

$$(ab)(\zeta) = (a_1 b_2 - a_2 b_1)\zeta_1 \zeta_2 + (a_2 b_3 - a_3 b_2)\zeta_2 \zeta_3 + (a_3 b_1 - a_1 b_3)\zeta_3 \zeta_1,$$

$$(abc)(\zeta) = \begin{vmatrix} a_1 & b_1 & c_1 \\ a_2 & b_2 & c_2 \\ a_3 & b_3 & c_3 \end{vmatrix} \zeta_1 \zeta_2 \zeta_3. \tag{2.25}$$

Usually the elements of the exterior algebra refer to some vectors or antisymmetric tensors in an n-dimensional space. The examples in (2.24) are 1-vectors with the products (ab) and (abc) being 2-vector and 3-vector, respectively.

Exterior algebra uses a grading different from that introduced in Sect. 2.2. Linear combinations of the monomials in (2.5) with fixed k are k-vectors. This set of k-vectors is denoted by $\mathscr{E}_k$. The grading decomposes the elements into the sets $\mathscr{E}_k$ for $k = 0 \ldots n$. Thus the vectors a and b in (2.5), (2.6) are k- and l-vectors, respectively. Generally, the product ab of $a \in \mathscr{E}_k$, $b \in \mathscr{E}_l$ is a $k + l$-product $ab \in \mathscr{E}_{k+l}$.

Often the basis vectors $\mathbf{e}_i$ are used instead of ζ_i. To indicate that the product is anticommutative one may use a wedge, $\wedge$, as the sign for multiplication and the product is called the *wedge product*, for example

$$a(\mathbf{e}) = a_1 \mathbf{e}_1 + a_2 \mathbf{e}_2 + a_3 \mathbf{e}_3, \tag{2.26}$$

$$(ab)(\mathbf{e}) = (a_1 b_2 - a_2 b_1)\mathbf{e}_1 \wedge \mathbf{e}_2 + (a_2 b_3 - a_3 b_2)\mathbf{e}_2 \wedge \mathbf{e}_3 + (a_3 b_1 - a_1 b_3)\mathbf{e}_3 \wedge \mathbf{e}_1.$$

Obviously the coefficients of the product ab are the coeffients of the cross product $a \times b$. The coefficient of $\zeta_1 \zeta_2 \zeta_3$ in the product abc is the triple product of the three vectors a, b, c, if $\mathbf{a} = \sum_i a_i \mathbf{e}_i$ with the orthonormal vectors $\mathbf{e}_i$ (similarly for b and c).

Quite generally, such products are called wedge products and the algebra is called an exterior algebra. Such ideas in general dimensions were the basis of Grassmann's extension calculus.

Problems

2.1 Calculate

$$\Big(\sum_{i=1}^{3} \sum_{j=i+1}^{4} a_{i,j} \xi_i \xi_j \Big)^2, \quad a_{i,j} \in \mathscr{A}_0, \quad \xi_i \in \mathscr{A}_1.$$

2.2 Solve $x^2 = a^2 + 2\xi_1 \xi_2$ for x with $a \in \mathbb{C}$, $\xi_1, \xi_2 \in \mathscr{A}_1$. Is there a solution for $a = 0$?

2.3 Solve $x^2 = \xi_1 \xi_2 \xi_3$ for x with $\xi_i \in \mathscr{A}_1$. The result shall depend only on the Grassmannians ξ_i, $i = 1..3$.

2.4 Substitute $x = z + \alpha \zeta_1 + \beta \zeta_2 + c \zeta_1 \zeta_2$, $x, z, c \in \mathscr{A}_0$, $\alpha, \beta, \zeta_i \in \mathscr{A}_1$ in $g(x, \zeta_1, \zeta_2) = g_{\emptyset}(z) + g_1(z)\gamma_1 \zeta_1 + g_2(z)\gamma_2 \zeta_2 + g_{12}(z)\zeta_1 \zeta_2$, $\gamma_i \in \mathscr{A}_1$, $g_{.}(z) \in \mathscr{A}_0$ and determine $f(z, \zeta_1, \zeta_2) = g(x, \zeta_1, \zeta_2)$ in the form $f(z, \zeta_1, \zeta_2) = f_{\emptyset}(z) + f_1(z)\zeta_1 + f_2(z)\zeta_2 + f_{12}(z)\zeta_1 \zeta_2$.

2.5 Calculate A^{-1} from $A = a_0 + \xi_1 \eta_1 + \xi_2 \eta_2 + a_2 \xi_1 \xi_2 \eta_1 \eta_2$, $a_0 \in \mathbb{C}$, $a_0 \neq 0$, $a_2 \in \mathscr{A}_0$, $\xi_i, \eta_i \in \mathscr{A}_1$.

References

[25]   F.A. Berezin, *Introduction to Superanalysis* (Reidel, Dordrecht, 1987)
[55]   B. DeWitt, *Supermanifolds* (Cambridge University Press, Cambridge, 1984)
[65]   K.B. Efetov, Supersymmetry and theory of disordered metals. Adv. Phys. **32**, 53 (1983)
[297] J. Zinn-Justin, *Quantum Field Theory and Critical Phenomena* (Clarendon Press, Oxford, 1993)

Chapter 3
Grassmann Analysis

Abstract Differentiation and integration of Grassmann variables are introduced. Application is made to Gauss integrals. A second part to exterior algebra follows.

3.1 Differentiation

The operation of differentiation of Grassmann variables is introduced.

A function, f, depending on $\zeta \in \mathscr{A}_1$ is linear in ζ,

$$f(\zeta) = f(0) + \zeta f_l = f(0) + f_r \zeta, \quad f_r = \mathscr{P}(f_l), \tag{3.1}$$

where $f(0)$, f_r and f_l do not depend on ζ (The parity operator $\mathscr{P}$ was defined in Sect. 2.2). The *left and right derivatives* are defined by

$$\frac{\mathrm{d}}{\mathrm{d}\zeta}f = f_l, \quad \mathrm{d}f/\mathrm{d}\zeta = f_r. \tag{3.2}$$

The differential reads

$$\mathrm{d}f(\zeta) = \mathrm{d}\zeta\,\frac{\mathrm{d}}{\mathrm{d}\zeta}f = \mathrm{d}f/\mathrm{d}\zeta \cdot \mathrm{d}\zeta. \tag{3.3}$$

Example 3.1.1 One obtains with $\zeta, \alpha \in \mathscr{A}_1$

$$\frac{\mathrm{d}}{\mathrm{d}\zeta}\zeta = 1, \quad \mathrm{d}\zeta/\mathrm{d}\zeta = 1, \quad \frac{\mathrm{d}}{\mathrm{d}\zeta}(\alpha\zeta) = -\alpha, \quad \mathrm{d}(\alpha\zeta)/\mathrm{d}\zeta = \alpha.$$

Grassmann variables and differentials anticommute

$$\mathrm{d}\zeta_i\,\zeta_j = -\zeta_j\,\mathrm{d}\zeta_i. \tag{3.4}$$

© Springer-Verlag Berlin Heidelberg 2016

F. Wegner, *Supermathematics and its Applications in Statistical Physics*,
Lecture Notes in Physics 920, DOI 10.1007/978-3-662-49170-6_3

In the following we will use both the left and right derivatives. According to (3.1) they can differ in sign. When using the literature, one should always check sign conventions.

The product rule reads

$$\frac{\partial}{\partial\zeta}(fg) = (\frac{\partial}{\partial\zeta}f)g + \mathscr{P}(f)\frac{\partial}{\partial\zeta}g, \tag{3.5}$$

$$\partial(fg)/\partial\zeta = \partial f/\partial\zeta\,\mathscr{P}(g) + f\partial g/\partial\zeta. \tag{3.6}$$

Multiple derivatives with respect to odd variables, $\zeta \in \mathscr{A}_1$ anticommute

$$\frac{\partial}{\partial\zeta_k}\frac{\partial}{\partial\zeta_l}f = -\frac{\partial}{\partial\zeta_l}\frac{\partial}{\partial\zeta_k}f. \tag{3.7}$$

From (2.10), (3.7), and the definition of $\frac{\partial}{\partial\zeta}$, with it acting to the right, one has

$$\zeta_k\zeta_l + \zeta_l\zeta_k = 0, \quad \frac{\partial}{\partial\zeta_k}\frac{\partial}{\partial\zeta_l} + \frac{\partial}{\partial\zeta_l}\frac{\partial}{\partial\zeta_k} = 0, \quad \frac{\partial}{\partial\zeta_k}\zeta_l + \zeta_l\frac{\partial}{\partial\zeta_k} = \delta_{k,l}. \tag{3.8}$$

Thus ζ and $\frac{\partial}{\partial\zeta}$ constitute a Clifford algebra.[1]

The derivative of a function $f(z,\zeta)$ with respect to an even variable $z \in \mathscr{A}_0$ is obtained by differentiating the components of the products of the ζs to z,

$$\frac{\partial(f_0(z) + f_1(z)\zeta)}{\partial z} = \frac{df_0(z)}{dz} + \frac{df_1(z)}{dz}\zeta. \tag{3.9}$$

There is no distinction between the right and left derivative with respect to even variables $z \in \mathscr{A}_0$. The conventional product rule applies

$$\frac{\partial(fg)}{\partial z} = \frac{\partial f}{\partial z}g + f\frac{\partial g}{\partial z}. \tag{3.10}$$

The derivatives with respect to $\zeta \in \mathscr{A}_1$ and $z \in \mathscr{A}_0$ commute,

$$\frac{\partial}{\partial z}\frac{\partial}{\partial\zeta}f = \frac{\partial}{\partial\zeta}\frac{\partial}{\partial z}f. \tag{3.11}$$

---

[1]A Clifford algebra is an associative algebra on the basis of elements η_i and their products obeying $\eta_i\eta_j + \eta_j\eta_i = u_{ij}$ with C-numbers $u_{ij} = u_{ji}$.

3.2 Integration

The integral with respect to Grassmann variables is introduced. It is equivalent to differentiation.

There is not really a range of values defined for Grassmann variables. Nevertheless, Berezin introduced the concept of an integral 'over the whole range of values' of a Grassmannian. Translational invariance is required for the integral over a Grassmannian,

$$\int d\zeta f(\zeta) = \int d\zeta f(\zeta + \alpha), \quad \zeta, \alpha \in \mathscr{A}_1 \tag{3.12}$$

in analogy to

$$\int_{-\infty}^{\infty} dx f(x) = \int_{-\infty}^{\infty} dx f(x + a), \quad x, a \in \mathscr{A}_0. \tag{3.13}$$

Equation (3.12) with $f(\zeta) = \zeta$ yields

$$\int d\zeta \alpha = 0. \tag{3.14}$$

Thus, one puts

$$\int d\zeta = 0, \quad \int d\zeta \zeta = 1, \tag{3.15}$$

where the last integral is arbitrarily normalized to 1.[2] Thus, integration with the differential $d\zeta$ to the left of the function coincides with left differentiation, i.e.

$$\int d\zeta f(\zeta) = \frac{\partial}{\partial \zeta} f(\zeta). \tag{3.16}$$

Since $\frac{\partial}{\partial \zeta} f(\zeta)$ does not depend on ζ, one obtains

$$\int d\zeta \frac{\partial}{\partial \zeta} f(\zeta) = 0, \tag{3.17}$$

---

[2]This agrees with the definitions of Zinn-Justin [297] and Salmhofer [227]. Berezin [25] and Efetov [65] use the opposite sign, which is related to the minus sign in (3.20). Sometimes other values are attributed to this integral.

which is analogous for a function $f(x)$, if $f(x \to \infty) = f(x \to -\infty) = 0$,

$$\int_{-\infty}^{\infty} \mathrm{d}x \frac{\mathrm{d}f(x)}{\mathrm{d}x} = 0. \tag{3.18}$$

Similarly, one introduces integration with the differential $\mathrm{d}\zeta$ to the right of the function. Since for $f \in \mathscr{A}_\nu$ and $\zeta \in \mathscr{A}_1$ one obtains

$$\int \mathrm{d}\zeta f(\zeta) = \frac{\partial}{\partial \zeta} f(\zeta) = (-)^{\nu-1} \partial f(\zeta)/\partial \zeta = (-)^{\nu} \int f(\zeta) \mathrm{d}\zeta, \tag{3.19}$$

and

$$\int f(\zeta) \mathrm{d}\zeta = -\partial f(\zeta)/\partial \zeta. \tag{3.20}$$

Note the minus sign.

Example 3.2.1 Compare with Example 3.1.1.

$$\int \mathrm{d}\zeta \zeta = 1, \quad \int \zeta \mathrm{d}\zeta = -1, \quad \int \mathrm{d}\zeta(\alpha\zeta) = -\alpha, \quad \int (\alpha\zeta)\mathrm{d}\zeta = -\alpha.$$

With (3.6), one obtains for partial integration

$$\int \mathrm{d}\zeta f(\zeta) (\frac{\partial}{\partial \zeta} g(\zeta)) = -\int \mathrm{d}\zeta (\frac{\partial}{\partial \zeta} f(\zeta)) \mathscr{P}(g(\zeta)),$$

$$\int \partial f(\zeta)/\partial \zeta \mathscr{P}(g(\zeta)) \mathrm{d}\zeta = -\int f(\zeta) \partial g(\zeta)/\partial \zeta. \tag{3.21}$$

Since derivatives anticommute, the same applies to differentials

$$\mathrm{d}\zeta_l \mathrm{d}\zeta_k = -\mathrm{d}\zeta_k \mathrm{d}\zeta_l. \tag{3.22}$$

3.3   Gauss Integrals I

Gauss integrals, that is, integrals over the exponential of a quadratic form of Grassmann variables, yield the determinant of the coefficient matrix. In contrast Gauss integrals over complex numbers yield the inverse of the determinant of the coefficient matrix.

Gauss integrals play an important role in applications of Grassmann variables and we will come back to them several times. Here, integrals over $2r$ Grassmann variables $\xi_1, \ldots \xi_r, \eta_1, \ldots \eta_r \in \mathscr{A}_1$ are considered,

$$I_- = \int \prod_{k=1}^{r} (\mathrm{d}\,\eta_k \mathrm{d}\,\xi_k) \exp(\sum_{k,l=1}^{r,r} \xi_k a_{kl} \eta_l), \tag{3.23}$$

with elements $a_{kl} \in \mathscr{A}_0$ which do not depend on the variables ξ and η, over which the integral is performed. The exponential function can be represented by a Taylor expansion. For $r = 1$, one obtains

$$\int \mathrm{d}\,\eta_1 \mathrm{d}\,\xi_1 (1 + \xi_1 a_{11} \eta_1) = \int \mathrm{d}\,\eta_1 a_{11} \eta_1 = a_{11}. \tag{3.24}$$

For larger r, we proceed recursively by integrating first over ξ_1

$$I_- = \int \prod_{k=2}^{r} (\mathrm{d}\,\eta_k \mathrm{d}\,\xi_k) \mathrm{d}\,\eta_1 \sum_{m=1}^{r} a_{1m} \eta_m \exp(\sum_{k=2,l=1}^{r,r} \xi_k a_{kl} \eta_l). \tag{3.25}$$

Next the integral over η_m is performed,

$$I_- = \sum_{m=1}^{r} (-)^{m-1} a_{1m} \int \prod_{k=1}^{m-1} (\mathrm{d}\,\eta_k \mathrm{d}\,\xi_{k+1}) \prod_{k=m+1}^{r} (\mathrm{d}\,\eta_k \mathrm{d}\,\xi_k) \exp(\sum_{k \neq 1, l \neq m} \xi_k a_{kl} \eta_l). \tag{3.26}$$

The factor $(-)^{m-1}$ is due to the rearrangement of the differentials $\mathrm{d}\,\xi$ und $\mathrm{d}\,\eta$, since $\mathrm{d}\,\eta_m \mathrm{d}\,\xi_{k+1} \mathrm{d}\,\eta_k = -\mathrm{d}\,\eta_k \mathrm{d}\,\xi_{k+1} \mathrm{d}\,\eta_m$ has been used for $k = m - 1$ down to 1 in order to obtain $\mathrm{d}\,\eta_m$ to the right of the other differentials. From (3.26), one sees that the integral I_- is a sum of r terms of the form $(-)^{m-1} a_{1m}$ times an integral of the same form depending on the $(r-1) \times (r-1)$ matrix that is obtained from a by cancelling the first line and the mth column. This is the recursion formula for determinants. Since the integral for $r = 1$, i.e. (3.24), yields the determinant of the 1×1 matrix, the integral evaluates to the determinant of the matrix a,

$$I_- = \int \prod_{k=1}^{r} (\mathrm{d}\,\eta_k \mathrm{d}\,\xi_k) \exp(\sum_{k,l=1}^{r,r} \xi_k a_{kl} \eta_l) = \det(a). \tag{3.27}$$

If we reverse the sign of a and exchange the sequence of $\mathrm{d}\,\eta$ and $\mathrm{d}\,\xi$, then

$$\int \prod_{k=1}^{r} (\mathrm{d}\,\xi_k \mathrm{d}\,\eta_k) \exp(-\sum_{k,l=1}^{r,r} \xi_k a_{kl} \eta_l) = \det(a). \tag{3.28}$$

In comparison the integral of the exponential function over the complex numbers x yields

$$I_+ = \int [D x] \exp(-\sum_{k,l=1}^{r,r} x_k^* a_{kl} x_l) = \frac{1}{\det(a)} \tag{3.29}$$

with

$$[D x] = \prod_{k=1}^{r} \frac{d\,\Re x_k d\,\Im x_k}{\pi}, \tag{3.30}$$

where x^* denotes the complex conjugate of x. Both real and imaginary part of x_k run from $-\infty$ to $+\infty$. The integral exists if the real part of all eigenvalues of a are positive. For the integration with complex numbers we use pairs of complex conjugate ones. One can do something similar for Grassmann variables, but is not forced to. The essential difference between Gauss integrals over Grassmann variables and over complex variables, is that in one case one obtains the determinant, in the other its inverse. Just as in the present case, the use of Grassmann variables frequently yields the inverse of the result obtained with complex variables.

Let us add linear terms in the exponent with $u, v \in \mathcal{A}_0$,

$$\begin{aligned} I_+(u^*, v) &= \int [D x] \exp(-\sum_{k,l=1}^{r,r} x_k^* a_{kl} x_l + \sum_k (u_k^* x_k + x_k^* v_k)) \\ &= \int [D x] \exp(-(x^\dagger - u^\dagger a^{-1}) a (x - a^{-1} v) + u^\dagger a^{-1} v) \\ &= \frac{1}{\det(a)} \exp(u^\dagger a^{-1} v). \end{aligned} \tag{3.31}$$

See Problem 3.6.

Here, and later on, we determine expectation values, in particular, moments for Gaussian densities, $d\rho$. We generalize the notion *density*, insofar as we do not require the densities to be positive, only to be normalized,

$$\int d\rho = 1. \tag{3.32}$$

The expectation value of A is defined by

$$\langle A \rangle = \int d\rho A. \tag{3.33}$$

Here the density is given by

$$d\rho = \det(a)[D x] \exp(-x^\dagger ax), \tag{3.34}$$

where x is a column vector and $x^\dagger$ its hermitian adjoint row vector. Second moments are obtained from

$$\frac{\partial^2 I_+(u^*, v)}{\partial u_q^* \partial v_p}\bigg|_{u=v=0} = \int [D x] x_p^* x_q \exp(-x^\dagger ax) = (a^{-1})_{q,p} \frac{1}{\det(a)}, \tag{3.35}$$

which yields

$$\langle x_q x_p^* \rangle = (a^{-1})_{q,p} \tag{3.36}$$

and

$$\langle \exp(u^\dagger x + x^\dagger v) \rangle = \exp(u^\dagger \langle xx^\dagger \rangle v). \tag{3.37}$$

Note: $\langle xx^\dagger \rangle$ is an $(r \times r)$-matrix. Generally, any derivative of $I_+(u^*, v)$ with respect to v_p yields a factor x_p^* under the integral and a derivative with respect to u_q^* yields a factor x_q under the integral. Thus, one obtains

$$\langle A(x, x^*) \rangle = A(\frac{\partial}{\partial u^*}, \frac{\partial}{\partial v}) \exp(u^\dagger \langle xx^\dagger \rangle v)|_{u=v=0}. \tag{3.38}$$

For higher moments one obtains

$$\langle x_{i_1}^* \ldots x_{i_n}^* x_{j_1} \ldots x_{j_n} \rangle = \sum_P \prod_{k=1}^n \langle x_{i_k}^* x_{j_{P(k)}} \rangle = \mathrm{per}_{k,l}(\langle x_{i_k}^* x_{j_l} \rangle), \tag{3.39}$$

where P runs over all permutations of the indices k of j. We denote the *permanent* as per. The permanent differs from the determinant insofar as it does not contain the sign factors associated to the permutation. It does not have as interesting properties as the determinant.

Equation (3.39) is obtained from (3.38) by first taking the n derivatives to v_{i_k},

$$\prod_{k=1}^n \frac{\partial}{\partial v_{i_k}} \exp(u^\dagger \langle xx^\dagger \rangle v) = \prod_{k=1}^n u^\dagger \langle xx_{i_k}^* \rangle \exp(u^\dagger \langle xx^\dagger \rangle v). \tag{3.40}$$

Then one sets $v = 0$, so that the exponential becomes one. The derivatives with respect to $u_{j_l}^*$ yield the terms in (3.39). If the number of factors x^* differs from the number of factors x, then the expectation value vanishes.

Similarly, one finds for I_- with $\alpha, \beta \in \mathscr{A}_1$

$$I_-(\alpha, \beta) = \int \prod_{k=1}^{r} (\mathrm{d}\xi_k \mathrm{d}\eta_k) \exp\left(-\sum_{k,l=1}^{r,r} \xi_k a_{kl} \eta_l + \sum_k (\xi_k \alpha_k + \beta_k \eta_k)\right)$$

$$= \int \prod_{k=1}^{r} (\mathrm{d}\xi_k \mathrm{d}\eta_k) \exp(-({}^t\xi - {}^t\beta a^{-1})a(\eta - a^{-1}\alpha) + {}^t\beta a^{-1}\alpha)$$

$$= \det(a) \exp({}^t\beta a^{-1}\alpha) \tag{3.41}$$

provided $\mathrm{ord}\,(\det(a)) \neq 0$. The transpose of the column array ξ is denoted by ${}^t\xi$, and similar for other arrays. Then

$$\partial^2 I_-(\alpha, \beta)/(\partial \alpha_p \partial \beta_q)\big|_{\alpha=\beta=0} = \int \prod_{k=1}^{r} (\mathrm{d}\xi_k \mathrm{d}\eta_k) \eta_q \xi_p \exp(-{}^t\xi a\eta)$$

$$= (a^{-1})_{q,p} \det(a). \tag{3.42}$$

Thus, for the Gaussian density

$$\mathrm{d}\rho = \det(a)^{-1} \prod (\mathrm{d}\xi \mathrm{d}\eta) \exp(-{}^t\xi a\eta), \tag{3.43}$$

we obtain the averages

$$\langle \eta_q \xi_p \rangle = (a^{-1})_{qp}, \quad \langle \exp({}^t\xi\alpha + {}^t\beta\eta) \rangle = \exp({}^t\beta\langle\eta\,{}^t\xi\rangle\alpha). \tag{3.44}$$

Similarly, as for (3.38), one obtains

$$\langle A(\xi, \eta) \rangle = A\left(-\frac{\partial}{\partial\alpha}, \frac{\partial}{\partial\beta}\right) \exp({}^t\beta\langle\eta\,{}^t\xi\rangle\alpha)\big|_{\alpha=\beta=0}. \tag{3.45}$$

For higher moments, one obtains

$$\langle \eta_{i_1}\xi_{j_1} \ldots \eta_{i_n}\xi_{j_n} \rangle = \sum_P (-)^P \prod_{k=1}^{n} \langle \eta_{i_k}\xi_{j_{P(k)}} \rangle = \det_{k,l}(\langle\eta_{i_k}\xi_{j_l}\rangle), \tag{3.46}$$

where P runs over all permutations of the indices k of j. Equation (3.46) is obtained from (3.45) similarly as (3.39) from (3.38). The derivatives with respect to v and u^* are replaced by those to α and β. Since we deal with Grassmann variables, we have to take care of the signs. Applying the derivatives in the sequence of the factors η and ξ on the l.h. side of (3.46) then the factors $\langle\eta_{i_k}\xi_{j_k}\rangle$ appear in this order, which yields the correct sign $(-)^P = 1$ for the identity. The other contributions are obtained by permutations of ξ and correspondingly by the derivatives $-\partial/\partial\alpha$. One obtains the factor $(-)^P$ in (3.46), since they anticommute. If the number of factors η differs from the number of factors ξ, then the expectation value vanishes.

3.4 Exterior Algebra II

We continue the treatment of Sect. 2.4 on exterior algebra. We introduce the Hodge dual and the inner product, the exterior and the interior differentials, and the Laplace-de Rham operator. Finally, Maxwell's equations are expressed in terms of differential forms.

Hodge Dual Continuing the considerations on exterior algebra we first introduce the Hodge dual or Hodge star operation

$$* a(\eta) = s_{n,k} \int d\zeta^1 d\zeta^2 \ldots d\zeta^n \exp \left(\sum_{i=1}^{n} g_i \eta^i \zeta^i \right) a(\zeta), \quad g_i = \pm 1. \tag{3.47}$$

The star operation transforms k-vectors into $n-k$-vectors, i.e. $a \in \mathscr{E}_k \mapsto *a \in \mathscr{E}_{n-k}$, where n is the dimension of the basis ξ and η, respectively. The factors g_i indicate the metric. We do not consider a general metric as e.g. necessary in general relativity, compare [188, 253], but will apply it below to Maxwell's equations in special relativity in the flat Minkowski space. Then factors $g_i = \pm 1$ are sufficient. One defines the sign $s_{n,k}$ so that

$$a(\zeta) = \zeta^1 \zeta^2 \ldots \zeta^k \quad \succ \quad *a(\zeta) = g_{k+1} \zeta^{k+1} g_{k+2} \zeta^{k+2} \ldots g_n \zeta^n. \tag{3.48}$$

Then

$$s_{n,k} = (-)^{k(n-k/2-1/2)} \tag{3.49}$$

and

$$a(\zeta) = \sum_{i_1 \ldots i_k} a_{i_1, i_2, \ldots i_k} \zeta^{i_1} \zeta^{i_2} \ldots \zeta^{i_k} / k! \tag{3.50}$$

with totally antisymmetrized $a_{i_1, i_2, \ldots i_k}$ yields

$$* a(\zeta) = \sum_{i_1 \ldots i_n} a_{i_1, i_2, \ldots i_k} \epsilon_{i_1, i_2, \ldots i_n} g_{i_{k+1}} \zeta^{i_{k+1}} \ldots g_{i_n} \zeta^{i_n} / (k!(n-k)!) \tag{3.51}$$

with the totally antisymmetric tensor ϵ with $\epsilon_{1,2,\ldots n} = +1$.

A different way to introduce the Hodge dual is to express it by means of derivatives,

$$* a(\xi) = (-)^{k(k-1)/2} a(\frac{\partial}{g.\partial\xi.})\omega(\xi), \quad \omega = g\xi^1\xi^2\ldots\xi^n, \quad g = \prod_{i=1}^{n} g_i. \quad (3.52)$$

The sign corresponds to the inversion of the order of the derivatives,

$$\xi^{i_k}\xi^{i_{k-1}}\ldots\xi^{i_1} = (-)^{k(k-1)/2}\xi^{i_1}\ldots\xi^{i_{k-1}}\xi^{i_k}. \quad (3.53)$$

Examples for the Hodge dual are the products in (2.25). They yield, with Euclidean metric $g_i = 1$,

$$* (a \wedge b)(\eta) = (a_1b_2 - a_2b_1)\eta^3 + (a_2b_3 - a_3b_2)\eta^1 + (a_3b_1 - a_1b_3)\eta^2,$$

$$*(a \wedge b \wedge c)(\eta) = \begin{vmatrix} a_1 & b_1 & c_1 \\ a_2 & b_2 & c_2 \\ a_3 & b_3 & c_3 \end{vmatrix}. \quad (3.54)$$

Double application of the Hodge dual maps the vector (maybe apart from the sign) into itself

$$* *a = s_{n,k}s_{n,n-k}(-)^{n(n-1)/2}ga = g(-)^{k(n-k)}a. \quad (3.55)$$

Inner Product The inner product of $a \in \mathscr{E}_k$ with $b \in \mathscr{E}_l$ is defined by

$$(a \cdot b)(\zeta) := (-)^{k(k-1)/2} a(\{g_i\frac{\partial}{\partial\zeta^i}\})b(\zeta)$$

$$= \sum_{i_1,\ldots,i_l} a_{i_1\ldots i_k}g_{i_1}\ldots g_{i_k}b_{i_1\ldots i_l}\zeta^{i_{k+1}}\ldots \zeta^{i_l}/(k!(l-k)!)$$

$$= (-)^{(n-l)(l-k)}g * (a \wedge *b)(\zeta) \in \mathscr{E}_{l-k} \quad (3.56)$$

with totally antisymmetrized $a_{i_1,\ldots,i_k}$ as in (3.50), similarly for $b_{i_1,\ldots,i_l}$. If $k = l$, then

$$(a \cdot b)(\zeta) = \sum_{i_1\ldots i_k} a_{i_1\ldots i_k}g_{i_1}\ldots g_{i_k}b_{i_1,\ldots i_k}/k! = \sum_{i_1<\ldots<i_k} a_{i_1,\ldots i_k}g_{i_1}\ldots g_{i_k}b_{i_1,\ldots i_k} \in \mathscr{E}_0.$$

$$(3.57)$$

Differential Forms Let us assume that a depends on coordinates x in an n-dimensional space. We introduce differential k-forms on the basis dx,

$$a(dx) = a_{i_1,i_2,\ldots i_k}(x)dx^{i_1}dx^{i_2}\ldots dx^{i_k}/k!, \quad (3.58)$$

where now the differentials dx^i are anticommuting and $a_{...}$ is totally antisymmetric. The exterior derivative is defined by

$$d := \sum_i \frac{\partial}{\partial x^i} dx^i, \tag{3.59}$$

which yields

$$d\,a(dx) = a_{i_1,i_2,\ldots i_k;i}(x)dx^i dx^{i_1} dx^{i_2} \ldots dx^{i_k}/k! \in \mathscr{E}_{k+1},$$

$$a_{i_1,i_2,\ldots i_k;i}(x) := \frac{\partial a_{i_1,i_2,\ldots i_k}(x)}{\partial x_i}. \tag{3.60}$$

The interior derivative is defined by

$$\delta := \sum_i \frac{\partial}{\partial x^i} g_i \frac{\partial}{\partial(dx^i)} \tag{3.61}$$

and thus acts as an inner product,

$$\delta a(dx) = \sum_{i_1 \ldots i_k} a_{i_1,\ldots i_k;i_1}(x)dx^{i_2} \ldots dx^{i_k}/(k-1)! = g(-)^{(n-k)(k-1)} * d * a(dx). \tag{3.62}$$

The squares of d and of δ vanish,

$$d^2 = 0, \quad \delta^2 = 0. \tag{3.63}$$

The Laplace-de Rham operator is given by

$$\square := (d + \delta)^2 = d\delta + \delta d = \sum_i g_i \frac{\partial^2}{\partial x^{i2}}, \tag{3.64}$$

which is easily deduced from (3.59), (3.61), (3.63). Applied on a k-form one finds

$$*d = (-)^k \delta*, \quad *\delta = (-)^{k+1}d * . \tag{3.65}$$

Examples In $n = 3$ dimensions with Cartesian coordinates ($g_i = 1$) one obtains for a scalar, a, the gradient $\operatorname{grad} a$,

$$d\,a = a_{;1}dx^1 + a_{;2}dx^2 + a_{;3}dx^3 = \operatorname{grad} a\, dx. \tag{3.66}$$

and $\delta a = 0$. For a vector, $a = a_i d x^i$, one obtains the curl and the divergence

$$d a = (a_{2;1} - a_{1;2}) d x^1 d x^2 + (a_{3;2} - a_{2;3}) d x^2 d x^3 + (a_{1;3} - a_{3;1}) d x^3 d x^1,$$

$$*d a = (a_{2;1} - a_{1;2}) d x^3 + (a_{3;2} - a_{2;3}) d x^1 + (a_{1;3} - a_{3;1}) d x^2 = \operatorname{curl} a\, d x,$$

$$*a = a_1 d x^2 d x^3 + a_2 d x^3 d x^1 + a_3 d x^1 d x^2,$$

$$d * a = (a_{1;1} + a_{2;2} + a_{3;3}) d x^1 d x^2 d x^3,$$

$$\delta a = *d * a = a_{1;1} + a_{2;2} + a_{3;3} = \operatorname{div} a. \tag{3.67}$$

Thus,

$$\Box a = \triangle a = (d \delta + \delta d) a = \begin{cases} \operatorname{div} \operatorname{grad} a, & a \in \mathscr{E}_0, \\ \operatorname{grad} \operatorname{div} a + \operatorname{curl} \operatorname{curl} a, & a \in \mathscr{E}_1. \end{cases} \tag{3.68}$$

Maxwell's Equations for Electrodynamics We choose the Minkowski metric $g_i = (1, -1, -1, -1)$ with $i = 0, ..3$ where $i = 0$ indicates the time coordinate and $i = 1, 2, 3$ the space coordinates, with $x^i = (t, x)$. We set light velocity $c = 1$. From the potential $A_i = (-\Phi, A)$, that is $A_0 = -\Phi$ and the three components $A_1, \ldots A_3$ of the vector potential,

$$A = \sum_i A_i d x^i \tag{3.69}$$

one obtains the electromagnetic field tensor

$$F = d A = \sum_{ij} F_{ij} d x^i d x^j / 2 \tag{3.70}$$

with

$$F_{ij} = A_{j;i} - A_{i;j}, \quad F_{0j} = -E_j, \quad F_{12} = B_3, \ldots \tag{3.71}$$

Obviously

$$d F = d^2 A = 0 \tag{3.72}$$

which yields the homogeneous Maxwell equations

$$\operatorname{div} B = 0, \quad \operatorname{curl} E + \frac{\partial B}{\partial t} = 0. \tag{3.73}$$

The inhomogeneous Maxwell equations connect the fields to the charge density and current density $j_i = (-\rho, j)$,

$$\delta F = j := \sum_i j_i d x^i. \tag{3.74}$$

Equivalently

$$\text{div} E = \rho, \quad \text{curl} B - \frac{\partial E}{\partial t} = j. \tag{3.75}$$

Conservation of charge is obtained from the equation of continuity

$$\delta^2 F = 0 = \delta j = \frac{\partial \rho}{\partial t} + \text{div } j. \tag{3.76}$$

The gauge transformation reads

$$A \to A + d \Lambda \tag{3.77}$$

with some scalar Λ. It leaves the tensor F unchanged, since $F \to F + d^2\Lambda = F$. If one chooses the Lorenz gauge $\delta A = 0$, then one obtains

$$\Box A = (\frac{\partial^2}{\partial t^2} - \triangle) A = (d\delta + \delta d)A = \delta d A = \delta F = j. \tag{3.78}$$

Generalized Stokes' Theorem   Stokes' theorem can be written in the general form

$$\int_\Omega d A = \int_{\partial\Omega} A \tag{3.79}$$

with $A \in \mathcal{E}_l$. The integral on the l.h.s. runs over the $l + 1$-dimensional volume Ω, and the integral on the r.h.s. over the l-dimensional boundary, $\partial\Omega$, of this volume. If $l + 1 = n$, then the l.h.s. yields immediately

$$\int_\Omega \sum \frac{\partial a_{i_1,i_2,\dots,i_l}}{\partial x^i} d x^i d x^{i_1} d x^{i_2} \dots d x^{i_l} = \int_{\partial\Omega} \sum a_{i_1,i_2,\dots,i_l}|_{x_<^i}^{x_>^i} d x^{i_1} d x^{i_2} \dots d x^{i_l}, \tag{3.80}$$

where (for a convex volume) $x_>^i$ and $x_<^i$ are the upper and lower limit of x_i, respectively, for given $x^{i_1}, x^{i_2}, \dots x^{i_l}$. For a non-convex volume there may be several $x_>$ and $x_<$. Their contributions have to be summed. The sums in (3.80) run over i and $i_1 < i_2 < \dots < i_l$.

If $l+1 < n$, one parametrizes $x^i = x^i(y^1, y^2, \ldots y^{l+1})$. Then

$$d\, a_{i_1, i_2, \ldots i_l} = \sum_r \frac{\partial a_{i_1, i_2, \ldots i_l}}{\partial y^r} d\, y^r,$$

$$d x^{i_1} d x^{i_2} \ldots d x^{i_l} = \sum \frac{\partial(x^{i_1}, x^{i_2}, \ldots x^{i_l})}{\partial(y^{r_1}, y^{r_2}, \ldots y^{r_l})} d y^{r_1} d y^{r_2} \ldots d y^{r_l}. \qquad (3.81)$$

The sum in the second equation runs over the $l+1$ contributions $r_1 < r_2 < \ldots < r_l$. Hence the integral can be rewritten

$$\int_\Omega \sum \frac{\partial}{\partial y^r} \left(a_{i_1, i_2, \ldots i_l} \frac{\partial(x^{i_1}, x^{i_2}, \ldots x^{i_l})}{\partial(y^{r_1}, y^{r_2}, \ldots y^{r_l})} \right) d y^r d y^{r_1} d y^{r_2} \ldots d y^{r_l}. \qquad (3.82)$$

The derivatives $\frac{\partial}{\partial y^r} \frac{\partial(x \ldots)}{\partial(y \ldots)}$ do not contribute since the double derivatives $\frac{\partial^2 x}{\partial y^r \partial y^{r_k}}$ appear pairwise for $r \neq r_k$ and cancel, whereas $d y^r d y^{r_k}$ vanishes for $r = r_k$. Then the y^r-integration can be performed, finally yielding

$$\int_{\partial\Omega} \sum a_{i_1, i_2, \ldots i_l} d x^{i_1} d x^{i_2} \ldots d x^{i_l}. \qquad (3.83)$$

Note that the generalized Stokes' theorem does not use any metric.

Why $d x^2 d x^1 = -d x^1 d x^2$?   Express the area A of a planar two-dimensional region by an integral over the boundary and use, for simplicity's sake $x^1 = x$ and $x^2 = y$. We may write

$$A = \int 1 d x d y = \int (x_>(y) - x_<(y)) d y = \oint_{\text{counter-clockw.}} x d y$$

$$= \int 1 d y d x = \int (y_>(x) - y_<(x)) d x = \oint_{\text{clockwise}} y d x, \qquad (3.84)$$

where in one case one has to circle the boundary counter-clockwise, and in the other case in a clockwise direction. Thus one obtains the same orientation for the integral along the boundary with the choice $d y d x = -d x d y$.

Integration is always performed by starting with the integration over the leftmost $d x$. The sign of the contributions of the boundary integral is given by the orientation of the normal pointing outward, multiplied by the following l differentials $d x$ compared to the product of the $l+1$ differentials $d x$ in the integral over the volume Ω.

Problems

3.1 Let

$$f = f(0) + \sum_i a_i \zeta b_i, \quad a_i \in \mathscr{A}_{k_i}, \quad b_i \in \mathscr{A}_{l_i},$$

where $f(0)$, a_i, b_i do not depend on ζ. Determine $\frac{\partial}{\partial \zeta} f$ and $\partial f / \partial \zeta$. Check the last Eq. (3.1).

3.2 Calculate $\int d\xi (\xi - \eta) f(\xi)$ for $\xi, \eta \in \mathscr{A}_1$. Thus what is the meaning of the function $(\xi - \eta)$?

3.3 Calculate the integral $\int d\xi e^{i\alpha\xi}$ with $\alpha, \xi \in \mathscr{A}_1$. Compare with $\int_{-\infty}^{\infty} e^{iax} dx = 2\pi\delta(a)$ for $a, x \in \mathscr{A}_0$.

3.4 Show the product rule, i.e. (3.6).

3.5 Expand the exponential function in (3.23) in the Grassmann variables for $r = 2$ and perform the integration.

3.6 Show that for $x, a, u, v \in \mathbb{C}$, $\Re a > 0$, the relation

$$\int \exp(-(x^* - u^*)a(x - v)) d\Re x \, d\Im x = \int \exp(-x^* ax) d\Re x \, d\Im x$$

holds. Hint: Separate x in its real and imaginary part.

3.7 Determine $*\omega$ to ω of (3.52) and $*1$.

References

[25] F.A. Berezin, *Introduction to Superanalysis* (Reidel, Dordrecht, 1987)
[65] K.B. Efetov, Supersymmetry and theory of disordered metals. Adv. Phys. **32**, 53 (1983)
[188] C.W. Misner, K.S. Thorne, J.A. Wheeler, *Gravitation* (Freeman, New York, 2008)
[227] M. Salmhofer, *Renormalization – An Introduction*. Texts and Monographs in Physics (Springer, Berlin, Heidelberg, 1998)
[253] W. Thirring, *A Course in Mathematical Physics. 2. Classical Field Theory* (Springer, New York, 1979, 1986); *Lehrbuch der mathematischen Physik. 2. Klassische Feldtheorie* (Springer, Wien, 1978, 1990)
[297] J. Zinn-Justin, *Quantum Field Theory and Critical Phenomena* (Clarendon Press, Oxford, 1993)

Chapter 4
Disordered Systems

Abstract The replica trick is a general tool to perform averages for quenched disordered systems. It is mathematically not exact, however. In contrast, exact calculations for quantum-mechanical particles in a random potential can be performed by simultaneous use of complex and anticommuting variables. A first application is the derivation of Wigner's semi-circle law for the density of states of the Gaussian unitary ensemble.

4.1 Introduction

The concept of quenched and annealed disorder is introduced.

Many solids are disordered. Periodic arrangements of atoms, as in crystals, are rare and even if the atoms are arranged in a periodic lattice, then it is likely that there exist more or less numerous defects. Thus, one is led to consider disordered systems with parameters characterizing the disorder.

Disordered systems are called quenched if these parameters are fixed and do not evolve with time. A quench refers to a rapid cooling, which prohibits the system to reach its equilibrium state. In contrast, if the disorder evolves in time, the system is called annealed.

Statistical averages for such systems are obtained by averaging over the disorder. In order to obtain the expectation value of an observable A one has to evaluate

$$\overline{\langle A \rangle} = \overline{\mathrm{tr}\,(A\mathrm{e}^{-\beta H})/\mathrm{tr}\,(\mathrm{e}^{-\beta H})}, \qquad (4.1)$$

where the overline denotes the average over the disorder in the Hamiltonian and $\langle A \rangle$ denotes the average over the density $\mathrm{e}^{-\beta H}/Z(H)$. The main obstacle in averaging the disorder is that the disorder appears both in the numerator and the denominator of (4.1).

There are several ways to overcome this difficulty. For free particles in a random potential, the use of superfields is possible and gives the correct result. In many other cases this is not possible. For such cases, often the replica trick is used. However, that trick does not always yield the correct result [262].

© Springer-Verlag Berlin Heidelberg 2016
F. Wegner, *Supermathematics and its Applications in Statistical Physics*,
Lecture Notes in Physics 920, DOI 10.1007/978-3-662-49170-6_4

4.2 Replica Trick

The replica trick, which is often used to determine expectation values for disordered systems, is explained.

The replica trick [53, 63] will be presented here. The basic idea is to introduce n replicas of the original system and to determine the result for general n. One then has the nth power of the partition function in the denominator. The trick is the extrapolation to $n = 0$. The denominator then degenerates to unity and it is sufficient to average $e^{-\beta H}$ in the numerator. There are two variants of the replica trick.

4.2.1 First Variant

One averages the disorder of the free energy. One starts from

$$\ln Z = \lim_{n \to 0} \frac{Z^n - 1}{n}. \tag{4.2}$$

One determines the average of Z^n for $n \in \mathbb{N}$ and uses the expression for $n \in \mathbb{R}$ to extrapolate to $n = 0$. Then the averaged free energy is given by

$$\overline{F} = -k_\mathrm{B} T \overline{\ln Z} = -k_\mathrm{B} T \lim_{n \to 0} \frac{\overline{Z^n} - 1}{n}. \tag{4.3}$$

An average over some quantity, $A(S)$, can be obtained from

$$Z(J) = \mathrm{tr}\,(e^{-\beta H + JA(S)}), \quad \overline{\langle A(S) \rangle} = \frac{\overline{d \ln Z(J)}}{dJ} = \lim_{n \to 0} \frac{d\,\overline{Z^n}}{n\,dJ}. \tag{4.4}$$

4.2.2 Second Variant

Here we really introduce replicas. For a given variable, S, one introduces n replicas, S_α, where α runs from 1 to n. The average of $A(S)$ for a particular realization of the disorder is

$$\langle A(S) \rangle = \langle A(S_1) \rangle = \frac{\int [D\,S_1][D\,S_2] \dots [D\,S_n] A(S_1) e^{-\beta(H(S_1)+H(S_2)+\dots H(S_n))}}{Z^n} \tag{4.5}$$

where Z is the partition function of one of the replicas. Since in general it is difficult to average this expression over the disorder, one averages the numerator for general

$n \in \mathbb{N}$ and finally extrapolates to $n = 0$,

$$\overline{\langle A(S) \rangle} = \int [\mathrm{D}\, S_1][\mathrm{D}\, S_2] \ldots [\mathrm{D}\, S_n] A(S_1) \overline{\mathrm{e}^{-\beta(H(S_1)+H(S_2)+\ldots H(S_n))}}\big|_{n=0}. \qquad (4.6)$$

The problem with the replica trick is that Z^n and the integral (4.6) can usually only be determined for natural n, but the result is assumed to apply for real n. But it may differ, for example by an additional function like $\sin(\pi n)/n$, yielding a different result in the limit $n = 0$.

4.3   Quantum Mechanical Particle in a Random Potential

The simultaneous use of complex and Grassmann variables, which allows the average over disorder without problems from the denominator, is explained for a model of quantum mechanical particles in a random potential.

Early theoretical investigations of the consequences of disorder are due to Dyson [59], who investigated the dynamics of a disordered linear chain, and due to Anderson [10], who considered the transition from spin waves to local spin excitations (Anderson localization). A different approach was used by Wigner [284], who described the interacting many-body system in terms of a large random matrix. The elements of that matrix are equally Gaussian-distributed random variables. Closer investigation by Dyson [60] revealed three different classes of such systems, now commonly called Wigner-Dyson classes. Both approaches can be combined by allowing for n orbitals at each lattice site [269], which can be solved exactly in the limit $n \to \infty$, but may also be used otherwise.

Consider a quantum-mechanical particle in a random potential, described by a tight-binding Hamiltonian

$$H = \sum_{r,r'} |r\rangle f_{r,r'} \langle r'|, \qquad (4.7)$$

where the kets $|r\rangle$ number the orthonormalized wavefunctions of orbitals. This r may be the lattice point around which an orbital is located or it may stay for some quantum numbers. Typically, one is interested in the Green's function

$$G(r, r', z) = \langle r| \frac{1}{z - H} |r'\rangle. \qquad (4.8)$$

For example, the density of states at orbital r is given by

$$\rho(r, E) = \lim_{\eta \to 0} (G(r, r, E - \mathrm{i}\eta) - G(r, r, E + \mathrm{i}\eta))/(2\pi\mathrm{i}). \qquad (4.9)$$

This Green's function can be expressed by (3.35) and (3.42), respectively.

$$G(r,r',z) = s \det(s(z-f)) \int x_r x_{r'}^* \exp(-\sum_{r,r'} x_r^* s(z\delta_{r,r'} - f_{r,r'}) x_{r'}) \prod_r \frac{\mathrm{d}\,\Re x_r \mathrm{d}\,\Im x_r}{\pi}$$

$$= s \det(s(z-f))^{-1} \int \eta_r \xi_{r'} \exp(-\sum_{r,r'} \xi_r s(z\delta_{r,r'} - f_{r,r'}) \eta_{r'}) \prod_r (\mathrm{d}\,\xi_r \mathrm{d}\,\eta_r).$$

$$(4.10)$$

The factor s is chosen so that the integral converges. Since f is hermitian, we choose s according to the sign of the imaginary part of z,

$$s = -\mathrm{i}\,\mathrm{sign}\Im z. \qquad (4.11)$$

The potential is random. A certain distribution of the matrix elements, f, is given. One has to average over this distribution. Now the determinants of $s(z-f)$ play the disturbing role of the partition function in (4.1, 4.5). Instead of using the replica trick, one performs the integral, both over the complex and the Grassmann variables,

$$G(r,r',z) = s \int \left\{ \begin{array}{c} x_r x_{r'}^* \\ \eta_r \xi_{r'} \end{array} \right\} \exp(-\sum_{r,r'} s(x_r^*(z\delta_{r,r'} - f_{r,r'}) x_{r'} + \xi_r(z\delta_{r,r'} - f_{r,r'}) \eta_{r'}))$$

$$\times \prod_r (\frac{\mathrm{d}\,\Re x_r \mathrm{d}\,\Im x_r}{\pi} \mathrm{d}\,\xi_r \mathrm{d}\,\eta_r). \qquad (4.12)$$

This is the root for the supersymmetric method. One may choose the upper or the lower expression in the curly bracket. If the matrix elements f are Gaussian distributed, the average can be easily performed. This has been used in [65, 261, 272].

4.4 Semicircle Law

Wigner's semicircle law for the density of states of a model described by a hermitian matrix, with equally Gaussian distributed matrix elements, is derived.

Consider a model of $N \times N$ Gaussian-distributed hermitian matrices f with probability density

$$P(f)\mathrm{d}f = \mathrm{const}\, \exp(-\frac{gN}{2}\,\mathrm{tr}f^2) \prod_{r=1}^N \mathrm{d}f_{rr} \prod_{1 \le r < r' \le N} \mathrm{d}\,\Re f_{rr'}\mathrm{d}\,\Im f_{rr'}. \qquad (4.13)$$

The factor N yields a useful limit for large N and fixed g. The distribution is invariant under unitary transformations, $P(U^\dagger f U) = P(f)$, for unitary matrices U. The matrix ensemble defined by (4.13) is called the Gaussian unitary ensemble (GUE). It is, in particular, invariant for the diagonal unitary matrix $U = \mathrm{diag}(e^{i\phi_r})$, which yields

$$G(r, r', z) = e^{i(\phi_r - \phi_{r'})} G(r, r', z) \tag{4.14}$$

for arbitrary phases ϕ_r and $\phi_{r'}$. Thus, G is diagonal in r and r'. The ensemble is also invariant under permutations of the indices r. Thus the averaged Green's function can be simply expressed by a function of z alone, $\overline{G(r, r', z)} = G(z)\delta_{r,r'}$. From (4.12) one obtains

$$G(z) = \frac{s}{N} \int x^\dagger x \exp\left(-sz(x^\dagger x + {}^t\xi\eta)\right) \overline{\exp\left(s(x^\dagger f x + {}^t\xi f \eta)\right)}$$

$$\times \prod_{r=1}^{N} \left(\frac{\mathrm{d}\,\Re x_r \mathrm{d}\,\Im x_r}{\pi} \mathrm{d}\,\xi_r \mathrm{d}\,\eta_r\right), \tag{4.15}$$

where x, ξ and η are column vectors. The transposed ${}^t\xi$ and $x^\dagger$ are row vectors. Thus $x^\dagger x$, ${}^t\xi\eta \in \mathscr{A}_0$, ${}^t\xi x, x^\dagger \eta \in \mathscr{A}_1$, whereas $xx^\dagger$ is an $N \times N$ matrix. The average yields

$$\overline{\exp\left(s(x^\dagger f x + {}^t\xi f \eta)\right)} = \exp\left(-\frac{1}{2gN} \sum_{r,r'} (x_r^* x_r + \xi_{r'}\eta_r)(x_r^* x_{r'} + \xi_r \eta_{r'})\right)$$

$$= \exp\left(-\frac{1}{2gN}[(x^\dagger x)^2 + 2({}^t\xi x)(x^\dagger \eta) - ({}^t\xi\eta)^2]\right). \tag{4.16}$$

In the next step we use the Hubbard-Stratonovich transformation

$$\exp\left(\frac{1}{2gN}({}^t\xi\eta)^2\right) = \left(\frac{gN}{2\pi}\right)^{1/2} \int \mathrm{d}\,u \exp\left(-u({}^t\xi\eta) - \frac{gN}{2}u^2\right). \tag{4.17}$$

In this way the quadrilinear Grassmannian term $({}^t\xi\eta)^2$ is expressed by a term containing only terms bilinear in the Grassmann variables, multiplied, however, with a new real variable, u. The integral over the Grassmann variables can then be performed,

$$\int \prod_r (\mathrm{d}\,\xi_r \mathrm{d}\,\eta_r) \exp(-{}^t\xi K \eta) = \det K, \quad K = (sz + u)\mathbf{1} + \frac{1}{gN}xx^\dagger,$$

$$\det K = (sz + u)^{N-1}(sz + u + \frac{1}{gN}x^\dagger x). \tag{4.18}$$

The determinant of K is obtained by using that the vectors perpendicular to x yield eigenvalues $sz + u$ of K, and x itself yields the last factor. Thus G reduces to

$$G(z) = \frac{s}{N}\left(\frac{gN}{2\pi}\right)^{1/2} \int d\,ux^\dagger x(sz+u)^{N-1}(sz+u+\frac{1}{gN}x^\dagger x)$$

$$\times \exp\left(-szx^\dagger x - \frac{1}{2gN}(x^\dagger x)^2 - \frac{gN}{2}u^2\right)\prod_r \frac{d\,\Re x_r d\,\Im x_r}{\pi}. \quad (4.19)$$

Since the integrand depends only on the square $x^\dagger x$, polar coordinates are introduced. This yields

$$\int \delta(x^\dagger x - v)\prod_r \frac{d\,\Re x_r d\,\Im x_r}{\pi} = \frac{v^{N-1}}{(N-1)!}. \quad (4.20)$$

Thus, the determination of $G(z)$ is reduced to an integral over two variables u and v,

$$G(z) = \frac{s}{N!}\left(\frac{gN}{2\pi}\right)^{1/2}\int_{-\infty}^{+\infty} d\,u \int_0^\infty d\,v \frac{sz+u+\frac{v}{gN}}{sz+u}h(z,u,v),$$

$$h = v^N(sz+u)^N \exp\left(-szv - \frac{v^2}{2gN} - \frac{gN}{2}u^2\right). \quad (4.21)$$

This is the formula for general N. We use the saddle point method in the limit $N \to \infty$. The saddle points u_0, v_0 of h are obtained from

$$\frac{\partial h}{\partial u} = N(\frac{1}{sz+u} - gu)h|_{u=u_0,v=v_0} = 0, \quad u_0 = -\frac{sz}{2} \pm \sqrt{\frac{1}{g} - \frac{z^2}{4}},$$

$$\frac{\partial h}{\partial v} = (\frac{N}{v} - sz - \frac{v}{gN})h|_{u=u_0,v=v_0} = 0, \quad \frac{v_0}{gN} = -\frac{sz}{2} \pm \sqrt{\frac{1}{g} - \frac{z^2}{4}}. \quad (4.22)$$

Also $u_0 = -sz$ and $v_0 = 0$ are saddle points. However, the integrand vanishes at these points. Moreover, the integrand vanishes if one takes the solutions (4.22) with opposite signs. Since the integral over v runs from 0 to $+\infty$, only the saddle point with positive real part of v contributes. Then h evaluates to $h(z,u_0,v_0) = N^N \exp(-N)$. The second derivatives at the saddle point are

$$\frac{\partial^2 \ln h}{\partial u^2} = -N\left(\frac{1}{(sz+u_0)^2} + g\right), \quad \frac{\partial^2 \ln h}{\partial v^2} = -\left(\frac{N}{v_0^2} + \frac{1}{gN}\right),$$

$$\frac{\partial^2 \ln h}{\partial u \partial v} = 0, \quad \frac{\partial^2 \ln h}{\partial u^2}\frac{\partial^2 \ln h}{\partial v^2} = 4 - gz^2. \quad (4.23)$$

This yields the Green's function

$$G(z) = \lim_{N \to \infty} \frac{N^{N+1/2} e^{-N}}{N!} \frac{s\sqrt{g}}{\sqrt{2\pi}} \frac{sz + u_0 + \frac{v_0}{gN}}{sz + u_0} \frac{2\pi}{\sqrt{\frac{\partial^2 \ln h}{\partial u^2} \frac{\partial^2 \ln h}{\partial v^2}}}$$

$$= \frac{g}{2}\left(z + s\sqrt{\frac{4}{g} - z^2}\right). \tag{4.24}$$

The imaginary part of G yields Wigner's semi-circle law for the density of states [285, 286],

$$\rho(E) = \frac{g}{2\pi}\sqrt{\frac{4}{g} - E^2}, \quad -\sqrt{\frac{4}{g}} \le E \le +\sqrt{\frac{4}{g}}. \tag{4.25}$$

This derivation is along the lines of volume 2 of [122]. Porter [214] contains a collection of early papers on systems with random matrices. Details of the spectra of such systems, in particular those of matrices with Gaussian distributed matrix elements, are found in the books by Mehta [182, 183].

We continue the calculation of this and similar models in Chap. 21.

Problem

4.1 Lloyd Model [169] Choose $H = \sum_r \epsilon_r |r\rangle\langle r| + \sum_{r,r'} t_{r,r'} |r\rangle\langle r'|$ with constant hopping matrix elements t and independently Lorentzian distributed diagonal matrix elements ϵ_r, $dP = \prod_r (\frac{\Gamma}{\pi(\Gamma^2 + \epsilon_r^2)} d\epsilon_r)$.

(i) How is the averaged Green's function $\overline{G(r, r', z)}$ related to the Green's function $G^{(0)}$ for the Hamiltonian with fixed $\epsilon_r = 0$? [Start from (4.12)].
(ii) Can the result be generalized to the average of products of Green's functions?

References

[10] P.W. Anderson, Absence of diffusion in certain random lattices. Phys. Rev. **109**, 1492 (1958)
[53] P.G. de Gennes, Exponents for the excluded volume problem as derived by the Wilson method. Phys. Lett. **38A**, 339 (1972)
[59] F.J. Dyson, The dynamics of a disordered linear chain. Phys. Rev. **92**, 1331 (1953)

 [60] F.J. Dyson, Statistical Theory of Energy Levels of Complex Systems. I, II,
      III. J. Math. Phys. **3**, 140, 157, 166 (1962)
 [63] S.F. Edwards, P.W. Anderson, Theory of spin glasses. J. Phys. F **5**, 965 (1975)
 [65] K.B. Efetov, Supersymmetry and theory of disordered metals. Adv. Phys. **32**,
      53 (1983)
[122] C. Itzykson, J.-M. Drouffe, *Statistical Field Theory*, 2 vols. (Cambridge
      University Press, Cambridge, 1989)
[169] P. Lloyd, Exactly solvable model of electronic states in a three-dimensional
      Hamiltonian: Non-existence of localized states. J. Phys. C **2**, 1717 (1969)
[182] M.L. Mehta, *Random Matrices and the Statistical Theory of Energy Levels*
      (Academic Press, New York, 1967)
[183] M.L. Mehta, *Random Matrices* (Academic Press, Boston, 1991)
[214] C.E. Porter, *Statistical Theories of Spectra* (Academic Press, New York,
      1965)
[261] J.J.M. Verbaarschot, H.A. Weidenmüller, M.R. Zirnbauer, Grassmann inte-
      gration in stochastic quantum physics: the case of compound-nucleus scat-
      tering. Phys. Repts. **129**, 367 (1985)
[262] J.J.M. Verbaarschot, M.R. Zirnbauer, Critique of the replica trick. J. Phys. A
      17, 1093 (1985)
[269] F. Wegner, Disordered systems with *n* Orbitals per site: $n = \infty$ limit. Phys.
      Rev. B **19**, 783 (1979)
[272] F. Wegner, Algebraic derivation of symmetry relations for disordered elec-
      tronic systems. Z. Phys. B **49**, 297 (1983)
[284] E.P. Wigner, On a class of analytic functions from the quantum theory of
      collisions. Ann. Math. **53**, 36 (1951)
[285] E.P. Wigner, Characteristic vectors of bordered matrices with infinite dimen-
      sions. Ann. Math. **62**, 548 (1955)
[286] E.P. Wigner, On the distribution of the roots of certain symmetric matrices.
      Ann. Math. **67**, 325 (1958)

Chapter 5
Substitution of Variables, Gauss Integrals II

Abstract A Gaussian integral over Grassmann variables yields a Pfaffian. Its
connection with the determinant is derived. The Jacobian for transformations of
Grassmann variables under the integral is presented.

5.1 Gauss Integrals II, Pfaffian Form

*The integral over the exponential of a form bilinear in Grassmann variables yields
a Pfaffian.*

In Chap. 3 we introduced the integral

$$I_- = \int \prod_{k=1}^{r} (\mathrm{d}\,\eta_k \mathrm{d}\,\xi_k) \exp(\sum_{k,l=1}^{r,r} \xi_k a_{kl} \eta_l) \tag{5.1}$$

over Grassmann variables ξ and η. The exponential was a sum of monomials, where
one factor ξ was multiplied by one factor η. This restriction is not necessary. We
consider

$$I_- = \int \mathrm{d}\,\xi_n \dots \mathrm{d}\,\xi_2 \mathrm{d}\,\xi_1 \exp(\tfrac{1}{2} \sum_{k,l=1}^{n} \xi_k a_{k,l} \xi_l), \quad a_{k,l} \in \mathscr{A}_0, \quad \xi_i \in \mathscr{A}_1, \tag{5.2}$$

where, without restriction of generality, it is assumed that the matrix a is antisymmetric, $a_{k,l} = -a_{l,k}$. The integral over ξ_1 yields

$$I_- = \int \mathrm{d}\,\xi_n \dots \mathrm{d}\,\xi_2 \sum_{j=2}^{n} a_{1,j}\xi_j \exp(\tfrac{1}{2} \sum_{k,l \neq 1,j} \xi_k a_{k,l} \xi_l)$$

$$= \sum_{j=2}^{n} (-)^j a_{1,j} \int \mathrm{d}\,\xi_n \dots \mathrm{d}\,\xi_{j+1} \mathrm{d}\,\xi_{j-1} \dots \mathrm{d}\,\xi_2 \exp(\tfrac{1}{2} \sum_{k,l \neq 1,j} \xi_k a_{k,l} \xi_l). \tag{5.3}$$

Thus we have obtained a recursion formula for I_-. One obtains the integral by
performing the corresponding integral after deleting the first and jth variable,

© Springer-Verlag Berlin Heidelberg 2016
F. Wegner, *Supermathematics and its Applications in Statistical Physics*,
Lecture Notes in Physics 920, DOI 10.1007/978-3-662-49170-6_5

multiplying it by $(-)^j a_{1,j}$ and summing over j. This recursion formula looks similar to that for a determinant, however, here, not only are the first row and jth column of a deleted, but also the jth row and first column. Hence the result is a polynomial in the as of order $n/2$. This expression is called the Pfaffian form of the antisymmetric matrix a. Thus one has

$$\mathrm{pf}(a) = \int \mathrm{d}\,\xi_n \ldots \mathrm{d}\,\xi_2 \mathrm{d}\,\xi_1 \exp(\tfrac{1}{2}\sum_{k,l=1}^{n} \xi_k a_{k,l}\xi_l). \tag{5.4}$$

In particular, one has

$$\begin{aligned}
\mathrm{pf}(a) &= 1 & n &= 0 \\
\mathrm{pf}(a) &= a_{1,2} & n &= 2 \\
\mathrm{pf}(a) &= a_{1,2}a_{3,4} - a_{1,3}a_{2,4} + a_{1,4}a_{2,3} & n &= 4 \\
\mathrm{pf}(a) &= 0 & n &\text{ odd}
\end{aligned} \tag{5.5}$$

Below we will see that the Pfaffian is closely connected to the determinant.

Let us add a source term to the integral

$$\begin{aligned}
I_-(\alpha) &= \int \mathrm{d}\,\xi_n \ldots \mathrm{d}\,\xi_1 \exp(\tfrac{1}{2}{}^t\xi a\xi + {}^t\xi\alpha) \\
&= \int \mathrm{d}\,\xi_n \ldots \mathrm{d}\,\xi_1 \exp\left(\tfrac{1}{2}({}^t\xi - {}^t\alpha a^{-1})a(\xi + a^{-1}\alpha) + \tfrac{1}{2}{}^t\alpha a^{-1}\alpha\right) \\
&= \mathrm{pf}(a) \exp\left(\tfrac{1}{2}{}^t\alpha a^{-1}\alpha\right), \quad \alpha_i \in \mathscr{A}_1. \tag{5.6}
\end{aligned}$$

Note that ${}^t(a^{-1}\alpha) = -{}^t\alpha a^{-1}$, since a and a^{-1} are antisymmetric. Expanding (5.6) up to second order in α, the density

$$\mathrm{d}\,\rho = \mathrm{pf}(a)^{-1}\mathrm{d}\,\xi_n \ldots \mathrm{d}\,\xi_1 \exp(\tfrac{1}{2}{}^t\xi a\xi) \tag{5.7}$$

yields

$$\langle \xi_k \xi_l \rangle = (a^{-1})_{lk}, \quad \langle \exp\left({}^t\xi\alpha\right)\rangle = \exp\left(-\tfrac{1}{2}{}^t\alpha\langle \xi\,{}^t\xi\rangle\alpha\right). \tag{5.8}$$

As before, (3.32), (3.33), $\mathrm{d}\,\rho$ is not a density in the usual sense, since it cannot be said to be positive, but the integral equals 1.

5.2   Variable Substitution I

The Jacobian for variable substitutions of Grassmann variables is the inverse determinant of the partial derivatives.

If we express variables $\zeta \in \mathscr{A}_1$ by variables $\eta \in \mathscr{A}_1$

$$\zeta = \zeta(\eta) = \zeta(0) + \frac{\partial \zeta}{\partial \eta}\eta, \tag{5.9}$$

then integration of

$$f(\zeta) = f(0) + \zeta \frac{\partial}{\partial \zeta}f, \tag{5.10}$$

yields

$$\int d\,\zeta f(\zeta) = \frac{\partial}{\partial \zeta}f, \quad \int d\,\eta f(\zeta) = \int d\,\eta (f(0) + \zeta \frac{\partial}{\partial \zeta}f) = \frac{\partial \zeta}{\partial \eta}\frac{\partial}{\partial \zeta}f. \tag{5.11}$$

Hence

$$\int d\,\zeta f(\zeta) = \int d\,\eta \left(\frac{\partial \zeta}{\partial \eta}\right)^{-1} f(\zeta(\eta)), \tag{5.12}$$

where we use that the partial derivative of ζ to η does not depend on η. Note that in comparison to variable substitutions of elements from $\mathscr{A}_0$ here the inverse of the derivative enters.

Replacing several odd variables, where we allow non-linear transformations, we obtain

$$\int d\,\zeta_m \ldots d\,\zeta_2 d\,\zeta_1 f(\zeta_1, \zeta_2, \ldots, \zeta_m)$$

$$= \int d\,\eta_m \ldots d\,\eta_2 d\,\eta_1 \frac{1}{\det(\partial \zeta / \partial \eta)} f(\zeta_1(\eta_1, \ldots), \ldots). \tag{5.13}$$

Thus the Jacobian for odd elements of the algebra is not the determinant $\det(\partial \zeta / \partial \eta)$, but its inverse. If we substitute the variables one after the other

$$(\eta_1, \eta_2, \ldots \eta_m) \rightarrow (\zeta_1, \eta_2, \ldots \eta_m) \rightarrow \ldots \rightarrow (\zeta_1, \zeta_2, \ldots \zeta_m), \tag{5.14}$$

then one obtains factors

$$D_i = \frac{\partial \zeta_i(\zeta_1, \ldots \zeta_{i-1}, \eta_i, \ldots \eta_m)}{\partial \eta_i} = \frac{\partial(\zeta_1, \ldots \zeta_{i-1}, \zeta_i, \eta_{i+1}, \ldots \eta_m)}{\partial(\zeta_1, \ldots \zeta_{i-1}, \eta_i, \eta_{i+1}, \ldots \eta_m)}. \tag{5.15}$$

The product of these factors yields

$$D_m D_{m-1} \ldots D_1 = \det(\partial \zeta / \partial \eta). \tag{5.16}$$

The matrix elements of $\partial\zeta/\partial\eta$ commute since they are even.[1]

Equation (5.16) can be shown by complete induction: One can express $D_i = \det(M_i)$ with

$$(M_i)_{kl} = \begin{cases} \delta_{kl} & k \neq i, \\ \frac{\partial\zeta_i(\zeta_1,\ldots\zeta_{i-1},\eta_i\ldots\eta_m)}{\partial\zeta_l} & k = i, l < i, \\ \frac{\partial\zeta_i(\zeta_1,\ldots\zeta_{i-1},\eta_i\ldots\eta_m)}{\partial\eta_l} & k = i, l \geq i. \end{cases} \tag{5.17}$$

All rows except row number i contains elements δ_{kl}. Thus, $\det(M_i) = D_i$. By induction in i it is shown that the product of the Ms obey,

$$(M_m\ldots M_i)_{kl} = \begin{cases} \delta_{kl} & k < i, \\ \frac{\partial\zeta_k(\zeta_1,\ldots\zeta_{i-1},\eta_i\ldots\eta_m)}{\partial\zeta_l} & k \geq i, l < i, \\ \frac{\partial\zeta_k(\zeta_1,\ldots\zeta_{i-1},\eta_i\ldots\eta_m)}{\partial\eta_l} & k \geq i, l \geq i. \end{cases} \tag{5.18}$$

Initial condition: Comparison of (5.18) with (5.17) shows that (5.18) holds for $i = m$. $M_m\ldots M_i = (M_m\ldots M_{i+1})M_i$ provides the induction from $i+1$ to i

$$(M_m\ldots M_i)_{kl} = \begin{cases} (M_i)_{kl} & k \leq i, \\ \frac{\partial\zeta_k(\ldots\zeta_i,\eta_{i+1}\ldots)}{\partial\zeta_l} + \frac{\partial\zeta_k(\ldots\zeta_i,\eta_{i+1}\ldots)}{\partial\zeta_i}\frac{\partial\zeta_i(\ldots\zeta_{i-1},\eta_i\ldots)}{\partial\zeta_l} & k > i, l < i, \\ \frac{\partial\zeta_k(\ldots\zeta_i,\eta_{i+1}\ldots)}{\partial\zeta_i}\frac{\partial\zeta_i(\ldots\zeta_{i-1},\eta_i\ldots)}{\partial\eta_i} & k > i, l = i, \\ \frac{\partial\zeta_k(\ldots\zeta_i,\eta_{i+1}\ldots)}{\partial\eta_l} + \frac{\partial\zeta_k(\ldots\zeta_i,\eta_{i+1}\ldots)}{\partial\zeta_i}\frac{\partial\zeta_i(\ldots\zeta_{i-1},\eta_i\ldots)}{\partial\eta_l} & k > i, l > i. \end{cases} \tag{5.19}$$

ζ_k in (5.19) for $k > i$ depends explicitly on $\zeta_1,\ldots\zeta_{i-1},\eta_{i+1},\ldots\eta_m$ and implicitly on these variables and η_i via $\zeta_i(\ldots\zeta_{i-1},\eta_i,\ldots)$. Thus, these expressions yield $(M_m\ldots M_i)_{kl}$ as given in (5.18). Thus by induction it holds for $i = 1$, too, which proves (5.16).

The multiplication theorem for determinants reads $\det(UV) = \det(U)\det(V)$. There is a corresponding multiplication theorem for Pfaffian forms. Substitute $\xi_k = \sum_l j_{k,l}\eta_l$ in (5.2). Then (5.2) and (5.13) yield

$$\int d\xi_r\ldots d\xi_2 d\xi_1 \exp(\tfrac{1}{2}\sum_{k,l}\xi_k a_{k,l}\xi_l) = pf(a)$$

$$= \int d\eta_r\ldots d\eta_2 d\eta_1 \exp(\tfrac{1}{2}\sum_{k,l}\eta_k({}^t jaj)_{k,l}\eta_l)/\det(j)$$

$$= pf({}^t jaj)/\det(j), \tag{5.20}$$

---

[1]The sequence of substitutions can be singular, even if $\det(\partial\zeta/\partial\eta) \neq 0$. Then one may use an infinitesimally close non-singular transformation and finally perform the limit. For the transformation $\zeta_1 = \eta_2, \zeta_2 = \eta_1$ one may use $\zeta_1 = \eta_2 + c\eta_1, \zeta_2 = \eta_1$ with $c \to 0$. We leave it to the reader to calculate D_1 and D_2 and consider the product.

where $^t j$ denotes the transposed of j. Hence

$$\mathrm{pf}(^t jaj) = \mathrm{pf}(a)\det(j). \tag{5.21}$$

Note that for an antisymmetric matrix a also $^t jaj$ is antisymmetric.

5.3 Gauss Integrals III, Pfaffian Form and Determinant

The square of a Pfaffian is the determinant of the antisymmetric matrix.
Now we derive the connection between the Pfaffian form and the determinant. We form the square of the Gauss integral (5.2)

$$I_-^2 = \int d\zeta_n \ldots d\zeta_2 d\zeta_1 d\xi_n \ldots d\xi_2 d\xi_1 \exp(\tfrac{1}{2}\sum_{k,l=1}^{n}(\xi_k a_{k,l}\xi_l + \zeta_k a_{k,l}\zeta_l)) \tag{5.22}$$

and perform the transformation

$$\xi_k = \frac{1}{\sqrt{2}}(\eta_k + \eta_k'), \quad \zeta_k = \frac{i}{\sqrt{2}}(\eta_k - \eta_k'). \tag{5.23}$$

One obtains with $\partial(\xi,\zeta)/\partial(\eta',\eta) = i$ and $\xi_k\xi_l + \zeta_k\zeta_l = \eta_k'\eta_l + \eta_k\eta_l'$

$$I_-^2 = i^{-n}(-)^{n(n-1)/2}\int \exp(-\sum_{k,l=1}^{n}\eta_k' a_{k,l}\eta_l)\prod_k^n(d\eta_k' d\eta_k) \tag{5.24}$$

The factor $(-)^{n(n-1)/2}$ comes from the rearrangement of the factors $d\zeta$ and $d\zeta'$. The prefactors yield in total $(-i)^{n^2}$, which for even n gives $+1$ and for odd n $-i$. I_- vanishes for odd n, so that this factor does not play a role. Thus, the left hand side of (5.24) yields $\mathrm{pf}^2(a)$ due to (5.4) and the right hand side $\det(a)$ due to (3.27). Hence

$$\mathrm{pf}^2(a) = \det(a). \tag{5.25}$$

Consider again the integral (3.23),

$$I_- = \int \prod_{k=1}^{r}(d\eta_k d\xi_k)\exp(\sum_{k,l=1}^{r,r}\xi_k a_{k,l}\eta_l). \tag{5.26}$$

The substitution $\zeta_k = \sum_l a_{k,l} \eta_l$ yields

$$I_- = \int \prod_{k=1}^{r} (\mathrm{d}\,\zeta_k \mathrm{d}\,\xi_k) \exp(\sum_k^{r} \xi_k \zeta_k) \det(a), \qquad (5.27)$$

where the ζ- and ξ-integrations are now trivial. One obtains the determinant.

We saw that the integral (5.4) led to the Pfaffian form and thus, to the square root of the determinant. This has its counterpart in the integrals over real variables

$$I_+ = \int_{-\infty}^{+\infty} \mathrm{d}\,x_r \ldots \mathrm{d}\,x_1 \exp(-\tfrac{1}{2} \sum_{k,l=1}^{r,r} x_k a_{k,l} x_l) = \frac{(2\pi)^{r/2}}{\sqrt{\det(a)}}. \qquad (5.28)$$

Similarly Eqs. (5.6) to (5.8) have their counterparts in

$$I_+(b) = \int_{-\infty}^{+\infty} \mathrm{d}\,x_r \ldots \mathrm{d}\,x_1 \exp(-\tfrac{1}{2}\,{}^{t}xax + {}^{t}xb)$$

$$= \frac{(2\pi)^{r/2}}{\sqrt{\det(a)}} \exp(\tfrac{1}{2}\,{}^{t}ba^{-1}b), \quad b_i \in \mathscr{A}_0, \qquad (5.29)$$

where the matrix a is symmetric and the real parts of its eigenvalues have to be positive. Then the density

$$\mathrm{d}\rho = \frac{\sqrt{\det(a)}}{(2\pi)^{r/2}} \mathrm{d}\,x_r \ldots \mathrm{d}\,x_1 \exp\left(-\tfrac{1}{2}\,{}^{t}xax\right) \qquad (5.30)$$

yields

$$\langle x_k x_l \rangle = (a^{-1})_{kl}, \quad \langle \exp({}^{t}xb) \rangle = \exp\left(\tfrac{1}{2}\,{}^{t}b\langle x\,{}^{t}x\rangle b\right). \qquad (5.31)$$

Problems

5.1 Show that the determinant of an antisymmetric matrix of odd dimension vanishes.

5.2 Check Eq. (5.21) for a 2×2 matrix explicitly.

5.3 Substitute

$$\zeta_1 = \alpha + \beta \eta_1 \eta_2 + a_{11} \eta_1 + a_{12} \eta_2, \quad \zeta_2 = \gamma + \delta \eta_1 \eta_2 + a_{21} \eta_1 + a_{22} \eta_2,$$

where Greek quantities are odd elements and Latin ones are even elements of the algebra. Determine $\partial(\zeta_1, \zeta_2)/\partial(\eta_1, \eta_2)$ and its inverse. Express the integrals over ζ_1 and ζ_2 (i) of a constant, (ii) of ζ_1 and ζ_2 and (iii) of $\zeta_1\zeta_2$ in terms of integrals over the ηs, and show that they yield the correct results.

Chapter 6
The Complex Conjugate

Abstract Two types of conjugation are introduced. They show a similar behavior to the antilinear operations of hermitian conjugation and of time-reversal.

6.1 Description

The properties of the two types of conjugation are given.
  We introduce the operation of conjugation. It turns out that there are two such operations, which are called the *conjugate of first and second kind*.

The Conjugate of First Kind It is denoted by * and obeys

$$1^* = 1, \quad a \in \mathscr{A}_i \to a^* \in \mathscr{A}_i, \quad (a+b)^* = a^* + b^*. \tag{6.1}$$

These three conditions hold for the conjugate of second kind, too. In addition, we must define what the application of conjugation yields twice and how the conjugate of a product is related to the product of the conjugate. For the conjugate of first kind, the requirement is

$$a^{**} = a, \quad (ab)^* = b^* a^*. \tag{6.2}$$

The Conjugate of Second Kind It is denoted by $^\times$ and, as in (6.1), one requires

$$1^\times = 1, \quad a \in \mathscr{A}_i \to a^\times \in \mathscr{A}_i, \quad (a+b)^\times = a^\times + b^\times. \tag{6.3}$$

However, instead of (6.2) one requires

$$a^{\times\times} = \mathscr{P}(a), \text{ thus } a^{\times\times} = (-)^\nu a \text{ for } a \in \mathscr{A}_\nu, \tag{6.4}$$

$$(ab)^\times = a^\times b^\times. \tag{6.5}$$

The two types of conjugation are described in [224].

© Springer-Verlag Berlin Heidelberg 2016
F. Wegner, *Supermathematics and its Applications in Statistical Physics*,
Lecture Notes in Physics 920, DOI 10.1007/978-3-662-49170-6_6

6.2   Similarity to Antilinear Operations in Quantum Mechanics

These two types of conjugation are compared with the two types of antilinear operations in quantum mechanics.

At first glance one may be surprised to learn that there are two types of antilinear mappings. In fact, there are two different antilinear mappings in quantum mechanics. One is the transformation to the hermitian adjoint and corresponds to the conjugate of first kind. For this operation the laws in (6.2) apply. In contrast, the antilinear operation of time reversal corresponds to the conjugate of second kind. Upon this operation, annihilation or creation operators transform into their time reversed operator without changing their sequence (6.5). The creation operator of a fermion (or a one-fermion state) transforms under double time reversal into its negative in agreement with (6.4). Thus, both operations of conjugation can appear simultaneously.

We note that for complex numbers the usual calculational rules hold and thus, both types of conjugation yield the same results.

For the conjugate of first kind there is always a decomposition in real and imaginary parts possible, since

$$(a \pm a^*)^* = \pm(a \pm a^*). \tag{6.6}$$

This applies also for the conjugate of second kind provided $a \in \mathscr{A}_0$. It is not possible for the conjugate of second kind, if $a \in \mathscr{A}_1$. There is no state of odd fermion number, which is time reversal invariant. Instead the time reversal state of odd fermion number is perpendicular to the original one.

We realize that for the Gauss integrals over the Grassmann variables ξ and η in Sect. 3.3 we can replace ξ by η^*. In this way the analogy between the integrals over Grassmann variables and over complex variables becomes more obvious. We have to assume that all variables η^* and η are independent.

Problem

6.1   Show $(a^*b)^* = b^*a$. When does $(a^\times b)^\times = b^\times a$ hold?

Reference

[224] V. Rittenberg, M. Scheunert, Elementary construction of graded Lie groups. J. Math. Phys. **19**, 709 (1978)

Chapter 7
Path Integrals for Fermions and Bosons

Abstract Path integrals are introduced for both bosons and fermions. The different statistics of these two types of particles is apparent in the different signs of the commutation relations and in the fact that bosonic states can be arbitrarily often occupied, whereas fermionic states allow only single occupation or no occupation. Consequently, bosons are described by commuting fields, and fermions by anticommuting fields. This formulation can be used for renormalization group calculations and serves to derive the perturbative expansion in the interaction, and Feynman diagrams.

7.1 Coherent States

Coherent states are defined and their representation by means of even and odd elements of $\mathscr{A}$ is given.

The creation and annihilation operators of bosons are denoted by $b^\dagger$ and b, respectively, and those of fermions by $f^\dagger$ and f, resp. The commutator and anticommutator relations read

$$[b_i^\dagger, b_j^\dagger] = [b_i, b_j] = 0, [b_i, b_j^\dagger] = \delta_{i,j}, \tag{7.1}$$

$$\{f_i^\dagger, f_j^\dagger\} = \{f_i, f_j\} = 0, \{f_i, f_j^\dagger\} = \delta_{i,j}, \tag{7.2}$$

where the squared bracket indicates the commutator $[A, B] := AB - BA$ and the curly brace the anticommutator $\{A, B\} := AB + BA$.

In order to obtain the path integral representation, one introduces coherent states $|c\rangle$ und $|\gamma\rangle$ with $c \in \mathscr{A}_0$ and $\gamma \in \mathscr{A}_1$. The states are defined by

$$b|c\rangle = c|c\rangle, \quad f|\gamma\rangle = \gamma|\gamma\rangle. \tag{7.3}$$

The states read

$$|c\rangle = \exp(cb^\dagger)|0\rangle, \quad |\gamma\rangle = \exp(-\gamma f^\dagger)|0\rangle = |0\rangle - \gamma f^\dagger|0\rangle = |0\rangle + f^\dagger|0\rangle\gamma. \tag{7.4}$$

© Springer-Verlag Berlin Heidelberg 2016
F. Wegner, *Supermathematics and its Applications in Statistical Physics*,
Lecture Notes in Physics 920, DOI 10.1007/978-3-662-49170-6_7

The vacuum is denoted by $|0\rangle$, thus $b|0\rangle = 0$ and $f|0\rangle = 0$. In addition to the kets one introduces the bras as hermitean conjugates of the kets,

$$\langle c^*| = \langle 0| \exp(c^*b), \quad \langle \gamma^*| = \langle 0| \exp(-f\gamma^*) = \langle 0| - \langle 0|f\gamma^* = \langle 0| + \gamma^*\langle 0|f. \tag{7.5}$$

One considers γ and γ^* anticommuting with f and $f^\dagger$, $f\gamma^* = -\gamma^*f$. γ^* is the complex conjugate of γ. At present it is only important that $\gamma^* \in \mathscr{A}_1$ is independent of γ. Then the bra-kets read

$$\langle c^*|c'\rangle = \exp(c^*c'), \quad \langle \gamma^*|\gamma'\rangle = \langle 0|0\rangle + \langle 0|f\gamma^*\gamma'f^\dagger|0\rangle = 1 + \gamma^*\gamma' = \exp(\gamma^*\gamma'). \tag{7.6}$$

This yields the identity for completeness

$$\mathbf{1} = \int \frac{\mathrm{d}\,\Re c\,\mathrm{d}\,\Im c}{\pi} |c\rangle\langle c^*| \exp(-cc^*) = \sum_{n=0}^{\infty} b^{\dagger n}|0\rangle \frac{1}{n!} \langle 0|b^n, \tag{7.7}$$

$$\mathbf{1} = \int \mathrm{d}\gamma^*\mathrm{d}\gamma |\gamma\rangle\langle \gamma^*| \exp(-\gamma^*\gamma) = |0\rangle\langle 0| + f^\dagger|0\rangle\langle 0|f. \tag{7.8}$$

The trace of an operator, C, yields

$$\mathrm{tr}\,(C) = \int \frac{\mathrm{d}\,\Re c\,\mathrm{d}\,\Im c}{\pi} \langle c^*|C|c\rangle \exp(-cc^*),$$

$$\mathrm{tr}\,(C) = \int \mathrm{d}\gamma^*\mathrm{d}\gamma \langle \gamma^*|C| - \gamma\rangle \exp(-\gamma^*\gamma). \tag{7.9}$$

A priori one would not have expected the minus sign in $| - \gamma\rangle$, since one would have multiplied the operator C with the representation of $\mathbf{1}$ and then one would have brought the ket $|\gamma\rangle$ under the trace to the right. This procedure gives the correct result for bosons. This does not apply for fermions. The reason is that γ^*, γ, f and $f^\dagger$ are considered odd elements of the algebra and thus $\langle \gamma^*|$ and $|\gamma\rangle$ as even elements. Thus, we have exchanged them under the trace without further thought. However, one has to calculate $\langle 0|C|0\rangle + \langle 0|fCf^\dagger|0\rangle$, where f or $f^\dagger$ has to be commuted with the vacuum states without changing a sign, although f and $f^\dagger$ are odd elements. The best thing to do is to check the equation for the fermions explicitly (Problem 7.2).

Further, one obtains with the definitions (7.3) and (7.5), and Eq. (7.6)

$$\langle c^*|(b^\dagger)^k b^l|c'\rangle = (c^*)^k c'^l \exp(c^*c'),$$

$$\langle \gamma^*|(f^\dagger)^k f^l|\gamma'\rangle = (\gamma^*)^k \gamma'^l \exp(\gamma^*\gamma'). \tag{7.10}$$

Hence each normal ordered operator C can be written

$$\langle c^*|C(b^\dagger,b)|c'\rangle = \exp(c^*c')C(c^*,c'),$$
$$\langle \gamma^*|C(f^\dagger,f)|\gamma'\rangle = \exp(\gamma^*\gamma')C(\gamma^*,\gamma'). \tag{7.11}$$

Normal ordered means that the creation operators are left of the annihilation operators.[1]

7.2 Path Integral Representation

The grand canonical partition function and Green's functions are expressed in terms of path integrals for imaginary times τ.

We have performed the preparations for the path integral representation of the grand canonical (g.c.) partition function Z defined by

$$Z = \mathrm{tr}\ (\exp(-\beta H)) = \lim_{n\to\infty}\ \mathrm{tr}\left((1-\frac{\beta}{n}H)^n\right) \tag{7.12}$$

with the Hamilton operator H expressed by creation and annihilation operators in normal ordering. For simplicities' sake we include the term $-\mu N$ in H. Insertion of the identities (7.7), (7.8) yields

$$\int \prod_{l=1}^{n}\frac{\mathrm{d}\,\Re c_l\mathrm{d}\,\Im c_l}{\pi}\langle c_n^*|1-\beta H/n|c_{n-1}\rangle\langle c_{n-1}^*|1-\beta H/n|c_{n-2}\rangle \ldots$$
$$\langle c_1^*|1-\beta H/n|c_n\rangle e^{-c_n^*c_n-c_{n-1}^*c_{n-1}-\ldots-c_1^*c_1}, \tag{7.13}$$

$$\int \prod_{l=1}^{n}(\mathrm{d}\,\gamma_l^*\mathrm{d}\,\gamma_l)\langle \gamma_n^*|1-\beta H/n|\gamma_{n-1}\rangle\langle \gamma_{n-1}^*|1-\beta H/n|\gamma_{n-2}\rangle \ldots$$
$$\langle \gamma_1^*|1-\beta H/n|-\gamma_n\rangle e^{-\gamma_n^*\gamma_n-\gamma_{n-1}^*\gamma_{n-1}-\ldots-\gamma_1^*\gamma_1}. \tag{7.14}$$

Now we insert the matrix elements using (7.11)

$$\langle c_l^*|1-\beta H(b^\dagger,b)/n|c_{l-1}\rangle = (1-\beta H(c_l^*,c_{l-1})/n)e^{c_l^*c_{l-1}} \tag{7.15}$$
$$= \exp(-\beta H(c_l^*,c_{l-1})/n + c_l^*c_{l-1}) + O(n^{-2}),$$
$$\langle \gamma_l^*|1-\beta H(f^\dagger,f)/n|\gamma_{l-1}\rangle = (1-\beta H(\gamma_l^*,\gamma_{l-1})/n)e^{\gamma_l^*\gamma_{l-1}} \tag{7.16}$$
$$= \exp(-\beta H(\gamma_l^*,\gamma_{l-1})/n + \gamma_l^*\gamma_{l-1}) + O(n^{-2}).$$

---

[1] Normal ordering can be introduced generally for any $\mathrm{e}^{-\beta H^{(0)}}$, where $H^{(0)}$ is bilinear in the creation and annihilation operators, including ground state limits $\beta \to \infty$.

This yields the expressions for the g.c. partition function

$$Z_B = \lim_{n \to \infty} \int \prod_{l=1}^{n} \frac{d \Re c_l d \Im c_l}{\pi} \exp\left(\sum_{l=1}^{n} (c_l^*(c_{l-1} - c_l) - \frac{\beta}{n} H(c_l^*, c_{l-1}))\right),$$

(7.17)

$$Z_F = \lim_{n \to \infty} \int \prod_{l=1}^{n} (d\gamma_l^* d\gamma_l) \exp\left(\sum_{l=1}^{n} (\gamma_l^*(\gamma_{l-1} - \gamma_l) - \frac{\beta}{n} H(\gamma_l^*, \gamma_{l-1}))\right).$$

(7.18)

with the boundary conditions $c_0 = c_n$, $\gamma_0 = -\gamma_n$. The calculation was performed with one operator b and f, respectively. It can be performed in the same way for operators of several states.

The formulation (7.17), (7.18) represents the discretized form of $\exp(-\int_0^\beta d\tau H)$. One may add source terms to the Hamiltonian, such that

$$H_{B,s} = H(b^\dagger, b) - b^\dagger J(\tau) - J^*(\tau)b,$$
$$H_{F,s} = H(f^\dagger, f) - f^\dagger \eta(\tau) - \eta^*(\tau)f,$$

(7.19)

which allow one to derive correlations of creation and annihilation operators depending on τ. Then the g.c. partition function becomes a function of $J, J^* \in \mathscr{A}_+$ and $\eta, \eta^* \in \mathscr{A}_-$. The η and η^* are independent for different τ. The g.c. partition function reads

$$Z_B(J^*, J) = \lim_{n \to \infty} \int \prod_{l=1}^{n} \frac{d \Re c_l \Im c_l}{\pi}$$

(7.20)

$$\exp\left(\sum_{l=1}^{n} (c_l^*(c_{l-1} - c_l) - \frac{\beta}{n}(H(c_l^*, c_{l-1}) - c_l^* J(\beta l/n) - J^*(\beta l/n)c_l))\right),$$

$$Z_F(\eta^*, \eta) = \lim_{n \to \infty} \int \prod_{l=1}^{n} (d\gamma_l^* d\gamma_l)$$

(7.21)

$$\exp\left(\sum_{l=1}^{n} (\gamma_l^*(\gamma_{l-1} - \gamma_l) - \frac{\beta}{n}(H(\gamma_l^*, \gamma_{l-1}) - \gamma_l^* \eta(\beta l/n) - \eta^*(\beta l/n)\gamma_l))\right).$$

With $\tau = \beta l/n$, one writes the continuum limit $n \to \infty$ of (7.20), (7.21)

$$Z_B(J^*, J) = C \int [D\,\Re c\, D\,\Im c]\exp\left(\int_0^\beta d\tau(-S_B(\tau) + J^*(\tau)c(\tau) + c^*(\tau)J(\tau))\right),$$

$$S_B(\tau) = c^*(\tau+0)\frac{dc(\tau)}{d\tau} + H(c^*(\tau+0), c(\tau)), \tag{7.22}$$

$$Z_F(\eta^*, \eta) = C \int [D\gamma^* D\gamma]\exp\left(\int_0^\beta d\tau(-S_F(\tau) + \gamma^*(\tau)\eta(\tau) + \eta^*(\tau)\gamma(\tau))\right),$$

$$S_F(\tau) = \gamma^*(\tau+0)\frac{d\gamma(\tau)}{d\tau} + H(\gamma^*(\tau+0), \gamma(\tau)). \tag{7.23}$$

Equations (7.22), (7.23) constitute the path integral representation of the g.c. partition function. The quantum mechanical problem in d dimensions is transformed into a classical one in $d+1$ dimensions, where τ appears as an extra dimension. It stands for the imaginary time $i\tau$, where τ runs from 0 to β (compare the operators in (7.25), (7.27)). Summation is performed over all paths parametrized by $(c^*(\tau), c(\tau))$ and $(\gamma^*(\tau), \gamma(\tau))$. This formulation is the starting point for renormalization calculations. See for example the book by Salmhofer [227]. Some care has to be taken to determine the normalization indicated by the factor C. One way is to return to Eqs. (7.20), (7.21). Care has also to be taken that the τ of the creation operators is infinitesimally larger than those of the annihilation operators.

One defines correlations (called Green's functions) of the annihilation and creation operators at different times

$$G_B(\tau_1, \tau_2) = -\langle T_\tau \bar{b}(\tau_1)b(\tau_2)\rangle = \begin{cases} -\langle \bar{b}(\tau_1)b(\tau_2)\rangle & \tau_1 \ge \tau_2 \\ -\langle b(\tau_2)\bar{b}(\tau_1)\rangle & \tau_2 > \tau_1 \end{cases} \tag{7.24}$$

$$\bar{b}(\tau) = e^{\tau H}b^\dagger e^{-\tau H}, b(\tau) = e^{\tau H}be^{-\tau H} \tag{7.25}$$

$$G_F(\tau_1, \tau_2) = \langle T_\tau \bar{f}(\tau_1)f(\tau_2)\rangle = \begin{cases} \langle \bar{f}(\tau_1)f(\tau_2)\rangle & \tau_1 \ge \tau_2 \\ -\langle f(\tau_2)\bar{f}(\tau_1)\rangle & \tau_2 > \tau_1 \end{cases} \tag{7.26}$$

$$\bar{f}(\tau) = e^{\tau H}f^\dagger e^{-\tau H}, f(\tau) = e^{\tau H}fe^{-\tau H}. \tag{7.27}$$

The τ-ordering operator T_τ orders all contributions with increasing τ from right to left. If fermionic operators are reordered, then the expression has to be multiplied by $(-)^P$, which indicates the parity of the permutation P.

The functions $\bar{b}(\tau)$ and $\bar{f}(\tau)$ are not the hermitian adjoints to $b(\tau)$ and $f(\tau)$. Using

$$\langle \bar{b}(\tau_1)b(\tau_2)\rangle = Z_B^{-1}\,\mathrm{tr}\,(e^{(\tau_1-\tau_2-\beta)H}b^\dagger e^{(\tau_2-\tau_1)H}b), \quad \tau_1 \ge \tau_2 \tag{7.28}$$

and similarly for the other expectation values in (7.24)–(7.27) one shows that the Green's functions depend only on the time difference

$$G_{B,F}(\tau_1, \tau_2) = G_{B,F}(\tau_2 - \tau_1). \tag{7.29}$$

The cyclic invariance under the trace yields

$$G_B(\tau - \beta) = G_B(\tau), \quad G_F(\tau - \beta) = -G_F(\tau), \quad 0 < \tau < \beta. \tag{7.30}$$

The Green's functions are discontinuous at $\tau = 0$,

$$G_B(-0) = -\langle b^\dagger b \rangle, \quad G_B(+0) = -\langle bb^\dagger \rangle, \quad G_B(-0) - G_B(+0) = 1, \quad (7.31)$$
$$G_F(-0) = \langle f^\dagger f \rangle, \quad G_F(+0) = -\langle ff^\dagger \rangle, \quad G_F(-0) - G_F(+0) = 1. \quad (7.32)$$

Correlations are obtained by functional derivatives of Z. The functional derivative is defined by

$$\delta Z_B = \int d\tau \frac{\delta Z_B}{\delta J(\tau)} \delta J(\tau) = \lim_{n \to \infty} \sum_l \frac{\beta}{n} \frac{\partial Z_B}{\partial J(\beta l/n)} \delta J(\beta l/n), \tag{7.33}$$

$$\delta Z_F = \int d\tau (\delta Z_F/\delta \eta(\tau)) \delta \eta(\tau) = \lim_{n \to \infty} \sum_l \frac{\beta}{n} \partial Z_F/\partial \eta(\beta l/n) \delta \eta(\beta l/n). \tag{7.34}$$

Similar terms have to be added in the middle and on the r.h.s. of these equations for the derivative with respect to J^* and η^*. The second derivative is introduced in a similar way. Then one obtains the functional derivatives

$$\frac{\delta Z_B}{\delta J(\tau)}\bigg|_{J=0} = \text{tr}\left(\exp(-\beta H)\bar{b}(\tau)\right) = \langle \bar{b}(\tau) \rangle Z_B, \tag{7.35}$$

$$-\frac{1}{Z_B} \frac{\delta^2 Z_B}{\delta J(\tau_1)\delta J^*(\tau_2)}\bigg|_{J=0} = G_B(\tau_1, \tau_2), \tag{7.36}$$

$$\delta Z_F/\delta \eta(\tau)\big|_{\eta=0} = \text{tr}\left(\exp(-\beta H)\bar{f}(\tau)\right) = \langle \bar{f}(\tau) \rangle Z_F, \tag{7.37}$$

$$-\frac{1}{Z_F} \delta^2 Z_F/\delta \eta(\tau_1)\delta \eta^*(\tau_2)\big|_{\eta=0} = G_F(\tau_1, \tau_2). \tag{7.38}$$

The fermionic sources are multiplied by Grassmann variables. If we would have used complex variables instead, then already the second derivatives would vanish.

The Green's functions (7.36), (7.38) depend on imaginary times $t = i\tau$.

7.3 Free Particles

Green's functions for free bosons and fermions are calculated as a function of imaginary time and as a function of the Matsubara frequencies.

7.3.1 Starting from Functions of τ

The g.c. partition function for a free boson and fermion with Hamiltonian $H^{(0)} = (\epsilon - \mu) b^\dagger b$ and $H^{(0)} = (\epsilon - \mu) f^\dagger f$, resp., is expressed by the determinant of the $n \times n$ matrix $A_{\mathrm{B,F}}$ using (7.17) and (7.18)

$$A_{\mathrm{B,F}} = \begin{pmatrix} 1 & -a & 0 & \cdots & \\ 0 & 1 & -a & 0 & \cdots \\ \cdots & & & & \\ & \cdots & 0 & 1 & -a \\ \mp a & 0 & \cdots & 0 & 1 \end{pmatrix} = 1 \mp a^n, \tag{7.39}$$

$$Z_{\mathrm{B}}^{(0)} = \frac{1}{\det(A_{\mathrm{B}})} = \frac{1}{1 - e^{-\beta(\epsilon - \mu)}},$$

$$Z_{\mathrm{F}}^{(0)} = \det(A_{\mathrm{F}}) = 1 + e^{-\beta(\epsilon - \mu)}. \tag{7.40}$$

with $a = 1 - \beta(\epsilon - \mu)/n$. The only contributions are obtained as products from the **main diagonal** and from the upper *side diagonal* including the element in the *lower left corner*. This later product carries an extra factor $(-)^{n-1}$.

Consider the Green's functions for $\tau_1 = l_1 \beta / n$, $\tau_2 = l_2 / n$. They are determined as the matrix elements $(A^{-1})_{l_2, l_1}$, which is $(-)^{l_1 - l_2} \det(A^{(l_2, l_1)}) / \det(A)$ where $A^{(l_2, l_1)}$ is the matrix A, where column l_1 and row l_2 are deleted. Examples of such matrices for $n = 7$ are

$$A^{(5,3)} = \begin{pmatrix} 1 & -a & 0 & 0 & 0 & 0 \\ 0 & 1 & 0 & 0 & 0 & 0 \\ 0 & 0 & -a & 0 & 0 & 0 \\ 0 & 0 & 1 & -a & 0 & 0 \\ 0 & 0 & 0 & 0 & 1 & -a \\ \mp a & 0 & 0 & 0 & 0 & 1 \end{pmatrix} \quad A^{(3,5)} = \begin{pmatrix} 1 & -a & 0 & 0 & 0 & 0 \\ 0 & 1 & -a & 0 & 0 & 0 \\ 0 & 0 & 0 & 1 & 0 & 0 \\ 0 & 0 & 0 & 0 & -a & 0 \\ 0 & 0 & 0 & 0 & 1 & -a \\ \mp a & 0 & 0 & 0 & 0 & 1 \end{pmatrix}. \tag{7.41}$$

In general, one obtains

$$\det(A^{(l_2, l_1)}) = \begin{array}{ll} (-a)^{l_2 - l_1} & l_2 \geq l_1, \\ \pm (-)^{l_2 - l_1} a^{n + l_2 - l_1} & l_2 < l_1. \end{array} \tag{7.42}$$

Thus, in the limit $n \to \infty$ one obtains the Green's functions for a free particle

$$G_{B,F}^{(0)}(\tau) = \begin{cases} \frac{-e^{-\tau(\epsilon-\mu)}}{1 \mp e^{-\beta(\epsilon-\mu)}} & \tau < 0 \\ \frac{\mp e^{-\tau(\epsilon-\mu)}}{e^{\beta(\epsilon-\mu)} \mp 1} & \tau > 0. \end{cases} \tag{7.43}$$

Here, and in the following, upper signs refer to bosons, lower signs to fermions. The g.c. partition function reads with sources

$$Z_B^{(0)}(J^*, J) = \frac{1}{1 - e^{-\beta(\epsilon-\mu)}} \exp\left(-\int d\tau_1 d\tau_2 J^*(\tau_2) G_B^{(0)}(\tau_2 - \tau_1) J(\tau_1)\right), \tag{7.44}$$

$$Z_F^{(0)}(\eta^*, \eta) = (1 + e^{-\beta(\epsilon-\mu)}) \exp\left(-\int d\tau_1 d\tau_2 \eta^*(\tau_2) G_F^{(0)}(\tau_2 - \tau_1) \eta(\tau_1)\right). \tag{7.45}$$

7.3.2   Matsubara Frequencies

Often the Fourier transformation is introduced instead of the functions of τ,

$$c(\tau) = \sqrt{T} \sum_k e^{-i\omega_k \tau} \hat{c}(\omega_k), \quad \text{similarly for } J, b, \gamma, \eta, f \tag{7.46}$$

$$c^*(\tau) = \sqrt{T} \sum_k e^{i\omega_k \tau} \hat{c}^*(\omega_k), \quad \text{similarly for } J^*, \bar{b}, \gamma^*, \eta^*, \bar{f}. \tag{7.47}$$

We put $k_B = 1$. Since $c(\beta) = c(0)$ for bosons and $\gamma(\beta) = -\gamma(0)$ for fermions is required, one obtains $e^{i\omega_k \beta} = \pm 1$. Thus the ω_k are given by

$$\omega_k = \begin{cases} 2k\pi T & \text{for bosons} \\ (2k + 1)\pi T & \text{for fermions} \end{cases} \quad k \in \mathbb{Z}. \tag{7.48}$$

These frequencies are called Matsubara frequencies.

The inverse transformations of (7.46), (7.47) are

$$\hat{c}(\omega_k) = \sqrt{T} \int_0^\beta d\tau e^{i\omega_k \tau} c(\tau), \quad \text{similarly for } \hat{J}, \hat{b}, \hat{\eta}, \hat{\gamma}, \hat{f}, \tag{7.49}$$

$$\hat{c}^*(\omega_k) = \sqrt{T} \int_0^\beta d\tau e^{-i\omega_k \tau} c^*(\tau), \quad \text{similarly for } \hat{J}^*\hat{\bar{b}}, \hat{\eta}^*, \hat{\gamma}^*, \hat{\bar{f}}. \tag{7.50}$$

The g.c. partition functions $Z_{B,F}^{(0)}$ obtained from (7.20), (7.21) read, in Fourier components,

$$Z_B^{(0)}(\hat{J}^*, \hat{J}) = \lim_{n\to\infty} C \prod_{k=[n/2]-n+1}^{[n/2]} \frac{d\,\Re\hat{c}(\omega_k)d\,\Im\hat{c}(\omega_k)}{\pi}$$

$$\times \exp\left(-p_k\hat{c}^*(\omega_k)\hat{c}(\omega_k) + \hat{J}^*(\omega_k)\hat{c}(\omega_k) + \hat{c}^*(\omega_k)\hat{J}(\omega_k)\right),$$

(7.51)

$$Z_F^{(0)}(\hat{\eta}^*, \hat{\eta}) = \lim_{n\to\infty} C^{-1} \prod_{k=[n/2]-n+1}^{[n/2]} d\,\hat{\gamma}^*(\omega_k)d\,\hat{\gamma}(\omega_k)$$

$$\times \exp\left(-p_k\hat{\gamma}^*(\omega_k)\hat{\gamma}(\omega_k) + \hat{\eta}^*(\omega_k)\hat{\gamma}(\omega_k) + \hat{\gamma}^*(\omega_k)\hat{\eta}(\omega_k)\right).$$

(7.52)

$[n/2]$ indicates the largest integer less than or equal to $n/2$. The product runs from $[n/2] - n + 1$ to $[n/2]$ to make sure that in the continuum limit both positive and negative k are included. C is the Jacobian for the Fourier transform of the components from τ to ω,

$$C = (nT)^n, \quad p_k = \begin{cases} nT(1 - e^{2k\pi i/n}a) & \text{for bosons,} \\ nT(1 - e^{(2k+1)\pi i/n}a) & \text{for fermions.} \end{cases}$$

(7.53)

Thus, one obtains

$$Z_B^{(0)}(\hat{J}^*, \hat{J}) = \lim \prod_k \left(\frac{nT}{p_k}\right) \exp\left(\sum_k \frac{\hat{J}^*(\omega_k)\hat{J}(\omega_k)}{p_k}\right),$$

(7.54)

$$Z_F^{(0)}(\hat{\eta}^*, \hat{\eta}) = \lim \prod_k \left(\frac{p_k}{nT}\right) \exp\left(\sum_k \frac{\hat{\eta}^*(\omega_k)\hat{\eta}(\omega_k)}{p_k}\right)$$

(7.55)

with

$$\prod_k \left(\frac{p_k}{nT}\right) = \begin{cases} \prod_k(1 - e^{2k\pi i/n}a) = 1 - a^n & \text{for bosons,} \\ \prod_k(1 - e^{(2k+1)\pi i/n}a) = 1 + a^n & \text{for fermioms.} \end{cases}$$

(7.56)

The limit $n \to \infty$ yields

$$p_k = -i\omega_k + \epsilon - \mu,$$

(7.57)

$$Z_B^{(0)}(\hat{J}^*, \hat{J}) = \frac{1}{1 - e^{-\beta(\epsilon-\mu)}} \exp\left(-\sum_k \hat{J}^*(\omega_k)\hat{G}_B^{(0)}(\omega_k)\hat{J}(\omega_k)\right),$$

(7.58)

$$\hat{G}_B^{(0)}(\omega_k) = \frac{1}{i\omega_k - \epsilon + \mu} = \langle T_\tau \hat{\bar{b}}(\omega_k)\hat{b}(\omega_k)\rangle, \tag{7.59}$$

$$Z_F^{(0)}(\hat{\eta}^*, \hat{\eta}) = (1 + e^{-\beta(\epsilon-\mu)})\exp\left(-\sum_k \hat{\eta}^*(\omega_k)G_F^{(0)}(\omega_k)\hat{\eta}(\omega_k)\right), \tag{7.60}$$

$$\hat{G}_F^{(0)}(\omega_k) = \frac{1}{i\omega_k - \epsilon + \mu} = \langle T_\tau \hat{\bar{f}}(\omega_k)\hat{f}(\omega_k)\rangle. \tag{7.61}$$

Generalizing to several bosonic and fermionic degrees of freedom is straightforward. The g.c. partition function for

$$H^{(0)} = \sum_{p,s}(\epsilon_{p,s} - \mu)\begin{cases} b_{p,s}^\dagger b_{p,s} \\ f_{p,s}^\dagger f_{p,s} \end{cases} \tag{7.62}$$

is obtained by multiplying the corresponding g.c. partition functions for the states (p, s), where p may stand for momentum and s for the z-component of the spin. The g.c. potential $F^{(0)}$ of this system reads

$$F_{B,F}^{(0)} = -T\ln Z_{B,F}^{(0)}(0,0) = \pm T\sum_{p,s}\ln(1 \mp e^{-\beta(\epsilon_{p,s}-\mu)}). \tag{7.63}$$

and the free action is given by

$$\mathscr{S}_B^{(0)} = \int_0^\beta d\tau \sum_{p,s} c_{p,s}^*(\tau+0)(\frac{dc_{p,s}(\tau)}{d\tau} + (\epsilon_{p,s}-\mu)c_{p,s}(\tau))$$
$$= \sum_{p,s,k} \hat{c}_{p,s}^*(\omega_k)(-i\omega_k + \epsilon_{p,s} - \mu)\hat{c}_{p,s}(\omega_k), \tag{7.64}$$

$$\mathscr{S}_F^{(0)} = \int_0^\beta d\tau \sum_{p,s} \gamma_{p,s}^*(\tau+0)(\frac{d\gamma_{p,s}(\tau)}{d\tau} + (\epsilon_{p,s}-\mu)\gamma_{p,s}(\tau))$$
$$= \sum_{p,s,k} \hat{\gamma}_{p,s}^*(\omega_k)(-i\omega_k + \epsilon_{p,s} - \mu)\hat{\gamma}_{p,s}(\omega_k). \tag{7.65}$$

For large systems the sum over p can be replaced by an integral, since the density of the momenta p increases proportionally to the volume $\mathscr{V}$,

$$F_{B,F}^{(0)} = \pm T\mathscr{V}\int \frac{d^3p}{(2\pi)^3}\sum_s \ln(1 \mp e^{-\beta(\epsilon_{p,s}-\mu)}). \tag{7.66}$$

We have used that in the thermodynamic limit one replaces

$$\sum_q = \mathscr{V} \int \frac{\mathrm{d}^d q}{(2\pi)^d}. \tag{7.67}$$

Compare the paragraphs on *Fourier transform* and on *Thermodynamic limit and continuum limit* in Sect. 22.3, in particular Eqs. (22.41), (22.42), (22.50).

7.4 Interacting Systems and Feynman Diagrams

Shortly the basic idea of perturbation and diagrammatic expansion of systems with interaction is presented. The linked cluster theorem for the grand canonical potential and for correlations is derived. The Dyson-equation for the one-particle Green's function and the Bethe-Salpeter equation for two-particle Green's functions are given.

The presentation given in the previous sections is very intricate for free particles. The Green's functions (7.43), (7.59), (7.61) and the g.c. potential (7.66) can be derived in a simpler way. But their representation is very useful for interacting systems. One advantage is that the free particles are described by bilinear actions $\mathscr{S}^{(0)}$, (7.64), (7.65). Thus, expectation values are given in terms of Gaussians, which are easily expressed in terms of the free Green's functions. The other advantage is that time-ordering is automatically taken into account by the action $\mathscr{S}^{(0)}$.

We write the Hamiltonian of an interacting system

$$H_B = H^{(0)} + H^{\mathrm{int}}(b^\dagger, b), \quad H_F = H^{(0)} + H^{\mathrm{int}}(f^\dagger, f). \tag{7.68}$$

Then the g.c. partition function reads

$$Z_B = Z_B^{(0)} \left\langle \exp\left(-\int_0^\beta \mathrm{d}\tau H^{\mathrm{int}}(\bar{c}(\tau+0), c(\tau))\right)\right\rangle_0, \tag{7.69}$$

$$Z_F = Z_F^{(0)} \left\langle \exp\left(-\int_0^\beta \mathrm{d}\tau H^{\mathrm{int}}(\bar{\gamma}(\tau+0), \gamma(\tau))\right)\right\rangle_0, \tag{7.70}$$

where the expectation values $\langle \ldots \rangle_0$ are performed with respect to the free action $\mathscr{S}^{(0)}$, Eqs. (7.64), (7.65).

Expansion of the exponential yields a perturbation expansion in the coupling of the interaction. For example we consider a two-particle interaction

$$H^{\mathrm{int}} = \frac{1}{2} \sum_{p_1, p_2, q_1, q_2} V_{p_1, p_2, q_1, q_2} \times \begin{cases} b_{p_1}^\dagger b_{p_2}^\dagger b_{q_2} b_{q_1}, \\ f_{p_1}^\dagger f_{p_2}^\dagger f_{q_2} f_{q_1} \end{cases} \tag{7.71}$$

It translates into the action

$$\mathscr{S}^{\mathrm{int}} = \frac{1}{2} \sum_{p_1,p_2,q_1,q_2} \int_0^\beta \mathrm{d}\tau V_{p_1,p_2,q_1,q_2} \times \begin{cases} c_{p_1}^\dagger(\tau+0)c_{p_2}^\dagger(\tau+0)c_{q_2}(\tau)c_{q_1}(\tau), \\ \gamma_{p_1}^\dagger(\tau+0)\gamma_{p_2}^\dagger(\tau+0)\gamma_{q_2}(\tau)\gamma_{q_1}(\tau) \end{cases}$$

(7.72)

Instead of performing the integrals over τ one can use the representation by means of the Matsubara frequencies with

$$\int_0^\beta \mathrm{d}\tau c_{p_1}^*(\tau)c_{p_2}^*(\tau)c_{q_2}(\tau)c_{q_1}(\tau)$$

$$= T \sum_{k_1,k_2,l_1,l_2} \hat{c}_{p_1}^*(\omega_{k_1})\hat{c}_{p_2}^*(\omega_{k_2})\hat{c}_{q_2}(\omega_{l_2})\hat{c}_{q_1}(\omega_{l_1})\delta_{k_1+k_2,l_1+l_2},$$

(7.73)

and similarly for the fermions.

Feynman Diagrams The evaluation of the expectation values of products of $\mathscr{S}^{\mathrm{int}}$ requires the determination of expectation values of products of factors c^*, c, and γ^*, γ, respectively. Since $\mathscr{S}^{(0)}$ is bilinear, we can apply the expressions (3.39) and (3.46), resp. Thus, we have to decompose the products in all possible ways into pairs, to multiply the expectation values of these pairs of factors, which are the Green's functions $G^{(0)}$ of the free particles, (7.43), (7.59), (7.61), and to sum over all decompositions. For fermions the appropriate signs have to be taken into account.

To perform this expansion it is often useful to draw diagrams, called Feynman diagrams. One draws a vertex for each factor V of the interaction. Since the expectation value is given by the sum of the product of all factors $\langle \bar{b}b \rangle_0$ and $\langle \bar{f}f \rangle_0$, resp. one connects the vertices by lines representing these Green's functions, so that for the interaction (7.71) two lines start at each vertex for $\bar{b}$ ($\bar{f}$) and two lines end at each vertex for the factors b (f) in all possible ways. One writes down the corresponding factors V and $G^{(0)}$ and performs the integrals over the variables τ and the independent momenta p and q, which yields the contribution of the diagram. Symmetric diagrams are overcounted. They have to be divided by the number of symmetry operations, which map the diagrams on themselves (For details see textbooks.). The g.c. partition function Z is the product of $Z^{(0)}$ multiplied by 1 plus the sum of the contributions of all diagrams.

Some of the diagrams are connected, some are disconnected. The diagrams (a) to (c) in Fig. 7.1 are connected. There is a diagram in second order in the interaction consisting of two parts (a), which is thus disconnected.

Linked Cluster Theorem for the Grand Canonical Potential *The only diagrammatic contributions to the grand canonical potential are those of the connected diagrams.* We expand the g.c. partition function

$$Z = Z^{(0)}\langle \exp(-\mathscr{S}^{\mathrm{int}})\rangle_0 = Z^{(0)} \sum_{n=0}^\infty \frac{1}{n!}\langle(-\mathscr{S}^{\mathrm{int}})^n\rangle_0,$$

(7.74)

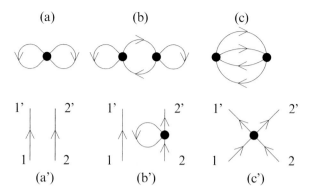

Fig. 7.1 Feynman diagrams. (**a**)–(**c**) are diagrams contributing to the g.c. potential in first and second order in the interaction; (**a'**)–(**c'**) are diagrams of a four point function in zeroth and first order of the interaction. There is a second diagram to (**a'**) with 1' and 2' exchanged. In diagram (**b'**) 1 and 2 and independently 1' and 2' can be exchanged. The *black dots* indicate interactions. The *lines* indicate the Green's functions $G^{(0)}$

The evaluation of $\langle(-\mathscr{S}^{\text{int}})^n\rangle_0$ yields various diagrams. Some of them are connected, that is all vertices V are connected by lines. We call the sum of these connected contributions

$$D_n = \frac{1}{n!}\langle(-\mathscr{S}^{\text{int}})^n\rangle_{\text{con}}. \tag{7.75}$$

Some diagrams of $\langle(-\mathscr{S}^{\text{int}})^n\rangle_0$ may consist of two connected diagrams with n_1 and n_2 factors $(-\mathscr{S}^{\text{int}})$, $n_1 + n_2 = n$. If $n_1 \neq n_2$, then there are $n!/(n_1!n_2!)$ ways to pick n_1 factors $(-\mathscr{S}^{\text{int}})$ out of the n factors. Thus they contribute

$$\frac{1}{n!}\frac{n!}{n_1!n_2!}\langle(-\mathscr{S}^{\text{int}})^{n_1}\rangle_{\text{con}}\langle(-\mathscr{S}^{\text{int}})^{n_2}\rangle_{\text{con}} = D_{n_1}D_{n_2}. \tag{7.76}$$

If $n_1 = n_2$, then by choosing n_1 factors $-\mathscr{S}^{\text{int}}$ also the other n_2 factors are a choice for the separation. Thus, in this case the contribution is $\frac{1}{2}D_{n_1}^2$. In general, we have decompositions in k_1 diagrams with n_1 factors $(-\mathscr{S}^{\text{int}})^{n_1}$, k_2 diagrams with n_2 factors $(-\mathscr{S}^{\text{int}})^{n_2}$, etc., then their contribution is

$$\frac{1}{n_1!^{k_1}k_1!n_2!^{k_2}k_2!\ldots}\langle(-\mathscr{S}^{\text{int}})^{n_1}\rangle_{\text{con}}^{k_1}\langle(-\mathscr{S}^{\text{int}})^{n_2}\rangle_{\text{con}}^{k_2}\cdots = \frac{1}{k_1!k_2!\ldots}D_{n_1}^{k_1}D_{n_2}^{k_2}\cdots \tag{7.77}$$

Summing all contributions, one obtains

$$Z = Z^{(0)}\exp(D_{\text{con}}), \quad D_{\text{con}} = \sum_{n=1}^{\infty}D_n. \tag{7.78}$$

The total g.c. potential is the g.c. potential of the free particles, $F^{(0)}$ minus temperature times the sum of all connected diagrams D_{con},

$$F = F^{(0)} - T \ln(Z/Z^{(0)}) = F^{(0)} - T D_{\text{con}}. \tag{7.79}$$

This is the linked cluster theorem for the g.c. potential.

Linked Cluster Theorem for Correlations *The correlation is obtained by summing only diagrams, where each part is connected to an external point.* In order to obtain the correlations we have to add the action $\mathscr{S}^{\text{source}}$ of the source terms to $\mathscr{S}^{\text{int}}$ and to replace $\mathscr{S}^{\text{int}}$ by $\mathscr{S}'$,

$$\mathscr{S}' = \mathscr{S}^{\text{int}} + \mathscr{S}^{\text{source}}, \tag{7.80}$$

$$\mathscr{S}_{\text{B}}^{\text{source}} = -\int_0^\beta d\tau \sum_{p,s} [J_{p,s}^*(\tau) c_{p,s}(\tau) + c_{p,s}^*(\tau) J_{p,s}(\tau)]$$

$$= -\sum_{p,s,k} [\hat{J}_{p,s}^*(\omega_k) \hat{c}_{p,s}(\omega_k) + \hat{c}_{p,s}^*(\omega_k) \hat{J}_{p,s}(\omega_k)], \tag{7.81}$$

$$\mathscr{S}_{\text{F}}^{\text{source}} = -\int_0^\beta d\tau \sum_{p,s} [\eta_{p,s}^*(\tau) \gamma_{p,s}(\tau) + \gamma_{p,s}^*(\tau) \eta_{p,s}(\tau)]$$

$$= -\sum_{p,s,k} [\hat{\eta}_{p,s}^*(\omega_k) \hat{\gamma}_{p,s}(\omega_k) + \hat{\gamma}_{p,s}^*(\omega_k) \hat{\eta}_{p,s}(\omega_k)], \tag{7.82}$$

The derivatives of the g.c. partition function with respect to the sources yield the g.c. partition function times expectation values. If the number of particles and translational invariance are conserved, then

$$\langle T_\tau \hat{b}_p(\omega) \hat{b}_q(\omega') \rangle = \frac{1}{Z_{\text{B}}} \frac{\partial^2 Z_{\text{B}}}{\partial J_{p,\omega} \partial J_{q,\omega'}^*} = -\delta_{p,q} \delta_{\omega,\omega'} \hat{G}_{\text{B}}^{(0)}(p, \omega) + \frac{\partial^2 D_{\text{con}}}{\partial J_{p,\omega} \partial J_{q,\omega'}^*}$$

$$= -\delta_{p,q} \delta_{\omega,\omega'} \hat{G}_{\text{B}}(p, \omega), \tag{7.83}$$

$$\langle T_\tau \hat{f}_p(\omega) \hat{f}_q(\omega') \rangle = \frac{1}{Z_{\text{F}}} \partial^2 Z_{\text{F}} / \partial \eta_p^*(\omega) \partial \eta_q(\omega')$$

$$= \delta_{p,q} \delta_{\omega,\omega'} \hat{G}_{\text{F}}^{(0)}(p, \omega) + \partial^2 D_{\text{con}} / \partial \eta_p^*(\omega) \partial \eta_q(\omega')$$

$$= \delta_{p,q} \delta_{\omega,\omega'} \hat{G}_{\text{F}}(p, \omega) \tag{7.84}$$

and

$$\hat{G}^{(2)}_{\mathrm{B,F}}(p_1,p_2,q_1,q_2;\omega_1,\omega_2,\omega_1',\omega_2')$$

$$:= \left.\begin{array}{c}\langle T_\tau \hat{\bar{b}}_{p_1}(\omega_1)\hat{\bar{b}}_{p_2}(\omega_2)\hat{b}_{q_2}(\omega_2')\hat{b}_{q_1}(\omega_1')\rangle \\ \langle T_\tau \hat{\bar{f}}_{p_1}(\omega_1)\hat{\bar{f}}_{p_2}(\omega_2)\hat{f}_{q_2}(\omega_2')\hat{f}_{q_1}(\omega_1')\rangle\end{array}\right\}$$

$$= \delta_{p_1,q_1}\delta_{\omega_1,\omega_1'}\hat{G}_{\mathrm{B,F}}(p_1,\omega_1)\delta_{p_2,q_2}\delta_{\omega_2,\omega_2'}\hat{G}_{\mathrm{B,F}}(p_2,\omega_2)$$

$$\pm\delta_{p_1,q_2}\delta_{\omega_1,\omega_2'}\hat{G}_{\mathrm{B,F}}(p_1,\omega_1)\delta_{p_2,q_1}\delta_{\omega_2,\omega_1'}\hat{G}_{\mathrm{B,F}}(p_2,\omega_2)$$

$$+\hat{C}_{\mathrm{B,F}}(p_1,p_2,q_1,q_2;\omega_1,\omega_2,\omega_1',\omega_2'). \tag{7.85}$$

The last term contains the cumulant $\hat{C}$ of the four-point function. It is obtained by taking the fourth derivative of D_{con} with respect to the four sources. The cumulant $\hat{C}$ of the four-point function has four external one-particle Green's functions. Taking these *legs* away one retains what is called the vertex $\hat{\Gamma}$ of the four-point function. Since the legs are taken away these vertices are also called amputated. Translational invariance in space and time yields δ-functions in momenta and frequencies. Then the cumulant reads

$$\hat{C}_{\mathrm{B,F}}(p_1,p_2,q_1,q_2;\omega_1,\omega_2,\omega_1',\omega_2')$$

$$= \delta_{p_1+p_2,q_1+q_2}\delta_{\omega_1+\omega_2,\omega_1'+\omega_2'}\hat{G}(p_1,\omega_1)\hat{G}(p_2,\omega_2)\hat{G}(q_1,\omega_1')\hat{G}(q_2,\omega_2')$$

$$\times\hat{\Gamma}_{\mathrm{B,F}}(p_1,p_2,q_1,q_2;\omega_1,\omega_2,\omega_1',\omega_2'). \tag{7.86}$$

Generally derivatives of Z, (7.79), are (apart from the contribution for free particles in $Z^{(0)}$) derivatives of D_{con} or products of such derivatives multiplied by the exponential $\exp(D_{\mathrm{con}})$, which cancels against the prefactor $1/Z$. This yields the linked cluster theorem for correlations.

A useful feature of Feynman diagrams is that they allow to present interaction processes in a pictorial way. They thus give a calculational procedure directly related to certain processes. Normally it is not possible to sum all diagrams. However, Feynman diagrams allow one to single out certain classes of diagrams, which can be summed, or to classify certain classes, which allow the derivation of certain useful relations. Of particular importance are the Dyson equation for the one-particle Green's function and the Bethe-Salpeter equation for two-particle Green's functions. They are shortly considered below.

Dyson Equation and Self-Energy Consider the diagrams of the one-particle Green's function, $\hat{G}(q,\omega)$. They can be cut into pieces by removing those propagators $\hat{G}^{(0)}(q,\omega)$, which decay the diagram into two pieces. What is left are the contributions of what is called the self-energy $\Sigma(q,\omega)$ (Fig. 7.2). This yields the

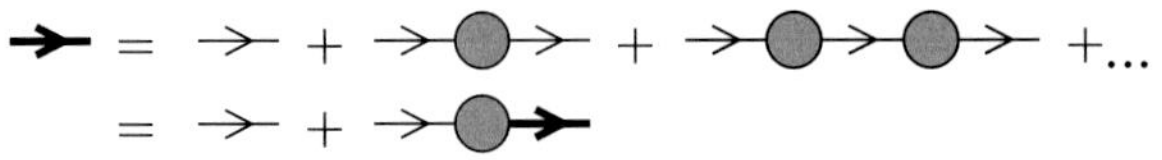

Fig. 7.2 Dyson equation. The full Green's function G is indicated by a *thick line*, $G^{(0)}$ is indicated by a *thin line*, the self-energy Σ by a *grey circle*

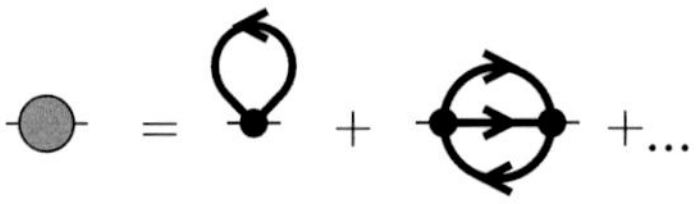

Fig. 7.3 Some contributions to the self-energy Σ. The self-energy itself is expressed by the full Green's function

Dyson equation,

$$
\begin{aligned}
\hat{G}(q,\omega) &= \hat{G}^{(0)} + \hat{G}^{(0)}\hat{\Sigma}\hat{G}^{(0)} + \hat{G}^{(0)}\hat{\Sigma}\hat{G}^{(0)}\hat{\Sigma}\hat{G}^{(0)} + \ldots \\
&= \hat{G}^{(0)}(q,\omega) + \hat{G}^{(0)}(q,\omega)\hat{\Sigma}(q,\omega)\hat{G}(q,\omega).
\end{aligned}
\tag{7.87}
$$

Division of this equation by $\hat{G}$ and $\hat{G}^{(0)}$ yields

$$
\hat{G}^{-1}(q,\omega) = \hat{G}^{(0)-1}(q,\omega) - \hat{\Sigma}(q,\omega)
\tag{7.88}
$$

and can be rewritten

$$
\hat{G}(q,\omega) = \frac{1}{i\omega - \epsilon_q + \mu - \hat{\Sigma}(q,\omega)}.
\tag{7.89}
$$

Thus the single-particle energy ϵ_q is corrected by the self-energy $\hat{\Sigma}(q,\omega)$ produced by the interaction. It is usually a much better approximation to use this Dyson-equation with an approximate self-energy than to determine the one-particle Green's function in some order of the interaction. We show two contributions to the self-energy in Fig. 7.3.

Bethe-Salpeter Equation Above we have introduced the self-energy as the sum of one-particle irreducible contributions to the one-particle Green's function. Similarly one introduces the two-particle irreducible vertex for the two-particle vertex. We rewrite the two-particle Green's function using (7.85), (7.86)

$$
\begin{aligned}
&\hat{G}^{(2)}(p_1 + q, p_2, p_1, p_2 + q; \omega_1 + \omega, \omega_2, \omega_1, \omega_2 + \omega) \\
&= \delta_{p_1,p_2}\delta_{\omega_1,\omega_2}\hat{G}(p_1 + q, \omega_1 + \omega)\hat{G}(p_1, \omega_1) \\
&\pm \delta_{q,0}\delta_{\omega,0}\hat{G}(p_1, \omega_1)\hat{G}(p_2, \omega_2) \\
&+ \hat{G}(p_1 + q, \omega_1 + \omega)\hat{G}(p_1, \omega_1)\hat{G}(p_2 + q, \omega_2 + \omega)\hat{G}(p_2, \omega_2) \\
&\times \hat{\Gamma}(p_1 + q, p_2, p_1, p_2 + q; \omega_1 + \omega, \omega_2, \omega_1, \omega_2 + \omega).
\end{aligned}
\tag{7.90}
$$

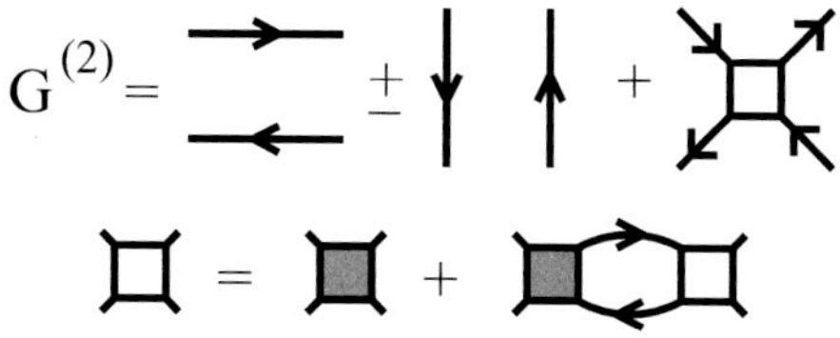

Fig. 7.4 Two-particle
Green's function and
Bethe-Salpeter equation. The
white square stands for the
vertex $\hat{\Gamma}$, the *grey square* for
the irreducible vertex $\hat{K}$

The vertex $\hat{\Gamma}$ can be expressed by the two-particle irreducible vertex $\hat{K}$

$$\hat{\Gamma}(p_1 + q, p_2, p_1, p_2 + q; \omega_1 + \omega, \omega_2, \omega_1, \omega_2 + \omega) \tag{7.91}$$
$$= \hat{K}(p_1 + q, p_2, p_1, p_2 + q; \omega_1 + \omega, \omega_2, \omega_1, \omega_2 + \omega)$$
$$+ \sum_{p_3, \omega_3} \hat{K}(p_1 + q, p_3, p_1, p_3 + q; \omega_1 + \omega, \omega_3, \omega_1, \omega_3 + \omega)$$
$$\times \hat{G}(p_3 + q, \omega_3 + \omega)\hat{G}(p_3, \omega_3)$$
$$\hat{\Gamma}(p_3 + q, p_2, p_3, p_2 + q; \omega_3 + \omega, \omega_2, \omega_3, \omega_2 + \omega).$$

These two equations are diagrammatically expressed in Fig. 7.4.

The simplest approximation is to insert the potential $\hat{V}_4$ for $\hat{K}$. Many systems conserve particle number, momentum, energy, and angular momentum. If one performs appropriate approximations, then these conservation laws are fulfilled. Such conserving approximations have been derived by Baym and Kadanoff [18, 128].

Here only two-particle interactions (7.71) were considered, which conserve the particle number. There are bosons like photons and phonons, which do not conserve the particle number. The same applies for a bose-condensate. In the case of superconductivity one considers terms generating or deleting pairs of electrons. One can easily generalize the path integrals to these cases.

Volume Dependence   We already found that translating the two-particle interaction from position space to momentum space gave rise to a factor $1/\mathcal{V}$. The same applies for the cumulants. We present the relation here. The commutators for operators b are written in momentum and position space as

$$[b_q, b_{q'}^\dagger] = \delta_{q,q'}, \quad [b(r), b^\dagger(r')] = \delta^d(r - r'). \tag{7.92}$$

This is consistent with the transformations

$$b^\dagger(r) = \frac{1}{\sqrt{\mathcal{V}}} \sum_q e^{iqr} b_q^\dagger, \quad b_q^\dagger = \frac{1}{\sqrt{\mathcal{V}}} \int d^d r \, e^{-iqr} b^\dagger(r). \tag{7.93}$$

in a periodicity volume $\mathcal{V}$. Then the interaction

$$H^{int} = \tfrac{1}{2} \int d^d r_1 d^d r_2 V_4(r_1 - r_2) b^\dagger(r_1) b^\dagger(r_2) b(r_1) b(r_2). \tag{7.94}$$

can be written

$$H^{\text{int}} = \frac{1}{2\mathcal{V}} \sum_{q_1,q_2,q} \hat{V}_q b^\dagger_{q_1+q} b^\dagger_{q_2-q} b_{q_2} b_{q_1} \tag{7.95}$$

with

$$\hat{V}_q = \int d^d r e^{iqr} V(r). \tag{7.96}$$

Consider now a connected diagram without external legs with n_k interactions with in total k creation and annihilation operators. In the example given above k was 4. Then the number n_G of Green's functions in the diagram is

$$n_G = \frac{1}{2} \sum_k k n_k. \tag{7.97}$$

The number n_{mom} of independent momenta is

$$n_{\text{mom}} = n_G - \sum_k n_k + 1. \tag{7.98}$$

Each vertex (interaction) carries a δ-function. Thus the sum over n_k is subtracted. Taking the conservation of the momenta in one interaction after the other into account, one finds that it is automatically fulfilled for the last one. Thus we have to add the one in (7.98). In the limit of large volume $\mathcal{V}$, the momenta become very dense and one replaces the sums over the momenta by integrals as in (7.66). Each interaction with k creation and annihilation operators contributes a power $1 - \frac{1}{2}k$ of the volume to the g.c. potential. Together with factors $\mathcal{V}$ from the integration over the momenta, one obtains the power

$$\sum_k (1 - \tfrac{1}{2}k) n_k + n_{\text{mom}} = 1, \tag{7.99}$$

indicating that the diagrammatic contributions to the g.c. potential are proportional to the volume (as expected).

We consider now connected diagrams with n_s external legs (source terms). Then the number of internal Green's functions decreases by $\frac{1}{2} n_s$. The number of constraints on the momenta remains $\sum_k n_k$. Consequently the power in $\mathcal{V}$ is $1 - \frac{1}{2} n_s$. The one-particle Green's function $\hat{G}$ does not depend on $\mathcal{V}$. However, the cumulant $\hat{C}$ is proportional to $1/\mathcal{V}$ in momentum space. This does not mean that it is negligible in the thermodynamic limit. See Problem 7.3.

Bibliographic Notes The Feynman diagrams [80] were first introduced for quantum electrodynamics. They were then generally used in particle physics and nuclear physics in a real time formalism for ground states, scattering processes, and excitations. Matsubara [177] showed that this technique can be also used in statistical physics at finite temperature using imaginary time τ. The real time Green's functions are uniquely determined by those along the imaginary axis, Baym and Mermin [19]. Keldysh [137] combined the real-time and imaginary-time formalism. See also the introduction to the Keldysh formalism by van Leeuwen et al. [257]. There are numerous textbooks on path integrals and Feynman diagrams. I mention only the books by Abrikosov, Gorkov, Dzyaloshinskii [2], Altland and Simons [7], Berezin [23, 24], Fetter and Walecka [79], Itzykson and Zuber [123], Kamenev [130], Kleinert [148], Negele and Orland [196], Popov [213], Salmhofer [227], and Zinn-Justin [297]. Candlin [44] and J.L. Martin [174, 175] were historically probably the first to use Grassmann variables for fermions.

Problems

7.1 Prove Eq. (7.9) for bosons with $C = e^{-kb^\dagger b}$.

7.2 Prove Eq. (7.9) for fermions with $C = a + cf^\dagger f$.

7.3 Express $\langle b^\dagger(r_1)b^\dagger(r_2)b(r_3)b(r_4)\rangle$ in terms of the Green's function $\hat{G}$ and the cumulant $\hat{C}$. Do these contributions depend on the volume $\mathcal{V}$?

References

[2] A.A. Abrikosov, L.P. Gorkov, I.E. Dzyaloshinski, *Methods of Quantum Field Theory in Statistical Physics* (Prentice-Hall, Englewood Cliffs, 1963)

[7] A. Altland, B. Simons, *Condensed Matter Field Theory* (Cambridge University Press, Cambridge, 2010)

[18] G. Baym, L.P. Kadanoff, Conservation laws and correlation functions. Phys. Rev. **124**, 287 (1961)

[19] G. Baym, N.D. Mermin, Determination of thermodynamic Green's functions. J. Math. Phys **2**, 232 (1961)

[23] F.A. Berezin, Canonical transformations in the representation of second quantization. Dok. Akad. Nauk SSSR **137**, 311 (1961)

[24] F.A. Berezin, *The Method of Second Quantization* (Academic Press, New York, 1966)

[44] D.J. Candlin, On sums over trajectories for systems with Fermi statistics. Nuovo Cim. **4**, 231 (1956)

[79] A.L. Fetter, J.D. Walecka, *Quantum Theory of Many-Particle Systems* (McGraw Hill, New York, 1971)

[80] R.P. Feynman, Space-time approach to quantum electrodynamics. Phys. Rev. **76**, 769 (1949)

[123] C. Itzykson, J.-B. Zuber, *Quantum Field Theory* (Mc-Graw Hill, New York, 1980)

[128] L.P. Kadanoff, G. Baym, *Quantum Statistical Mechanics* (Benjamin, New York, 1962)

[130] A. Kamenev, *Field Theory of Non-equilibrium Systems* (Cambridge University Press, Cambridge, 2011)

[137] L.V. Keldysh, Diagram technique for nonequilibrium processes. Zh. Eksp. Teor. Fiz. **47**, 1515 (1964); Sovj. Phys. JETP **20**, 1018 (1965)

[148] H. Kleinert, *Path Integrals in Quantum Mechanics, Statistics and Polymer Physics* (World Scientific, Singapore, 1990); *Pfadintegrale in der Quantenmechanik, Statistik und Polymerphysik* (BI Wissenschaftsverlag Mannheim, 1993)

[174] J.L. Martin, General classical dynamics, and the 'classical analogue' of a Fermi Oscillator. Proc. R. Soc. A **251**, 536 (1959)

[175] J.L. Martin, The Feynman principle for a Fermi system. Proc. R. Soc. A **251**, 543 (1959)

[177] T. Matsubara, A new approach to quantum-statistical mechanics. Prog. Theor. Phys. **14**, 351 (1955)

[196] J.W. Negele, H. Orland, *Quantum Many-Particle Systems*, 5th edn. (Westview Press, Reading, 1998)

[213] V.N. Popov, *Functional Integrals in Quantum Field Theory and Statistical Physics* (Reidel, Dordrecht, 1983)

[227] M. Salmhofer, *Renormalization: An Introduction*. Texts and Monographs in Physics (Springer, Berlin/Heidelberg, 1998)

[257] R. van Leeuwen, N.E. Dahlen, G. Stefanucci, C.-O. Almbladh, U. von Barth, Introduction to the Keldysh formalism, in *Time-Dependent Density Functional Theory* ed. by M.A.L. Marques et al. Lecture Notes in Physics, vol. 706 (Springer, Berlin, 2006), pp. 33–59

[297] J. Zinn-Justin, *Quantum Field Theory and Critical Phenomena* (Clarendon Press, Oxford, 1993)

Chapter 8
Dimers in Two Dimensions

Abstract Given a lattice, whose vertices are connected by edges, for example a square lattice, whose nearest neighbours are connected by edges in horizontal and vertical directions. The question is: In how many different ways can the edges be covered by dimers so that, at each lattice point, exactly one edge is covered by a dimer. This dimer problem can be solved for a plain lattice, if no edges cross each other. As examples, we consider the square lattice and the honeycomb lattice.

8.1 General Considerations

The number of configurations of dimers can be solved by Grassmann variables and be expressed in terms of a Pfaffian.

One starts out from a lattice, that is, from vertices (points) which are connected by edges (lines). A number of edges are covered by what is called a dimer in such a way, that each vertex is connected to a dimer. An example is shown in Fig. 8.1 for a small (4×4) square lattice. The dimer problem consists of the determination of the number of possible dimer configurations $\mathcal{N}$.

The solution can be found for example in [133, 134, 179, 251]. An interesting introduction is given in [140], and also [138, 139]. One introduces, at each lattice site x, a Grassmann variable ξ_x and the exponential function

$$\exp(\tfrac{1}{2} \sum_{x,x'} b_{x,x'} \xi_x \xi_{x'}), \quad b_{x,x'} = -b_{x',x}. \tag{8.1}$$

One puts $b_{x,x'} = \pm 1$, if x and x' are connected by an edge, otherwise $b_{x,x'} = 0$. Then integration of the exponential function over all ξs yields a contribution ± 1 from each dimer configuration. The problem is to choose the signs of the bs so that all contributions are positive. If this is possible, then the problem is reduced to the calculation of a Pfaffian.

To find the signs of the bs, let us compare two arbitrary dimer configurations by putting both on the lattice. At each lattice point two dimers end, one of each configuration. Thus, this double configuration consists of closed loops. On each loop the dimers belong alternately to one and the other configuration. Denote the

© Springer-Verlag Berlin Heidelberg 2016

F. Wegner, *Supermathematics and its Applications in Statistical Physics*,
Lecture Notes in Physics 920, DOI 10.1007/978-3-662-49170-6_8

Fig. 8.1 Square lattice and
dimers. **(a)** vertices and
edges, **(b)** complete dimer
covering of the lattice. The
dimers are indicated by *thick
lines*

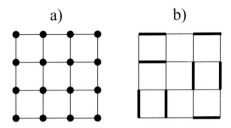

points along the loop by $x_1, x_2, \ldots x_{2n}$. Then, for one configuration, one obtains the
product

$$(b_{x_1,x_2}\xi_{x_1}\xi_{x_2})(b_{x_3,x_4}\xi_{x_3}\xi_{x_4})\ldots(b_{x_{2n-1},x_{2n}}\xi_{x_{2n-1}}\xi_{x_{2n}}),\tag{8.2}$$

and for the other one

$$(b_{x_{2n},x_1}\xi_{x_{2n}}\xi_{x_1})(b_{x_2,x_3}\xi_{x_2}\xi_{x_3})\ldots(b_{x_{2n-2},x_{2n-1}}\xi_{x_{2n-2}}\xi_{x_{2n-1}}).\tag{8.3}$$

If in the last expression ξ_{2n} is shifted to the end, then the sign changes due to the
$2n - 1$ transpositions of ξ_{2n} with the $2n - 1$ factors ξ_i. Thus, the product of all bs
along a loop has to obey

$$\prod_{i=1}^{2n} b_{x_i,x_{i+1}} = -1.\tag{8.4}$$

Suppose the lattice can be constructed by adding one plaquette after the other so
that at least one edge is new. We can then assign to one new edge a factor b, so that
the product of the bs for plaquettes with an even (odd) number of vertices equals -1
($s_0 = \pm 1$). These products should be equal for all plaquettes with an odd number of
vertices, if circled in the same direction. Obviously the product changes sign with
a change of direction: If this construction is provided, then the product of bs of a
non-overlapping loop equals -1, if the loop has an even number n of edges and, if
it includes an even number of vertices n_{vi}. The last condition is necessary to cover
the region inside the loop completely by dimers.

Consider any non-overlapping loop including all the plaquettes inside. Denote
the number of vertices of this partial lattice by n_v, the number of plaquettes by n_p
and the number of edges by n_e. Euler's theorem yields $n_v + n_p = n_e + 1$, since the
outside area is not counted as a plaquette. Further, denote the number of plaquettes
with an even number of edges (vertices) n_{pe}, those with an odd number n_{po}. Then
$n_p = n_{pe} + n_{po}$. Further, call the number of edges inside the loop n_{ei}. The number
of vertices on the loop equals the number n of edges on the loop. So, $n_e = n_{ei} + n$,
$n_v = n_{vi} + n$. Euler's theorem now reads $n_{vi} + n_{pe} + n_{po} = n_{ei} + 1$. If the number of
plaquettes with k edges is n_k, then one obtains, for the number of edges, $\sum_k k n_k =
2n_{ei} + n$. This relation yields $n_{po} \equiv n \bmod 2$.

One obtains the product s, of bs along the loop by multiplying the products of the bs of all plaquettes and a factor -1 for each internal edge. Since b carries in one direction $+1$, in the opposite one -1, $s = (-)^{n_{pe}} s_{po}^{n_{po}} (-)^{n_{ei}}$. Elimination of n_{ei} by means of Euler's relation and the relation between n_{po} and n yields $s = (-)^{n_{vi}+1} (-s_{po})^n$. Since n and n_{vi} are even, one obtains $s = -1$.

8.2 Square Lattice

An explicit expression for the number of dimer coverings on a square lattice is given.

For the square lattice one may choose

$$b_{k,l;k+1,l} = (-)^l = -b_{k+1,l;k,l}, \quad b_{k,l;k,l+1} = 1 = -b_{k,l+1;k,l}. \tag{8.5}$$

One obtains for a plaquette

$$b_{k,l;k+1,l} b_{k+1,l;k+1,l+1} b_{k+1,l+1;k,l+1} b_{k,l+1;k,l} = -1. \tag{8.6}$$

Now we evaluate the number of dimer configurations $\mathcal{N}$,

$$\mathcal{N} = \int \prod_{k,l} d\xi_{k,l} \exp(\tfrac{1}{2} \sum_{k,l,k',l'} (\delta_{k,k'} M_{l,l'} + M_{k,k'} D_{l,l'}) \xi_{k,l} \xi_{k',l'}) = \mathrm{pf}(\mathbf{1}_h \otimes M_v + M_h \otimes D_v) \tag{8.7}$$

with

$$M = \begin{pmatrix} 0 & 1 & & & 0 \\ -1 & 0 & 1 & & \\ & -1 & & \cdots & \\ & & \cdots & & 1 \\ 0 & & & -1 & 0 \end{pmatrix}, \tag{8.8}$$

$$D = \begin{pmatrix} -1 & & & 0 \\ & 1 & & \\ & & -1 & \\ & & & \cdots \\ 0 & & & \pm 1 \end{pmatrix}. \tag{8.9}$$

The indices $_h$ and $_v$ at the matrices indicate that these are matrices of dimension $L_h \times L_h$ and $L_v \times L_v$ for dimers on a rectangle of size $L_h \times L_v$. D and M anticommute. Thus, one obtains

$$\mathcal{N}^4 = \mathrm{pf}(\mathbf{1}_h \otimes M_v + M_h \otimes D_v)^4 = \det(\mathbf{1}_h \otimes M_v + M_h \otimes D_v)^2 = \det(\mathbf{1}_h \otimes M_v^2 + M_h^2 \otimes \mathbf{1}_v). \tag{8.10}$$

We did not define the sequence of the differentials $d\xi$ and thus the sign of the integral. Since the fourth power is taken and since the result is non-negative, the sign of $\mathcal{N}$ is obvious. The matrix in the last expression of (8.10) can be diagonalized. Denoting the eigenvalues of M_h and M_v by μ_m and ν_n yields

$$\mathcal{N}^4 = \prod_{m=1}^{L_h}\prod_{n=1}^{L_v}(\mu_m^2 + \nu_n^2). \tag{8.11}$$

The eigenfunctions $\psi_m(r)$, $r = 1, .., L_h$, of M_h and their eigenvalues are

$$\psi_m(r) = i^r \sin(\frac{\pi r m}{L_h + 1}), \quad \mu_m = 2i\cos(\frac{\pi m}{L_h + 1}). \tag{8.12}$$

The factors i are to be squared upon insertion in (8.10). One has $L_v L_h$ of these factors in the product. The total sign is positive, since one obtains dimer configurations only for even $L_v L_h$. Thus, one obtains

$$\mathcal{N}^4 = \prod_{m=1}^{L_h}\prod_{n=1}^{L_v}\left(4\cos^2(\frac{\pi m}{L_h + 1}) + 4\cos^2(\frac{\pi n}{L_v + 1})\right). \tag{8.13}$$

Finally one may multiply the horizontal edges by g_h and the vertical ones by g_v. Then one obtains, instead of (8.13),

$$\mathcal{N}^4(g_h, g_v) = \prod_{m=1}^{L_h}\prod_{n=1}^{L_v}\left(4g_h^2\cos^2(\frac{\pi m}{L_h + 1}) + 4g_v^2\cos^2(\frac{\pi n}{L_v + 1})\right). \tag{8.14}$$

$\mathcal{N}$ yields a homogeneous polynomial in g_h and g_v of order $L_h L_v/2$. The coefficient of $g_h^{k_h} g_v^{k_v}$ specifies the number of dimer configurations with k_h horizontal and k_v vertical dimers.

Three cases are to be distinguished:

(1) If both L_h and L_v are odd, then the factor for $m = (L_h + 1)/2$ und $n = (L_v + 1)/2$ vanishes in (8.12). Since the number of vertices is odd, there are no dimer configurations, that is $\mathcal{N} = 0$.
(2) If both L_h and L_v are even, then by using $\mu_m = -\mu_{L_h+1-m}$ and $\nu_n = -\nu_{L_v+1-n}$ the product can be reduced to half of the factors for both m and n, yielding

$$\mathcal{N}(g_h, g_v) = \prod_{m=1}^{L_h/2}\prod_{n=1}^{L_v/2}\left(4g_h^2\cos^2(\frac{\pi m}{L_h + 1}) + 4g_v^2\cos^2(\frac{\pi n}{L_v + 1})\right). \tag{8.15}$$

We mention, without proof, that for $L_h = L_v = 2n$ one obtains $\mathcal{N} = 2^n r_n^2$, with integer r_n, with $r_1 = 1$, $r_3 = 29$, $r_5 = 89893$. For r_2 and r_4 see Problems 8.2

and 8.3. Jokusch [125] has shown that the number of dimer configurations of square lattices with fourfold symmetry is a square or twice a square.

(3) For even L_h and odd L_v, one obtains, for $n = (L_v + 1)/2$, the factors $\cos^2(\pi m/(L_h + 1))$, which yields

$$\mathcal{N}(g_h, g_v) = \prod_{m=1}^{L_h/2} \left(2g_h \cos(\frac{\pi m}{L_h + 1}) \right)$$

$$\times \prod_{m=1}^{L_h/2} \prod_{n=1}^{(L_v-1)/2} \left(4g_h^2 \cos^2(\frac{\pi m}{L_h + 1}) + 4g_v^2 \cos^2(\frac{\pi n}{L_v + 1}) \right).$$

(8.16)

An analogous result is obtained for odd L_h and even L_v.

For $L_h = 2$ and $L_v = 3$, one obtains, for example,

$$\mathcal{N}(g_h, g_v) = g_h(g_h^2 + 2g_v^2).$$

(8.17)

There is one configuration with three horizontal dimers and two configurations with one horizontal and two vertical dimers.

One may also consider the limit of arbitrarily large systems by expressing the logarithm of $\mathcal{N}$ as a sum. After introduction of the wave vectors $(\pi m/(L_h + 1), \pi n/(L_v + 1))$ one performs the continuum limit. This yields [85, 133]

$$\lim_{L_h, L_v \to \infty} \mathcal{N}^{2/(L_h L_v)} = \exp(2C/\pi) = 1.791622\ldots$$

(8.18)

for $g_h = g_v = 1$ with the Catalan constant $C = 1-1/9+1/25-\ldots = 0.915965\ldots$. Thus, the number of configurations grows exponentially with the number of vertices.

8.3 Dimers and Tilings

Dimers lead to tilings of the plain. This is expressed in terms of the lattice dual to the original one. The dimers of the square lattice yield a domino tiling. The dimers of the hexagonal lattice yield a rhomboid tiling, which may be interpreted as projection of a three dimensional surface consisting of squares.

In Fig. 8.2a we show dimers on a (black) square lattice. These dimers are imbedded in rectangles in the shape of dominoes. This yields a domino tiling of the checkerboard.

Another interesting lattice is the hexagonal lattice. We show an example in Fig. 8.2b. One may embed each dimer with a rhomboid, also called lozenge, which yields a tiling of the plane in rhomboids. These rhomboids can be considered a

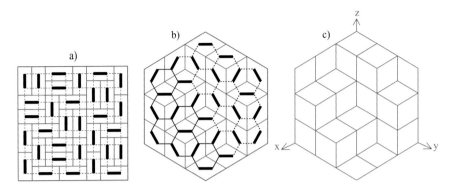

Fig. 8.2 Dimers and tilings. (**a**) checkerboard and dominoes, (**b**) honeycomb lattice and rhomboids, (**c**) rhomboid tiling again. Dimers are shown as *thick black lines*, the other parts of the lattice as *black thin dotted lines*. The tiling is shown in *red thin lines*

projection of a surface in three dimensions onto a plane, which consists of squares parallel to the xy-, xz-, and yz-plains. Then the ensemble of dimer configurations yields a class of random surfaces. These surfaces, which are more clearly seen in Fig. 8.2c, have the property that z decreases or stays constant with increasing x and increasing y.

Generally the tilings are obtained from the lattice dual to the original one. The dual lattice is obtained by putting a vertex in each plaquette of the original lattice. The vertices on adjacent plaquettes are connected by edges. These vertices and new edges constitute the dual lattice. The dual lattice of the square lattice is again a square lattice, whereas the dual lattice of the honeycomb lattice is a triangular lattice, see Fig. 9.4. Finally the edges crossing the dimers have to be removed from the dual lattice. This yields the domino rectangles and the rhomboids.

Problems

8.1 The double product for $\mathcal{N}$ can be reduced to a simple product by use of $a^n - 1 = \prod_{m=0}^{n-1}(a-\exp(2\pi mi/n))$ and $b^2+1+2b\cos(\phi) = (b+e^{i\phi})(b+e^{-i\phi})$. Perform the transformation.

8.2 Calculate the number of dimer configurations for a 4×4 lattice.

8.3 Calculate numerically how many ways 32 dominoes can be placed on a checkerboard.

8.4 Show that the number of dimer configurations of $2 \times n$ square lattices is given by the Fibonacci numbers [146].

References

[85] M.E. Fisher, Statistical mechanics of Dimers on a plane lattice. Phys. Rev. **124**, 1664 (1961)

[125] W. Jokusch, Perfect matchings and perfect squares. J. Combin. Theory A **67**, 100 (1994)

[133] P.W. Kasteleyn, The statistics of dimers on a lattice, the number of dimer arrangements on a quadratic lattice. Physica **27**, 1209 (1961)

[134] P.W. Kasteleyn, Dimer statistics and phase transitions. J. Math. Phys. **4**, 287 (1963)

[138] R. Kenyon, Dimer problems, in *Encyclopedia of Mathematical Physics*, ed. J.-P. Françoise, G.L. Naber, T.S. Tsun, (Academic Press, Amsterdam, 2006)

[139] R. Kenyon, Lectures on Dimers. arXiv:0910.3129v1 (2009)

[140] R. Kenyon, A. Okounkov, What is a dimer? Not. Am. Math. Soc. **52**, 342 (2005)

[146] D. Klarner, J. Pollack, Domino tilings of rectangles with fixed width. Discret. Math. **32**, 44 (1980)

[179] B. McCoy, T.T. Wu, *The Two-Dimensional Ising Model* (Harvard University Press, Cambridge, 1973)

[251] H.N.V. Temperley, M.E. Fisher, Dimer problem in statistical mechanics - an exact result. Phil. Mag. **6**, 1061 (1961)

Chapter 9
Two-Dimensional Ising Model

Abstract The solution of the two-dimensional Ising model on the square lattice is presented. The logarithmic divergence of the specific heat at the critical point is derived. The boundary tension in the ordered phase is determined. Duality arguments allow the determination of the exponential decay of the spin-spin correlation in the paramagnetic phase.

9.1 The Ising Model

The Ising model is introduced. Its importance for the theory of critical phenomena is pointed out.

9.1.1 The Model

We consider how the partition function of the two-dimensional Ising model on a square lattice can be calculated by means of Grassmann variables [86, 122, 179, 213, 228, 229].

The model consists of Ising spins $S(k, l) = \pm 1$ on lattice points (k, l), with k, l integer. The interaction between the spins is given by

$$H = -I_\mathrm{h} \sum_{k,l} S(k, l) S(k + 1, l) - I_\mathrm{v} \sum_{k,l} S(k, l) S(k, l + 1). \tag{9.1}$$

The partition function reads

$$Z = \sum_{\{S\}} \exp(-\beta H)$$

$$= \sum_{\{S\}} \prod_{k,l} (\cosh(\beta I_\mathrm{h})(1 + t_\mathrm{h} S(k, l) S(k + 1, l)))$$

© Springer-Verlag Berlin Heidelberg 2016
F. Wegner, *Supermathematics and its Applications in Statistical Physics*,
Lecture Notes in Physics 920, DOI 10.1007/978-3-662-49170-6_9

$$\times \prod_{k,l}(\cosh(\beta I_{\mathrm{v}})(1 + t_{\mathrm{v}}S(k,l)S(k,l+1))),$$

$$t_{\mathrm{h}} = \tanh(\beta I_{\mathrm{h}}), \quad t_{\mathrm{v}} = \tanh(\beta I_{\mathrm{v}}). \tag{9.2}$$

$S(k,l)$ has to be summed at each lattice point independently over ± 1. The expression under the sum is a polynomial in $S(k,l)$. Summation over S yields a non vanishing contribution only, if at each lattice site an even power of $S(k,l)$ appears. If one introduces for each factor tSS a corresponding edge in the lattice, then the partition function is a sum of contributions $Ct_{\mathrm{h}}^{k_{\mathrm{h}}}t_{\mathrm{v}}^{k_{\mathrm{v}}}$ over all closed graphs on the lattice. Closed means that an even number of edges ends at each lattice site. k_{h} and k_{v} are the numbers of horizontal and vertical edges in the lattice. Thus, the partition function is given by

$$Z = C \sum_{\mathrm{closed\ graphs}} t_{\mathrm{h}}^{k_{\mathrm{h}}}t_{\mathrm{v}}^{k_{\mathrm{v}}}, \tag{9.3}$$

with

$$C = 2^{N}(\cosh(\beta I_{\mathrm{h}}))^{N}(\cosh(\beta I_{\mathrm{v}}))^{N}, \tag{9.4}$$

where N is the number of spins, which equals the number of lattice points. We assume periodic and antiperiodic boundary conditions.

9.1.2   Phases and Singularities

For a long time, the only models describing transitions from disordered to ordered phases by breaking a symmetry were various types of molecular field models. The most prominent were Van der Waals theory (1873) for the gas-liquid transition, Weiss mean-field theory (1907) for magnetic phase transitions, and Landau theory (1937) for phase transition in general. They all gave a jump in the specific heat at the critical temperature T_{c} and predicted, for the order parameter, a square root law, $\sqrt{T_{\mathrm{c}} - T}$. This order parameter is the difference between the density of the liquid ρ_{l} and the vapor phase ρ_{v}, below the critical temperature. In a magnetic system, it is the spontaneous magnetization in the ferromagnetic phase. The compressibility and susceptibility respectively, diverge like $|T_{\mathrm{c}} - T|^{-1}$ in these theories.

    However, already in 1900 it was clear (see the review by Levelt-Sengers [165]) that experiments showed a different behaviour. For example Verschaffelt and Young found that in isopentane $\rho_{\mathrm{l}} - \rho_{\mathrm{v}}$ is proportional to $(T_{\mathrm{c}} - T)^{0.3434}$. Many more experiments supported that molecular field theory did not describe the critical behaviour correctly. Therefore the exact solution of the two-dimensional Ising model in 1944 and in 1949 by Onsager [200, 201], which gave a logarithmic singularity of the specific heat and a power law for the spontaneous magnetization,

$m \propto (T_c - T)^{1/8}$, showed that exact solutions could indeed give results different from molecular field approximation.

The calculation of the free energy was simplified by Kaufman [135] and a complete derivation of the spontaneous magnetization was given by Yang [292]. The representation of the Ising model by Grassmannians was given by Samuel [228, 229]. It is also given in Chap. 2 of the book by Itzykson and Drouffe [122]. The book by McCoy and Wu [179] contains a general summary and further references.

As a function of the parameters t_h and t_v we will find that the specific heat diverges logarithmically, if one of the four Pfaffians $2 - (1 \pm t_v)(1 \pm t_h)$ in Eq. (9.31) vanishes. These singularities indicate phase transitions. The transition lines are shown in Fig. 9.1. At high temperatures, i.e. for small values of t, the system is in the disordered paramagnetic region. At low temperatures, that is for t close to ± 1 the system is ordered. For ferromagnetic couplings I, that is positive t_h and t_v, the system becomes ferromagnetic with spins preferably lined up in one direction. The system approaches antiferromagnetic order in the three other corners of the figure. A direct calculation of order parameter correlations goes beyond the scope of this book and we refer the interested reader to [122, 179]. Using duality arguments we will give the decay of the spin-spin correlation in the paramagnetic region. It will be shown that there is long-range order in the regions indicated by f and af. This is done in the ferromagnetic case by introducing antiperiodic boundary conditions. Thus, the system is forced to have a domain wall with a cost in free energy proportional to the length of this wall in the ordered phase, whereas, in the paramagnetic region, this cost of energy becomes negligible as soon as the size of the system exceeds the correlation length of the spins. Similarly one finds that in the antiferromagnetic phases, with odd length L and periodic boundary conditions in the direction of the staggered magnetization again this cost in free energy shows up. In a final section, the duality transformation is presented. This transformation allowed Kramers and Wannier [153] to determine the critical temperature before Onsager's celebrated solution [200] of the Ising model. We also give some indications, how

Fig. 9.1 Regions of various phases in t_h-t_v space. p indicates the paramagnetic region, f the ferromagnetic one, af three types of antiferromagnetic order: af_1 staggered in horizontal direction, af_2 staggered in vertical direction, and af_3 staggered in both directions. The *phase transition lines* are determined by the vanishing Pfaffians (9.31)

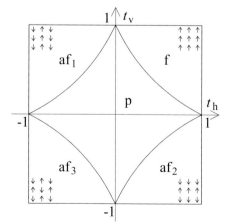

the solution can be generalized to Ising models on other two-dimensional lattices like the honeycomb lattice and the triangular lattice.

9.2 Representation by Grassmann Variables

The partition function of the Ising model on the square lattice is expressed as integral over Grassmann variables.

At each lattice site four Grassmann variables are introduced, shifted from the lattice site a little bit **up**, **down**, to the **right** and **left**, denoted by $\zeta_u(k, l)$, $\zeta_d(k, l)$, $\zeta_r(k, l)$ and $\zeta_l(k, l)$, resp. For each factor $1 + t_h S(k, l) S(k + 1, l)$ in (9.2) we introduce a factor

$$1 + t_h a_h \zeta_r(k, l) \zeta_l(k + 1, l) = \exp(t_h a_h \zeta_r(k, l) \zeta_l(k + 1, l)) \tag{9.5}$$

The lower indices indicate that the edge runs from (k, l) to the **right** and from $(k + 1, l)$ to the **left**. a_h is a sign factor, which is later chosen so that all contributions under the sum are positive (Fig. 9.2). Correspondingly, the factors

$$1 + t_v a_v \zeta_u(k, l) \zeta_d(k, l + 1) = \exp(t_v a_v \zeta_u(k, l) \zeta_d(k, l + 1)) \tag{9.6}$$

are introduced for the vertical edges. At each point, an integration is performed over all four ζs. With this prescription we would obtain only contributions, if all four edges would end at the point. Thus, we introduce an additional factor $\exp(R(k, l))$ with

$$R(k, l) = \tfrac{1}{2} \sum_{d, d'} a_{d, d'} \zeta_d(k, l) \zeta_{d'}(k, l), \tag{9.7}$$

Fig. 9.2 Location of the four Grassmann variables around lattice sites

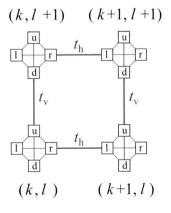

where d and d' stand for all four directions and a is antisymmetric. The integrals

$$I(f) = \int \exp(R) f(\zeta) \, d\zeta_u d\zeta_d d\zeta_r d\zeta_l \qquad (9.8)$$

yield

f	$I(f)$
1	$a_{ud}a_{rl} - a_{ur}a_{dl} + a_{ul}a_{dr}$
$\zeta_u\zeta_d$	a_{rl}
$\zeta_u\zeta_r$	$-a_{dl}$
$\zeta_u\zeta_l$	a_{dr}
$\zeta_d\zeta_r$	a_{ul}
$\zeta_d\zeta_l$	$-a_{ur}$
$\zeta_r\zeta_l$	a_{ud}
$\zeta_u\zeta_d\zeta_r\zeta_l$	1

$$(9.9)$$

If no edge ends at a lattice point, then we require that the factor equals one,

$$a_{ud}a_{rl} - a_{ur}a_{dl} + a_{ul}a_{dr} = 1. \qquad (9.10)$$

If we are able to choose the factors a so that each closed graph contributes with positive sign, then

$$Z = C \int \prod_{k,l} d\,\zeta_u d\,\zeta_d d\,\zeta_r d\,\zeta_l \qquad (9.11)$$

$$\times \exp\Big(\sum_{k,l} (t_h a_h \zeta_r(k,l)\zeta_l(k+1,l) + t_v a_v \zeta_u(k,l)\zeta_d(k,l+1) + R(k,l))\Big).$$

Now the signs a have to be determined. To do this we dissect the graph into closed not overlapping loops, which are traversed in left. We postpone the case where four edges meet at a point. Then, one loop contributes the factors

$$(t_1 a_1 \zeta_{d_1}(x_1)\zeta_{d'_2}(x_2))(t_2 a_2 \zeta_{d_2}(x_2)\zeta_{d'_3}(x_3)) \ldots (t_n a_n \zeta_{d_n}(x_n)\zeta_{d'_1}(x_1))$$

$$= -\prod_i (t_i a_i)(\zeta_{d'_1}(x_1)\zeta_{d_1}(x_1))(\zeta_{d'_2}(x_2)\zeta_{d_2}(x_2)) \ldots (\zeta_{d'_n}(x_n)\zeta_{d_n}(x_n)).$$

$$(9.12)$$

The factor $\zeta_{d'_1}(x_1)$ has passed $2n - 1$ factors ζ, which yields the minus sign. After this rearrangement there are always two factors $\zeta_{d'}\zeta_d$ at the same lattice point. Thus these pairs commute. We denote the number of factors $\zeta_{d'}\zeta_d$ by $n_{d',d}$. It counts the number of sites at which the loops enter from d' and leave in direction d. Denoting the number of links, which go in direction d by k_d, one obtains

$$k_u - n_{du} = n_{dr} + n_{dl} = n_{ru} + n_{lu},$$

$$k_d - n_{ud} = n_{ur} + n_{ul} = n_{rd} + n_{ld},$$

$$k_r - n_{lr} = n_{lu} + n_{ld} = n_{ur} + n_{dr},$$

$$k_l - n_{rl} = n_{ru} + n_{rd} = n_{ul} + n_{dl}. \tag{9.13}$$

k_u is given both by the number of bonds entering from below the sites and by the number of bonds leaving the sites in the upward direction. We proceed similarly for the other directions. Circling through a loop in left rotation, the number of turns to the left exceeds the number of turns to the right by four. For l loops, one obtains

$$4l = n_{ur} + n_{rd} + n_{dl} + n_{lu} - n_{ru} - n_{dr} - n_{ld} - n_{ul} \tag{9.14}$$

Equations (9.13) and (9.14) yield

$$n_{ur} = l + n_{ld},$$

$$n_{dl} = l + n_{ru},$$

$$n_{rd} = l + n_{ul},$$

$$n_{lu} = l + n_{dr}. \tag{9.15}$$

One is left with the following product, which for integer n and l has to yield $+1$,

$$1 = (-)^l a_v^{k_u} (-a_v)^{k_d} a_h^{k_r} (-a_h)^{k_l} \prod_{d,d'} I(\zeta_d \zeta_{d'})^{n_{d,d'}} \tag{9.16}$$

$$= (a_v a_h a_{dl} a_{ur})^{n_{ld}+n_{ru}+l} (-a_v a_h a_{dr} a_{ul})^{n_{ul}+n_{dr}+l} (-a_v a_{rl})^{n_{ud}+n_{du}} (-a_h a_{ud})^{n_{rl}+n_{lr}}.$$

This is fulfilled for

$$a_v a_h a_{dl} a_{ur} = 1, \quad a_v a_h a_{dr} a_{ul} = -1, \tag{9.17}$$

$$a_v a_{rl} = -1, \quad a_h a_{ud} = -1. \tag{9.18}$$

Until now we have not considered the case that four edges meet at a lattice point. In this case we may either connect left with down and right with up or we may connect left with up and right with down. Then $I(\zeta_d \zeta_l) I(\zeta_u \zeta_r)$ has to be replaced by $I(\zeta_d \zeta_l \zeta_u \zeta_r)$ or $I(\zeta_d \zeta_r) I(\zeta_u \zeta_l)$ has to be replaced by $I(\zeta_d \zeta_r \zeta_u \zeta_l)$. This yields the conditions

$$I(\zeta_d \zeta_l) I(\zeta_u \zeta_r) = I(\zeta_d \zeta_l \zeta_u \zeta_r), \quad a_{ur} a_{dl} = -1 \tag{9.19}$$

$$I(\zeta_d \zeta_r) I(\zeta_u \zeta_l) = I(\zeta_d \zeta_r \zeta_u \zeta_l), \quad a_{ul} a_{dr} = 1. \tag{9.20}$$

From Eqs. (9.10), (9.19), and (9.20) one obtains

$$a_{ud}a_{rl} = -1, \qquad (9.21)$$

The solution of Eqs. (9.17)–(9.21) is not unique. We use the solution

$$-a_v = a_h = a_{rl} = -a_{ud} = a_{ur} = -a_{dl} = a_{ul} = a_{dr} = 1. \qquad (9.22)$$

Thus, we have obtained the partition function in the form of a Gauss integral over Grassmann variables

$$Z = C \int \prod_{k,l} (d\,\zeta_u d\,\zeta_d d\,\zeta_r d\,\zeta_l) \exp(G), \qquad (9.23)$$

$$G = \sum_{k,l} (t_h \zeta_r(k,l) \zeta_l(k+1,l) - t_v \zeta_u(k,l) \zeta_d(k,l+1) + R(k,l)). \qquad (9.24)$$

9.3 Evaluation of the Partition Function

The partition function of the Ising model on the square lattice is expressed as a product.

We consider the model on a lattice with periodic boundary conditions and the extensions L_h and L_v. These boundary conditions correspond to the model on a torus. The evaluation given here is not precise, since there are graphs with one or several loops winding around the torus, for which Eq. (9.14) no longer holds. In Sect. 9.4 the necessary corrections are given. A Fourier transform of the variables ζ brings G in block-diagonal form,

$$\zeta_d(k,l) = \frac{1}{\sqrt{L_h L_v}} \sum_{m=1}^{L_h} \sum_{n=1}^{L_v} \exp(ikp(m) + ilq(n))\xi_d(p,q), \qquad (9.25)$$

$$p(m) = \frac{2m\pi}{L_h}, \quad q(n) = \frac{2n\pi}{L_v}.$$

G then reads

$$G = \sum_{p,q} (t_h \exp(-ip)\xi_r(p,q)\xi_l(-p,-q) - t_v \exp(-iq)\xi_u(p,q)\xi_d(-p,-q)$$

$$+ \frac{1}{2} \sum_{d,d'} a_{d,d'}\xi_d(p,q)\xi_{d'}(-p,-q)). \qquad (9.26)$$

It is sufficient to integrate over the pairs $\xi(p,q)$ and $\xi(-p,-q)$, and to multiply the products of these determinants

$$Z = C \prod_{p,q}{}' \det(A(p,q)), \tag{9.27}$$

$$\det(A(p,q)) = \int \prod_d (\mathrm{d}\,\xi_d(p,q)\mathrm{d}\,\xi_d(-p,-q))$$

$$\times \exp(\sum_{d,d'} \xi_d(p,q)A_{d,d'}(p,q)\xi_{d'}(-p,-q)) \tag{9.28}$$

$$A(p,q) = \begin{pmatrix} 0 & -1-t_v e^{-iq} & 1 & 1 \\ 1+t_v e^{iq} & 0 & 1 & -1 \\ -1 & -1 & 0 & 1+t_h e^{-ip} \\ -1 & 1 & -1-t_h e^{ip} & 0 \end{pmatrix}, \tag{9.29}$$

$$\det(A(p,q)) = (1 - t_v - t_h - t_v t_h)^2 + 2t_v(1-t_h^2)(1-\cos q)$$
$$+2t_h(1-t_v^2)(1-\cos p). \tag{9.30}$$

Obviously $\det(A(p,q)) \geq 0$. The prime indicates that multiplication extends only over one half of the wave-vectors (p,q), so that $(-p,-q)$ does not appear besides (p,q). If p and q are integer multiples of π, then $\xi_d(p,q) = \xi_d(-p,-q)$, so that instead of $\det(A(p,q))$ one has to insert the corresponding Pfaffian $\mathrm{pf}(A(p,q))$,

$$\mathrm{pf}(A(0,0)) = 1 - t_v - t_h - t_v t_h, \ \ \mathrm{pf}(A(0,\pi)) = 1 + t_v - t_h + t_v t_h,$$
$$\mathrm{pf}(A(\pi,0)) = 1 - t_v + t_h + t_v t_h, \ \ \mathrm{pf}(A(\pi,\pi)) = 1 + t_v + t_h - t_v t_h. \tag{9.31}$$

The Pfaffians are, apart from perhaps the sign, the square roots of the corresponding determinants.

We did not explicitly introduce the Jacobian for the Fourier transform, ζ to ξ. For $t_h = t_v = 0$, both the ζ integral and the ξ integral yield 1. Thus the Jacobian equals 1, since it is independent of t_h and t_v.

9.4 Loops Winding Around the Torus

Taking periodic boundary conditions into account one finds that the partition function of the Ising model is a sum of four determinants, which are expressed as products as before.

Finally, we consider how the calculation has to be modified for closed loops winding around the torus. For such loops, one has, instead of (9.14),

$$0 = n_{ur} + n_{rd} + n_{dl} + n_{lu} - n_{ru} - n_{dr} - n_{ld} - n_{ul}, \tag{9.32}$$

since for these loops the number of right turns equals the number of left turns. Thus, l counts only the loops not winding around the torus. The minus sign, (9.12), which appears by commuting the factor ζ with the others, has to be taken care of differently. The torus can be cut open in both directions by changing the sign of the factors a_h and a_v, respectively, at the border. More precisely, we replace a_h, in the terms $\zeta_r(L_h, l)\zeta_l(1, l)$ by $-a_h$. If the loop runs once in the horizontal direction around the torus, then one obtains the desired factor -1. This does not change anything for loops, which do not wind around the torus, since for them one obtains the factor -1 by crossing the border to the right as well as to the left. Instead of replacing a_h by $-a_h$, one may use antiperiodic boundary conditions $\zeta_d(k + L_h, l) = -\zeta_d(k, l)$. One proceeds analogously in the vertical direction. We can now distinguish four cases, since the boundary conditions can be chosen independently in both directions, and denote the corresponding expressions (9.27) by $Z_{\pm,\pm}$, where the first sign indicates a_h at the horizontal border and the second sign a_v at the vertical border. We claim

$$Z = \tfrac{1}{2}(-Z_{++} + Z_{+-} + Z_{-+} + Z_{--}). \tag{9.33}$$

To verify this expression first denote the number of windings of a loop in the vertical and horizontal directions by w_h und w_v, respectively. Several of such loops must agree in the number of windings, since otherwise they would intersect. If the number of loops is even, then each choice of the boundary conditions yields the correct result, so they are correctly counted in (9.33). If the number n_w of such loops is odd, then it is essential to realize that w_h and w_v are prime to each other. If they were not prime, then the loops would intersect each other. This means that w_h and w_v cannot both be even. In all other cases, the sign by which the contribution enters is given by

n_w	w_h	w_v	Z_{++}	Z_{+-}	Z_{-+}	Z_{--}
even	any	any	1	1	1	1
any	0	0	1	1	1	1
odd	even	odd	-1	1	-1	1
odd	odd	even	-1	-1	1	1
odd	odd	odd	-1	1	1	-1

$$\tag{9.34}$$

One observes that in all cases the sum on the right-hand side of (9.33) yields contributions with weight $+1$. The boundary conditions change p and q in (9.25) to

$$p(m) = \begin{cases} \frac{2m\pi}{L_h} & \text{for } Z_{+,s'} \\ \frac{(2m+1)\pi}{L_h} & \text{for } Z_{-,s'} \end{cases}, \tag{9.35}$$

$$q(n) = \begin{cases} \frac{2n\pi}{L_v} & \text{for } Z_{s,+} \\ \frac{(2n+1)\pi}{L_v} & \text{for } Z_{s,-} \end{cases}. \tag{9.36}$$

One observes that the sign of the $Z_{s,s'}$ depends on the signs of the Pfaffians. One obtains

$$Z_{s,s'} = \prod_{p \in \mu(s,L_h)} \prod_{q \in \mu(s',L_v)} \text{sign}(\text{pf}(A(p,q)))|Z_{s,s'}|. \qquad (9.37)$$

with the sets

$$\mu(+,L) = \begin{cases} (0,\pi) & \text{even } L, \\ (0) & \text{odd } L \end{cases}, \quad \mu(-,L) = \begin{cases} \emptyset & \text{even } L, \\ (\pi) & \text{odd } L \end{cases}. \qquad (9.38)$$

We have now obtained the exact result for the partition function of the Ising model on the square lattice with periodic boundary conditions.

9.5 Divergence of the Specific Heat

The logarithmic divergence of the specific heat at the critical point is obtained.

Provided that none of the Pfaffians (9.31) vanishes, we will show below that for all $|Z_{s,s'}|$ the thermodynamic limit is given by the expressions one obtains from (9.27) and (9.30) in the continuum limit,

$$\ln Z_{\text{cont}} := \lim_{L_v, L_h \to \infty} \frac{1}{L_h L_v} \ln |Z_{s,s'}|$$

$$= \ln(2 \cosh(\beta I_h) \cosh(\beta I_v)) \qquad (9.39)$$

$$+ \frac{1}{2(2\pi)^2} \int_{-\pi}^{\pi} dp \int_{-\pi}^{\pi} dq \ln(\det(A(p,q))).$$

The partition function Z, Eq. (9.33), is a linear combination of the $Z_{s,s'}$. Using the limit Z_{cont} and the signs of the Pfaffians (9.31), one finds, that depending on the phase and the boundary conditions, Z may be 2 or 1 or 0 times $Z_{\text{cont}}^{L_h L_v}$. But even when the factor 0 applies, $\ln Z$, which is of order $L_h L_v$, differs from $L_h L_v \ln Z_{\text{cont}}$ only by terms of order L_h or L_v as we will see in Sect. 9.7. Thus,

$$\lim_{L_v, L_h \to \infty} \frac{1}{L_h L_v} \ln Z = \ln Z_{\text{cont}} \qquad (9.40)$$

also holds. Z_{cont} is the partition function per lattice site in the thermodynamic limit.

The partition function is non-analytic where the determinant of the integrand (9.30) respectively the Pfaffian (9.31) vanishes. For ferromagnetic couplings $I_v > 0$, $I_h > 0$ this happens at $1 = t_v + t_h + t_v t_h$. We will later see that this equation determines the temperature of the phase transition. In order to investigate the non-analyticity, it is useful to consider the derivative with respect to

β and to investigate the behaviour of the integrand at $p = q = 0$. With

$$\kappa = \text{pf}(A(0,0)), \quad c_v = t_v(1 - t_h^2), \quad c_h = t_h(1 - t_v^2) \tag{9.41}$$

one obtains for the derivative of the singular part of the partition function

$$\frac{d}{d\beta}\left(\frac{1}{L_h L_v} \ln Z_{\text{sing}}\right) = \frac{d\kappa}{d\beta} \frac{\kappa}{(2\pi)^2} \int \frac{dp\, dq}{\kappa^2 + c_v q^2 + c_h p^2} \tag{9.42}$$

The integral has the non-analytic part

$$-\frac{\pi}{\sqrt{c_v c_h}} \ln(\kappa^2), \tag{9.43}$$

which yields a non-analyticity for the energy per lattice site of

$$\frac{1}{L_h L_v}\langle H\rangle_{\text{sing}} = \frac{\kappa}{4\pi\sqrt{c_v c_h}} \frac{d\kappa}{d\beta} \ln(\kappa^2). \tag{9.44}$$

where we have used $\langle H\rangle = -d \ln Z/d\beta$. Note that $\ln(\kappa^2)$ is negative. From (9.44) one deduces the logarithmic singularity of the specific heat

$$\frac{1}{L_h L_v} \frac{d\langle H\rangle}{dT}\bigg|_{\text{sing}} = \frac{\beta^2}{4\pi\sqrt{c_v c_h}}\left(\frac{d\kappa}{d\beta}\right)^2 |\ln(\kappa^2)|. \tag{9.45}$$

Similar singularities show up, when one of the three other Pfaffians (9.31) vanishes.

9.6 Other Lattices

Starting from the square lattice some couplings can be chosen to vanish or to be infinite. In this way the partition function of other planar lattices can be obtained. Examples are the triangular and the honeycomb lattice.

Can the procedure to evaluate the partition function be used for Ising models on other lattices? As far as it is known the partition function of all two-dimensional Ising models can be expressed by a Pfaffian, provided the edges of the interactions do not cross. Two possibilities to construct the partition function of other two-dimensional lattices, starting out from the square lattice, are:

(1) One may attach to some edges an infinite coupling, which corresponds to the choice $t = 1$. The connected spins are then forced to be parallel. Thus, they act as one spin. An example is shown in Fig. 9.3a, where the thick lines indicate $t = 1$. Thus, two spins merge to one spin, which is now connected to six different spins, which yields a triangular lattice, as shown in figure a'.

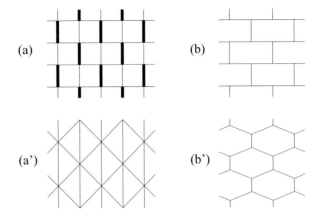

Fig. 9.3 Lattices obtained from the square lattice. Choices $t = 1$ and $t = 0$ for some bonds yield triangular and hexagonal lattices. See text

(2) One may eliminate some edges, which corresponds to the choice $t = 0$, as shown in Fig. 9.3b. This reduces the coordination number from four to three. A slight distortion of the brickwall lattice yields the honeycomb lattice shown in figure b'.

M.E. Fisher [86] relates the Ising model on a planar lattice to the dimer problem. He starts with lattices of coordination number 3 and proceeds to those with larger coordination numbers.

In an attempt to transfer this procedure to a three dimensional lattice one has the problem, as also there the factor $(-)^l$ for l loops appears, but that there is no relation analogous to (9.14). Thus, the transfer to three dimensions fails. However, the partition function of the three-dimensional Ising model can be given as integral over an exponential function, which does not only contain bilinear terms but also terms of fourth order in the Grassmann variables [121]. There is no exact solution known for three-dimensional Ising models.

9.7 Phases and Boundary Tension

By comparing the partition function for periodic and antiperiodic boundary conditions the boundary tension and thus the ordered regions are determined.

One expects that at high temperatures, that is for small β and thus small t_v and t_h, the system is in the paramagnetic phase, in which the Ising spins show only short-range correlations. At low temperatures, that is for t_v and t_h close to ± 1, the system is expected to be in the ordered phase. The logarithmic singularity in the specific heat, which appears at vanishing Pfaffians in (9.31) shows that there are strong fluctuations in the energy-energy correlations, since the specific heat can be expressed by the sum of these correlations over the whole system. One guesses that the phase transition between the disordered paramagnetic phase indicated by p and the ordered ferro- and antiferromagnetic phases indicated by f and af occur along these lines in the t_h-t_v diagram Fig. 9.1. For positive couplings, I_v and I_h, the spins tend to line up in parallel approaching ferromagnetic order. If one or both couplings are negative, then there is the tendency for a staggered order, either in the horizontal or vertical or both directions, yielding three types of antiferromagnetic order.

A calculation of the spontaneous magnetization is beyond the scope of this book. We refer the reader to [122, 179]. However, we may observe the long-range order in a different way. Let us consider the ferromagnetic and paramagnetic phases, with the argument for the antiferromagnetic phases running similarly. We consider the system with positive couplings I and, thus, positive ts. However, we change the periodic boundary conditions to antiperiodic ones, $S(k + L_h, l) = -S(k, l)$. This means that in Eq. (9.33), all $Z_{s,s'}$ have to be changed to $Z_{-s,s'}$ yielding

$$Z_{a,p} = \tfrac{1}{2}(-Z_{-+} + Z_{--} + Z_{++} + Z_{+-}), \tag{9.46}$$

where p stands for periodic and a for antiperiodic boundary conditions. Similarly we may introduce antiperiodic boundary conditions in both directions. Then $Z_{s,s'}$ in (9.33) changes to $Z_{-s,-s'}$,

$$Z_{a,a} = \tfrac{1}{2}(-Z_{--} + Z_{-+} + Z_{+-} + Z_{++}). \tag{9.47}$$

The dependence of $Z_{s,s'}$ as a function of s and s' is analyzed in Section 9.7.1. At leading order, one obtains

$$|Z_{s,s'}| = Z_{\text{cont}}^{L_h L_v} \left(1 - s\chi_{1,0} - s'\chi_{0,1} - 2ss'\chi_{1,1}\right), \tag{9.48}$$

where the χ are functions decaying exponentially with L,

$$\chi_{m,n} \propto \text{sign}(c_h^{mL_h})\text{sign}(c_v^{nL_v})(\lambda_{h\,m,n})^{mL_h}(\lambda_{v\,m,n})^{nL_v} \tag{9.49}$$

with

$$\lambda_{h\,1,0} = \left(\frac{1 - |t_v|}{|t_h|(1 + |t_v|)}\right)^{\pm 1}, \quad \lambda_{v\,0,1} = \left(\frac{1 - |t_h|}{|t_v|(1 + |t_h|)}\right)^{\pm 1} \tag{9.50}$$

and for $L_h = L_v$,

$$\lambda_{h\,1,1}\lambda_{v\,1,1} = \left(\frac{(1 - t_v^2)(1 - t_h^2)}{4|t_v t_h|}\right)^{\pm 1}. \tag{9.51}$$

The plus sign of the exponents applies in the f and af regions, the minus sign in the p region. The λs equal 1, when one of the Pfaffians, in (9.31), vanishes, otherwise they are less than 1.

Paramagnetism In the region p, all Pfaffians and thus all $Z_{s,s'}$, are positive. Thus, one obtains, for all boundary conditions, $Z = Z_{\text{cont}}^{L_h L_v}$ and the corrections by the χs yield corrections to the free energy, which decay exponentially with the size of the system.

Ferromagnetism In region f the situation is different. There, $pf(A(0,0))$ and thus, $Z_{+,+}$ is negative. For periodic boundary conditions in both directions one obtains $Z_{p,p} = Z_{cont}^{L_h L_v}$. If one or both boundary conditions are antiperiodic, then the leading contributions $Z_{cont}^{L_h L_v}$ cancel and one obtains

$$Z_{a,p} = Z_{p,p}\chi_{0,1}, \quad Z_{a,a} = 2Z_{p,p}\chi_{1,1}. \tag{9.52}$$

Thus, the antiperiodic boundary conditions make the free energies increase

$$F_{a,p} - F_{p,p} = -k_B T L_v \ln \lambda_{v\,0,1}, \tag{9.53}$$

$$F_{a,a} - F_{p,p} = -k_B T (L_v \ln \lambda_{v\,1,1} + L_h \ln \lambda_{h\,1,1}). \tag{9.54}$$

The reason is, going around the torus, the system has to have a domain wall somewhere. This domain wall costs energy (and yields some entropy). The cost in free energy is proportional to the length of the domain wall. This change in free energy per area (in three dimensions) and per length (in two dimensions) is called surface tension and boundary tension, resp. We remember that λ is less than one. Thus, one really has an increase in free energy. λ approaches 1 as one approaches the critical line. Thus, the surface tension goes to zero at the critical line. An explicit calculation of the boundary tension is given in Section 9.7.1. It was first determined by Avron et al. [13] from the spin-spin correlation function by means of the duality transformation (see Sect. 9.8).

Antiferromagnetism If t_v or t_h is negative, then the system tends to antiferromagnetism. The system can be easily transformed into one with ferromagnetic coupling, since the interaction remains invariant under any of the transformations (9.55) and (9.56),

$$S(k,l) \to (-)^k S(k,l), \quad t_h \to -t_h, \tag{9.55}$$

$$S(k,l) \to (-)^l S(k,l), \quad t_v \to -t_v. \tag{9.56}$$

If t_h is negative, then the transformation (9.55) is performed. If L_h is even, then periodic boundary conditions, as well as antiperiodic boundary conditions, are preserved under the transformation. If L_h is odd, however, then periodic boundary conditions are transformed into antiperiodic ones and vice versa. Similar arguments hold for t_v and L_v. Independently, one may use Eqs. (9.33), (9.46), (9.47) together with the signs of the Pfaffians (9.31), (9.37), (9.38) and the signs of the χs, (9.49), to evaluate the free energies, which yield the same results.

The differences in the free energies (9.53), (9.54), which can be obtained analogously for antiferromagnetic couplings, and which grow proportional to the lengths of domain walls, clearly indicate that the system is in an ordered phase with long-range correlations in these regions. In the region indicated by p, these differences decay exponentially with increasing size of the system, which indicates absence of long-range order. The length over which this correlation decays is inversely proportional to the product P of the Pfaffians (9.70), which in turn is proportional to the difference of the temperature T from the critical temperature T_c. Thus, the correlation length ξ obeys

$$\xi(T) \propto \frac{1}{T - T_c} \tag{9.57}$$

in contrast to $\xi(T) \propto (T - T_c)^{-1/2}$ in the molecular field approximation.

9.7.1 Appendix

The derivation of $\ln Z$ as sum of χs is given in this appendix. With

$$\cosh(\beta I) = \frac{1}{\sqrt{1 - t^2}}, \quad \det(A(p, q)) = c_0 - 2c_v \cos q - 2c_h \cos p, \tag{9.58}$$

$$c_0 = (1 + t_v^2)(1 + t_h^2), \quad \hat{c}_v = \frac{c_v}{c_0}, \quad \hat{c}_h = \frac{c_h}{c_0} \tag{9.59}$$

one obtains

$$\ln |Z_{s,s'}| = \frac{L_h L_v}{2} \ln \left(4 \frac{1 + t_v^2}{1 - t_v^2} \frac{1 + t_h^2}{1 - t_h^2} \right)$$
$$+ \frac{1}{2} \sum_{p,q} \ln(1 - 2\hat{c}_v \cos q - 2\hat{c}_h \cos p). \tag{9.60}$$

We evaluate

$$\sum_{p,q} \ln(1 - 2\hat{c}_v \cos q - 2\hat{c}_h \cos p) = \sum_{p,q} \ln(1 - \hat{c}_v e^{iq} - \hat{c}_v e^{-iq} - \hat{c}_h e^{ip} - \hat{c}_h e^{-ip})$$

$$= - \sum_{p,q,m_1,m_2,n_1,n_2} \frac{(m_1 + m_2 + n_1 + n_2 - 1)!}{m_1! m_2! n_1! n_2!}$$

$$\hat{c}_v^{n_1 + n_2} \hat{c}_h^{m_1 + m_2} e^{i(n_2 - n_1)q} e^{i(m_2 - m_1)p}. \tag{9.61}$$

With

$$\sum_p e^{i(m_2-m_1)p} = \begin{cases} s^m L_h & m_2 - m_1 = mL_h, \\ 0 & \text{otherwise} \end{cases} \tag{9.62}$$

with integer m one obtains

$$\sum_{p,q} \ln(1 - 2\hat{c}_v \cos q - 2\hat{c}_h \cos p) = - \sum_{m=0,n=0}^{\infty} (2 - \delta_{m,0})(2 - \delta_{n,0})s^m s'^n \chi_{m,n},$$

$$\chi_{m,n} = L_v L_h \sum_{m_1,n_1} \frac{(2m_1 + 2n_1 + mL_h + nL_v - 1)!}{m_1!(m_1 + mL_h)!n_1!(n_1 + nL_v)!}$$

$$\hat{c}_v^{2n_1 + nL_v} \hat{c}_h^{2m_1 + mL_h}. \tag{9.63}$$

The term $m_1 = n_1 = 0$ is excluded for $m = n = 0$. The term $\chi_{0,0}/L_v L_h$ does not depend on L. This term together with the first term in (9.60) yields Z_{cont}

$$\ln Z_{\text{cont}} = \tfrac{1}{2} \ln \left(4 \frac{1 + t_v^2}{1 - t_v^2} \frac{1 + t_h^2}{1 - t_h^2} \right) - \frac{1}{2L_v L_h} \chi_{0,0}. \tag{9.64}$$

In total, one obtains

$$\ln |Z_{s,s'}| = L_v L_h \ln Z_{\text{cont}} - \chi^{\text{e,e}} - s\chi^{\text{o,e}} - s'\chi^{\text{e,o}} - ss'\chi^{\text{o,o}} \tag{9.65}$$

with

$$\chi^{\text{e,e}} = \sum_{m,n} (2 - \delta_{m,0} - \delta_{n,0})\chi_{2m,2n},$$

$$\chi^{\text{o,e}} = \sum_{m,n} (2 - \delta_{n,0})\chi_{2m+1,2n},$$

$$\chi^{\text{e,o}} = \sum_{m,n} (2 - \delta_{m,0})\chi_{2m,2n+1},$$

$$\chi^{\text{o,o}} = \sum_{m,n} 2\chi_{2m+1,2n+1}. \tag{9.66}$$

The terms $\chi_{m,n}$ can be estimated for $m > 0$ or $n > 0$ by looking for the maximal term in the sum (9.63): One obtains

$$\frac{4(\hat{m} + \hat{n})^2}{\hat{m}^2 - a^2} \hat{c}_h^2 = 1 = \frac{4(\hat{m} + \hat{n})^2}{\hat{n}^2 - b^2} \hat{c}_v^2 \tag{9.67}$$

with

$$\hat{m} := m_1 + a, \quad a := \tfrac{1}{2}mL_{\mathrm{h}}, \quad \hat{n} := n_1 + b, \quad b := \tfrac{1}{2}nL_{\mathrm{v}}. \tag{9.68}$$

The solution of (9.67) reads

$$\hat{m} = \frac{4t_{\mathrm{h}}^2(1 + t_{\mathrm{v}}^2)a' + (1 - t_{\mathrm{v}}^2)^2(1 + t_{\mathrm{h}}^2)b'}{|P|},$$

$$\hat{n} = \frac{(1 - t_{\mathrm{h}}^2)^2(1 + t_{\mathrm{v}}^2)a' + 4t_{\mathrm{v}}^2(1 + t_{\mathrm{h}}^2)b'}{|P|} \tag{9.69}$$

with

$$P = \mathrm{pf}(A(0,0))\mathrm{pf}(A(0,\pi))\mathrm{pf}(A(\pi,0))\mathrm{pf}(A(\pi,\pi)), \tag{9.70}$$

$$a' = \sqrt{4t_{\mathrm{v}}^2a^2 + (1 - t_{\mathrm{v}}^2)^2b^2} = \frac{1}{1 + t_{\mathrm{h}}^2}\sqrt{\tfrac{1}{2}(B + P(b^2 - a^2))}, \tag{9.71}$$

$$b' = \sqrt{(1 - t_{\mathrm{h}}^2)^2a^2 + 4t_{\mathrm{h}}^2b^2} = \frac{1}{1 + t_{\mathrm{v}}^2}\sqrt{\tfrac{1}{2}(B + P(a^2 - b^2))}, \tag{9.72}$$

$$B = (c_0^2 - 4c_{\mathrm{h}}^2 + 4c_{\mathrm{v}}^2)a^2 + (c_0^2 + 4c_{\mathrm{h}}^2 - 4c_{\mathrm{v}}^2)b^2. \tag{9.73}$$

We mention

$$\frac{(\hat{m} + \hat{n})^2}{c_0^2} = \frac{\hat{m}^2 - a^2}{4c_{\mathrm{h}}^2} = \frac{\hat{n}^2 - b^2}{4c_{\mathrm{v}}^2}$$

$$= \frac{[(1 + t_{\mathrm{h}}^2)a' + (1 + t_{\mathrm{v}}^2)b']^2}{P^2} = \frac{B + 2Rc_0}{P^2} \tag{9.74}$$

with

$$R = \sqrt{4c_{\mathrm{v}}^2a^4 + (c_0^2 - 4c_{\mathrm{v}}^2 - 4c_{\mathrm{h}}^2)a^2b^2 + 4c_{\mathrm{h}}^2b^4} = a'b'. \tag{9.75}$$

We use Stirling's formula $a! \approx \sqrt{2\pi}a^{a+1/2}\mathrm{e}^{-a}$ and approximate the expression under the sum by a Gaussian. This yields

$$\chi_{m,n} \approx c_{m,n}\frac{(2\hat{m} + 2\hat{n})^{2\hat{m}+2\hat{n}}}{(\hat{m} + a)^{\hat{m}+a}(\hat{m} - a)^{\hat{m}-a}(\hat{n} + b)^{\hat{n}+b}(\hat{n} - b)^{\hat{n}-b}}\hat{c}_{\mathrm{h}}^{2\hat{m}}\hat{c}_{\mathrm{v}}^{2\hat{n}}$$

$$= c_{m,n}\left(\frac{\hat{m} - a}{\hat{m} + a}\right)^a\left(\frac{\hat{n} - b}{\hat{n} + b}\right)^b\underbrace{\left(\frac{4\hat{c}_{\mathrm{h}}^2(\hat{m} + \hat{n})^2}{\hat{m}^2 - a^2}\right)^{\hat{m}}}_{1}\underbrace{\left(\frac{4\hat{c}_{\mathrm{v}}^2(\hat{m} + \hat{n})^2}{\hat{n}^2 - b^2}\right)^{\hat{n}}}_{1},$$

$$\tag{9.76}$$

$$c_{m,n} = \frac{\text{sign}(c_{\text{h}}^{mL_{\text{h}}})\text{sign}(c_{\text{v}}^{nL_{\text{v}}})L_{\text{v}}L_{\text{h}}}{\sqrt{32\pi(\hat{n}a^2 + \hat{m}b^2)}} \tag{9.77}$$

and thus,

$$\chi_{m,n} \approx c_{m,n}\lambda_{\text{h}\,m,n}^{mL_{\text{h}}}\lambda_{\text{v}\,m,n}^{nL_{\text{v}}}, \quad \lambda_{\text{h}\,m,n} = \sqrt{\frac{\hat{m}-a}{\hat{m}+a}}, \quad \lambda_{\text{v}\,m,n} = \sqrt{\frac{\hat{n}-b}{\hat{n}+b}}. \tag{9.78}$$

The signs in expression (9.77) for $c_{m,n}$ come from the signs of $\hat{c}_{\text{v}}$ and $\hat{c}_{\text{h}}$ in (9.63). The λs approach 1 at the critical lines due to $|P| \to 0$ in the denominator for $\hat{m}$ and $\hat{n}$ in (9.69).

The coefficient $c_{m,n}$ is of order $L^{1/2}$ and decays close to the critical lines proportional to $|P|^{1/2}$. Similarly $-\ln\lambda \propto |P|$ close to criticality. Further

$$|\chi_{m,n}| \le |\chi_{m',n'}|, \quad \text{if } |m| \ge |m'| \text{ and } |n| \ge |n'|, \tag{9.79}$$

where the equal sign holds only, if $|m| = |m'|$ and $|n| = |n'|$. This can be easily seen, since the coefficients of the powers of c_{v} and c_{h} in (9.63) are smaller or equal, if the conditions in (9.79) hold. Thus, at leading order one has

$$\chi^{\text{o,e}} = \chi_{1,0}, \quad \chi^{\text{e,o}} = \chi_{0,1}, \quad \chi^{\text{o,o}} = \chi_{1,1}. \tag{9.80}$$

Several explicit values for $a = 0$, $b = 0$, and $a = b$ are listed in Table 9.1. Neglecting the prefactors $c_{m,n}$ in $\chi_{m,n}$, one can show that

$$|\chi_{m_1+m_2,n_1+n_2}| \ge |\chi_{m_1,n_1}\chi_{m_2,n_2}|, \tag{9.81}$$

where the equal sign holds only if $m_1/n_1 = m_2/n_2$. This can be seen by showing that

$$d(a,b) := -\ln|\chi| = a\ln\left(\frac{\hat{m}+a}{\hat{m}-a}\right) + b\ln\left(\frac{\hat{n}+b}{\hat{n}-b}\right). \tag{9.82}$$

has the *property of a distance*,

$$d(a_1,b_1) + d(a_2,b_2) \ge d(a_1 + a_2, b_1 + b_2). \tag{9.83}$$

Table 9.1 $\hat{m}$, $\hat{n}$, λ_{h}, and λ_{v} for $a = 0$, $b = 0$, and $a = b$

a,b	$\hat{m}$	$\hat{n}$	$\lambda_{\text{h}}^{-\text{sign}(P)}$	$\lambda_{\text{v}}^{-\text{sign}(P)}$
$b=0$	$\frac{t_{\text{h}}^2(1+\lvert t_{\text{v}}\rvert)^2+(1-\lvert t_{\text{v}}\rvert)^2}{\lvert t_{\text{h}}^2(1+\lvert t_{\text{v}}\rvert)^2-(1-\lvert t_{\text{v}}\rvert)^2\rvert}a$	$\frac{2\lvert t_{\text{v}}\rvert(1-t_{\text{h}}^2)}{\lvert t_{\text{h}}^2(1+\lvert t_{\text{v}}\rvert)^2-(1-\lvert t_{\text{v}}\rvert)^2\rvert}a$	$\frac{1-\lvert t_{\text{v}}\rvert}{\lvert t_{\text{h}}\rvert(1+\lvert t_{\text{v}}\rvert)}$	1
$a=0$	$\frac{2\lvert t_{\text{h}}\rvert(1-t_{\text{v}}^2)}{\lvert t_{\text{v}}^2(1+\lvert t_{\text{h}}\rvert)^2-(1-\lvert t_{\text{h}}\rvert)^2\rvert}b$	$\frac{t_{\text{v}}^2(1+\lvert t_{\text{h}}\rvert)^2+(1-\lvert t_{\text{h}}\rvert)^2}{\lvert t_{\text{v}}^2(1+\lvert t_{\text{h}}\rvert)^2-(1-\lvert t_{\text{h}}\rvert)^2\rvert}b$	1	$\frac{1-\lvert t_{\text{h}}\rvert}{\lvert t_{\text{v}}\rvert(1+\lvert t_{\text{h}}\rvert)}$
$a=b$	$\frac{4t_{\text{h}}^2(1+t_{\text{v}}^2)^2+(1+t_{\text{v}}^2)^2(1-t_{\text{v}}^2)^2}{\lvert(1-t_{\text{h}}^2)(1-t_{\text{v}}^2)-16t_{\text{h}}^2t_{\text{v}}^2\rvert}a$	$\frac{4t_{\text{v}}^2(1+t_{\text{h}}^2)^2+(1+t_{\text{v}}^2)^2(1-t_{\text{h}}^2)^2}{\lvert(1-t_{\text{h}}^2)(1-t_{\text{v}}^2)-16t_{\text{h}}^2t_{\text{v}}^2\rvert}a$	$\frac{(1+t_{\text{h}}^2)(1-t_{\text{v}}^2)}{2\lvert t_{\text{h}}\rvert(1+t_{\text{v}}^2)}$	$\frac{(1+t_{\text{v}}^2)(1-t_{\text{h}}^2)}{2\lvert t_{\text{v}}\rvert(1+t_{\text{h}}^2)}$

For $0 < q < 1$ one obtains, after some calculation,

$$d(qa + \delta a, qb + \delta b) + d((1-q)a - \delta a, (1-q)b - \delta b) \tag{9.84}$$
$$= d(a,b) + \frac{|P|}{q(1-q)a'b'((1+t_h^2)a' + (1+t_v^2)b')}(a\delta b - b\delta a)^2 + O(\delta a, \delta b)^3.$$

Noting that d is homogeneous in a and b of order 1, $d(a,b) = b\hat{d}(a/b)$, one obtains

$$d(qa + \delta a, qb + \delta b) + d((1-q)a - \delta a, (1-q)b - \delta b)$$
$$= d(a,b) + \frac{1}{2q(1-q)b^3}(a\delta b - b\delta a)^2 \hat{d}'' + O(\delta a, \delta b)^3. \tag{9.85}$$

Comparison of (9.84) and (9.85) shows that the second derivative of $\hat{d}(a/b)$ is positive and thus (9.81) holds. By exponentiating $\ln Z$, (9.65), products of χs can be neglected at leading order due to (9.81). For the discussion of the various boundary conditions it is sufficient to use (9.48). The correction $\chi^{e,e}$ is negligible.

9.8 Duality Transformation

The duality transformation connects properties of two-dimensional Ising models on dual lattices at high and low temperatures. It yields relations between the partition functions and allows for the determination of the critical temperature for self-dual lattices. It also connects the spin-spin correlation with the change of free energy for the dual system, in which the sign of the couplings along a corresponding line is changed.

Consider a two-dimensional lattice consisting of vertices, at which the spins are located. The vertices are connected by bonds between the interacting pairs of spins. If these bonds do not cross, then the lattice is called planar. In the following only planar lattices will be considered. Then the partition function can be represented as sum over all unions of closed graphs U, where each bond in the graph contributes a factor $t_b = \tanh(K_b)$ with $K_b = \beta I_b$.

$$Z\{K\} = C\{K\} \sum_U \prod_{b \in U} t_b, \quad C\{K\} = 2^N \prod_b \cosh(K_b). \tag{9.86}$$

Another expansion in terms of unions of closed graphs is obtained in the following way: Introduce domain walls between neighboring spins pointing in opposite direction. These domain walls form closed paths, too. The energy of such a configuration is given by

$$E = E_{\parallel} + 2\sum_b I_b \tag{9.87}$$

where $E_\parallel$ is the energy of the two states with all spins parallel aligned and the sum runs over all bonds b crossing the domain wall. These domain walls are located on the edges of the dual lattice. This dual lattice is obtained by putting a vertex in each plaquette of the original lattice. Vertices on adjacent plaquettes are connected by bonds $\hat{b}$. Thus there is a bond $\hat{b}$ of the dual lattice crossing a bond b of the original lattice. Similarly there is one vertex of the original lattice in each plaquette of the dual lattice and one vertex of the dual lattice in each original plaquette.

Thus the calculation of the partition function from (9.87) yields

$$Z_{p,p}\{K\} = 2\tilde{C}\{K\}\sum_{\hat{U}}\prod_{\hat{b}\in\hat{U}}\exp(-2K_{b(\hat{b})}), \quad \tilde{C}\{K\} = \prod_b \exp(K_b). \tag{9.88}$$

since the domain walls are unions of closed graphs $\hat{U}$ on the dual lattice. By reversing all spins one obtains the same domain walls, which yields the factor 2 in the first Eq. (9.88).

One obtains (apart from the prefactors $C, \tilde{C}$) the same partition function for the Ising model on the dual lattice

$$\hat{Z}\{\hat{K}\} = C\{\hat{K}\}\sum_{\hat{U}}\prod_{\hat{b}\in\hat{U}}\hat{t}_{\hat{b}}, \quad C\{\hat{K}\} = 2^{\hat{N}}\prod_{\hat{b}}\cosh(\hat{K}_{\hat{b}}), \tag{9.89}$$

if one chooses $\hat{t}_{\hat{b}} = \tanh(\hat{K}_{b(\hat{b})}) = \exp(-2K_{b(\hat{b})})$, which yields the relation

$$t_b + \hat{t}_{\hat{b}} + t_b\hat{t}_{\hat{b}} = 1. \tag{9.90}$$

Thus one obtains the duality relation

$$\frac{1}{C\{\hat{K}\}}\hat{Z}\{\hat{K}\} = \frac{1}{\tilde{C}\{K\}}Z\{K\}. \tag{9.91}$$

A lattice on the torus is not a planar lattice. But the partition function can be determined in a similar way. Going around the torus one crosses the domain walls an even number of times. Thus only loops going around the whole torus are to be counted, if they appear pairwise. An odd number of such loops is not allowed. From (9.34) one observes that now

$$Z_{p,p} = \tfrac{1}{2}(\hat{Z}_{+,+} + \hat{Z}_{+,-} + \hat{Z}_{-,+} + \hat{Z}_{-,-}). \tag{9.92}$$

The ratio between $Z_{p,p}$ and $\hat{Z}_{+,+}$ lies between $1/2$ and 2. In the following it can be neglected.

Euler's topological relation reads for the torus

$$N + \hat{N} = N_b, \tag{9.93}$$

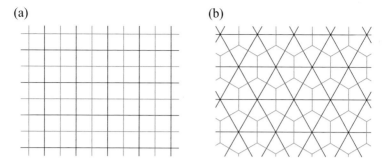

Fig. 9.4 Dual lattices. (**a**) Two square lattices are dual. (**b**) Triangular lattice and honeycomb lattice are dual

with N_b the number of bonds. One obtains for the coefficients C

$$\frac{C\{\hat{K}\}}{\tilde{C}\{K\}} = \frac{\tilde{C}\{\hat{K}\}}{C\{K\}} = 2^{\hat{N}-N_b/2} \prod_b (\sinh(2K_b))^{-1/2}$$

$$= \left(2^{N-N_b/2} \prod_b \left(\sinh(2\hat{K}_b)\right)^{-1/2}\right)^{-1}. \tag{9.94}$$

where $2t/(1-t^2) = \sinh(2K)$ has been used.

As shown in Fig. 9.4 the square lattice is dual to the square lattice and the triangular lattice is dual to the honeycomb lattice on the torus. For the square lattice the duals of the horizontal bonds are the vertical bonds and vice versa. Thus

$$\hat{t}_h = \exp(-2K_v) = \frac{1-t_v}{1+t_v}, \quad \hat{t}_v = \exp(-2K_h) = \frac{1-t_h}{1+t_h}. \tag{9.95}$$

Kramers and Wannier [153] used this fact to map the high-temperature behavior expressed by the expansion in t to the low-temperature behavior expressed by the expansion in powers of $\hat{t}$. This mapping is called duality transformation, since performing the map $t \rightarrow \hat{t}$ twice leads back to the original t. Assuming that the free energy has one critical temperature at which it becomes non-analytic, they determined the critical temperature of the Ising model on the square lattice from $t = \hat{t}$.

9.8.1 Order and Disorder Operators

The concept of duality allows connection between spin correlations in the dual lattice and the change of the free energy due to walls of opposite interactions in the original lattice (Kadanoff and Ceva [129]). The disorder is introduced in the dual

Fig. 9.5 Dual lattices. The
gray dots indicate the
location of the spin operators
on the dual lattice and those
of the disorder operators on
the original lattice

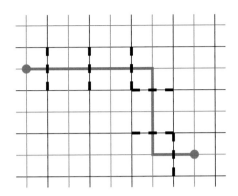

lattice in the following way: Choose two points on the dual lattice: Connect them by a line (both in gray in Fig. 9.5). This line cuts interaction lines on the original lattice indicated by thick dashed lines. One reverses the interaction along these lines. This introduces some disorder in the interaction. If the interaction was originally attractive, then across the line the interaction prefers pairs of spins with opposite signs at the ends of these dashed lines, which introduces a domain wall. The cost of free energy depends only on the locations of the two gray points. One can easily convince oneself that a different line between these two points yields the same free energy. Thus disorder variables $\mu(r)$ and $\mu(r')$ are introduced and the change in free energy (divided by $k_B T$) is denoted as correlation of the disorder variables

$$\exp(-\beta(F_{\text{dis}} - F_{\text{ord}})) = \langle \mu(r)\mu(r') \rangle \{K\} = \frac{Z\{K'\}}{Z\{K\}}, \tag{9.96}$$

where $K' = -K$ on the bonds with reversed interaction, otherwise $K' = K$.

Let us consider correlations of the spins $\langle S(r)S(r') \rangle$, first on the original lattice. For this purpose we write

$$S(r)S(r') = (S(r)S(r_1))(S(r_1)S(r_2)) \ldots (S(r_{n-1})S(r')), \tag{9.97}$$

where the spins apart from $S(r)$ and $S(r')$ appear pairwise and are grouped into pairs of neighboring spins. In order to evaluate the correlation one has to replace $\exp(KS(r_m)S(r_{m+1})) = \cosh K + \sinh KS(r_m)S(r_{m+1})$ by

$$S(r_m)S(r_{m+1}) \exp(KS(r_m)S(r_{m+1})) = \sinh K + \cosh KS(r_m)S(r_{m+1})$$

$$= -i(\cosh(K + \frac{i\pi}{2}) + \sinh(K + \frac{i\pi}{2})S(r_m)S(r_{m+1})) \tag{9.98}$$

$$= -i\exp\left((K + \frac{i\pi}{2})S(r_m)S(r_{m+1})\right). \tag{9.99}$$

Thus K has to be replaced by $K + \frac{i\pi}{2}$ for all bonds appearing in the product (9.97). We now consider the correlation on the dual lattice with the dual couplings $\hat{K}' = \hat{K}$ and $\hat{K} + \frac{i\pi}{2}$,

$$\langle S(r)S(r')\rangle_{\text{dual}}\{\hat{K}\} = (-i)^n \frac{\hat{Z}\{\hat{K}'\}}{\hat{Z}\{\hat{K}\}}. \tag{9.100}$$

We realize that the replacements $K \to -K$ and $\hat{K} \to \hat{K} + \frac{i\pi}{2}$ are performed at the associated bonds and dual bonds. Since $\hat{K}$ is the dual of K, it follows that $\hat{K} + \frac{i\pi}{2}$ is the dual of $-K$. Using (9.91) and

$$\frac{\hat{Z}\{\hat{K}'\}}{Z\{K'\}} = \frac{C\{\hat{K}\}}{\tilde{C}\{K\}} i^n \tag{9.101}$$

one obtains

$$\langle \mu(r)\mu(r')\rangle\{K\} = \langle S(r)S(r')\rangle_{\text{dual}}\{\hat{K}\}. \tag{9.102}$$

The duality transformation connects correlations between systems below and above the transition [129].

First consider the situation with $T > T_c$ on the original lattice, $\hat{T} < T_c$ on the dual lattice. Then the correlation $\langle S(r)S(r')\rangle_{\text{dual}}$ is long ranged. For large distances the correlation approaches $\langle S(r)\rangle^2$, that is the square of the spontaneous magnetization. Above T_c disorder has only a local effect, since the spins are correlated only over small distances. Thus $\lim_{|r-r'|\to\infty}\langle\mu(r)\mu(r')\rangle = \text{const.}$

Secondly consider the case $T < T_c$ on the original lattice, $\hat{T} > T_c$ on the dual lattice. Then the spin correlations are short ranged and decay. On the original lattice $T < T_c$ yields long range order. Therefore the change in free energy described by $F_{\text{a,a}} - F_{\text{p,p}}$ in (9.54) and by $F_{\text{dis}} - F_{\text{ord}}$ in (9.96) yields apart from local effects around r and r', which are absent in the antiperiodic boundary conditions the linear growth with distance $|r - r'|$. Thus the exponential decay of the spin correlation above T_c can be concluded from the surface tension below T_c,

$$\lim_{|r-r'|\to\infty} \frac{\ln(\langle S(r)S(r')\rangle)}{|r - r'|} = -\frac{1}{\xi}, \quad \xi = -|\cos\phi|\ln\lambda_{\text{h}11} - |\sin\phi|\ln\lambda_{\text{v}11}, \tag{9.103}$$

where

$$r - r' = |r - r'|(\cos\phi, \sin\phi), \quad b/a = \tan\phi. \tag{9.104}$$

Again the correlation length ξ diverges close to T_c as given in (9.57). Avron et al. [13] went the other way round. They calculated the surface tension from the correlation function.

Spontaneous Magnetization We do not calculate the spontaneous magnetization in this book. We mention only Onsager's announcement of the result as a discussion remark [201] in 1949: "Mathematically, the composition-temperature curve in a solid solution presents the same problem as the degree of order in a ferromagnetic with scalar spin. B. Kaufman and I have recently solved the latter problem (unpublished) for a two-dimensional rectangular net with interaction energies J, J'. If we write $\sinh(2J/kT)\sinh(2J'/kT) = 1/k$, then the degree of order, for $k < 1$, is simply $(1 - k^2)^{1/8}$." The solution was published by Yang [292] in 1952.

Problems

9.1 The condition for criticality is usually given by $\sinh(2\beta I_v)\sinh(2\beta I_h) = 1$. Derive this relation. Hint: Start from $\mathrm{pf}(A(0,0))\mathrm{pf}(A(\pi,\pi)) = 0$. What is obtained from $\mathrm{pf}(A(0,\pi))\mathrm{pf}(A(\pi,0)) = 0$?

9.2

(i) Convert the double products for the partition functions $Z_{s,s'}$ into a simple product over p and q, resp. by means of

$$\prod_{m=0}^{N-1}(a - be^{2\pi im/N}) = a^N - b^N. \tag{9.105}$$

(ii) Use these expressions to express $Z_{+,s'}/Z_{-,s'}$ and $Z_{s,+}/Z_{s,-}$ and determine the boundary tension in horizontal and vertical direction.

9.3

(i) As shown in Sect. 9.6 one transforms from the square to the triangular lattice by introducing bonds with $t = 1$. Perform the corresponding integrals over $\zeta_d(2)$ and $\zeta_u(1)$ in Fig. 9.6a.
(ii) Similarly one transforms to the honeycomb lattice by setting $t = 0$ and performing the same integrals.

9.4

(i) In order to determine the partition function of the honeycomb and triangular lattice one can first consider a square lattice, where from every second site one has four different couplings to the nearest neighbors as shown in Fig. 9.6b. One first integrates over the ζs of every second site, in the figure the central site.
(ii) Then one performs the Fourier transform for the other sites and determines the determinants $A(p,q)$ and the pfaffians, which may show up. Finally one may set t_4 to 0 (honeycomb case) or to 1 (triangular case).

Fig. 9.6 (a) Two
neighboring lattice sites. (b)
A central site surrounded by
the neighboring four sites

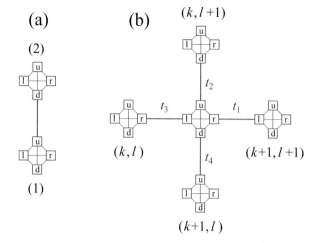

9.5

(i) Star-triangle transformation. One can eliminate every second spin on the honeycomb lattice by performing the sum in the partition function. $\sum_{S_0} e^{\beta I_h S_0 (S_1 + S_2 + S_3)} = c' e^{\beta I_t (S_1 S_2 + S_1 S_3 + S_2 S_3)}$. Determine the relation between βI_h and βI_t.

(ii) Starting from the triangular lattice and performing the duality transformation one obtains a model on the honeycomb lattice. Then applying the star-triangle transformation one is back on the triangular lattice. Determine the critical βI_t and βI_h.

9.6 Consider duality on the square lattice. How do κ, c_h, c_v, the Pfaffians (9.31) and the partition functions $Z_{\pm,\pm}$ transform under the duality transformation?

References

[13] J.E. Avron, H. van Beijeren, L.S. Schulman, R.K.P. Zia, Roughening transition, surface tension and equilibrium droplet shapes in a two-dimensional Ising system. J. Phys. A **15**, L81 (1982)

[86] M.E. Fisher, On the dimer solution of planar Ising models. J. Math. Phys. **7**, 1776 (1966)

[121] C. Itzykson, Ising fermions (II). Three dimensions. Nucl. Phys. B **210**, 477 (1982)

[122] C. Itzykson, J.-M. Drouffe, *Statistical Field Theory*, 2 vols. (Cambridge University Press, Cambridge, 1989)

[129] L.P. Kadanoff, H. Ceva, Determination of an operator algebra for the two-dimensional Ising model. Phys. Rev B **3**, 3918 (1971)

[135] B. Kaufman, Crystal statistics. II. Partition function evaluated by Spinor analysis. Phys. Rev. **76**, 1232 (1949)

[153] H.A. Kramers, G.H. Wannier, Statistics of the two-dimensional ferromagnet. Phys. Rev. **60**, 252–262 (1941)

[165] J.M.H. Levelt-Sengers, From van der Waals' equation to the scaling laws. Physica **73**, 73 (1974)

[179] B. McCoy, T.T. Wu, *The Two-Dimensional Ising Model* (Harvard University Press, Cambridge, 1973)

[200] L. Onsager, Crystal statistics. I. A two-dimensional model with an order-disorder transition. Phys. Rev. **65**, 117 (1944)

[201] L. Onsager, Discussion remark on p. 261 in G.S. Rushbrooke. On the theory of regular solutions. Nuovo Cimento (series 9) **6**, Suppl. 251 (1949)

[213] V.N. Popov, *Functional Integrals in Quantum Field Theory and Statistical Physics* (Reidel, Dordrecht, 1983)

[228] S. Samuel, The use of anticommuting variable integrals in statistical mechanics. I. The computation of partition functions. J. Math. Phys. **21**, 2806 (1980)

[229] S. Samuel, The use of anticommuting variable integrals in statistical mechanics. II. The computation of correlation functions. J. Math. Phys. **21**, 2815 (1980)

[292] C.N. Yang, The spontaneous magnetization of a two-dimensional Ising model. Phys. Rev. **85**, 808 (1952)

Part II
Supermathematics

Supermathematics is the combination of commuting and anticommuting variables on an equal footing. In particular, vectors and matrices with both even and odd components are introduced. The mathematics of complex vectors and matrices are generalized to these supervectors and supermatrices in Chaps. 10–14.

This includes two types of transpositions of matrices, two types of adjoint, the definition of real and symmetric matrices. Two types of superunitary groups are related to the two types of adjoint.

We have seen in part one that integrals over complex and real variables may cancel those over Grassmann variables. Here we will find that integrals over supervectors and supermatrices invariant under superunitary transformations have interesting cancellation properties which are derived in Chaps. 15 and 16. Chapter 17 contains a few more results on supermatrices.

The mathematics developed in this part will be used for the physical applications in Part III.

Chapter 10
Supermatrices

Abstract Until now we only considered the algebra of Grassmann-valued expressions and differentiation and integration with respect to Grassmann variables. We kept away from transformations between even and odd variables. This will now be done. In fact, symmetries will be observed, which include both types of variables. To begin with we use the differential, which immediately leads to matrices with elements of both even and odd elements. We generalize the transposition, the determinant, and the trace to its superforms.

10.1 Differential, Matrices, Transposition

Supermatrices are introduced as differentials. Functions of functions yield the multiplication law. Supertransposition is introduced.

We consider functions Z_k of variables Y_k, where $Z_1, \ldots Z_{n_1} \in \mathscr{A}_0$, $Z_{n_1+1}, \ldots Z_{n_1+m_1} \in \mathscr{A}_1$ and $Y_1, \ldots Y_{n_2} \in \mathscr{A}_0$, $Y_{n_2+1}, \ldots Y_{n_2+m_2} \in \mathscr{A}_1$. If the functions are differentiable, then the differential can be written in terms of the right-derivatives

$$\mathrm{d}Z_k = \sum_l W_{kl}\mathrm{d}Y_l, \quad W_{kl} = \partial Z_k/\partial Y_l. \tag{10.1}$$

The supermatrix W consists of four blocks

$$W = \begin{pmatrix} a & \alpha \\ \beta & b \end{pmatrix} \qquad \begin{matrix} a & \alpha \\ \beta & b \end{matrix} \begin{matrix} \}n_1 \\ \}m_1 \end{matrix} \tag{10.2}$$
$$\underbrace{}_{n_2} \underbrace{}_{m_2}$$

with matrices a and b, whose elements are in $\mathscr{A}_0$, and matrices α and β with elements in $\mathscr{A}_1$. a and α have n_1 rows, with b and β having m_1. Also, a and β have n_2 columns, with b and α having m_2. The set of supermatrices (10.2) is denoted by $\mathscr{M}(n_1, m_1, n_2, m_2)$. Often a is called the bosonic block and b the fermionic block. These two blocks, which contain even elements, are glued together by the blocks α and β of odd elements. We will often term supermatrices simply by matrices, but

© Springer-Verlag Berlin Heidelberg 2016
F. Wegner, *Supermathematics and its Applications in Statistical Physics*,
Lecture Notes in Physics 920, DOI 10.1007/978-3-662-49170-6_10

one should keep in mind that, generally, supermatrices are meant. The blocks are also called sectors.

Instead of (10.1), one may express the differentials by the left-derivatives

$$d Z_k = \sum_l d Y_l\, {}^{\mathrm{T}}W_{lk}, \qquad {}^{\mathrm{T}}W_{lk} = \frac{\partial}{\partial Y_l} Z_k. \tag{10.3}$$

${}^{\mathrm{T}}W$ is called the supertransposed of W,

$$
{}^{\mathrm{T}}W = \begin{pmatrix} {}^{\mathrm{t}}a & {}^{\mathrm{t}}\beta \\ -{}^{\mathrm{t}}\alpha & {}^{\mathrm{t}}b \end{pmatrix} \in \mathscr{M}(n_2, m_2, n_1, m_1). \tag{10.4}
$$

Depending on how ${}^{\mathrm{T}}W$ is introduced, the minus sign may appear in front of β instead in front of α. One should always check the sign convention. The index ${}^{\mathrm{t}}$ denotes the transposed of the matrix, e.g. ${}^{\mathrm{t}}\beta_{kl} = \beta_{lk}$. The minus sign in (10.4) comes from the operator $\mathscr{P}$ in (3.1). One notes that

$$
{}^{\mathrm{TT}}W = \begin{pmatrix} a & -\alpha \\ -\beta & b \end{pmatrix} = \mathscr{P}(W) = \sigma^{(n_1, m_1)} W \sigma^{(n_2, m_2)}, \qquad \sigma^{(n,m)} = \begin{pmatrix} \mathbf{1}_n & 0 \\ 0 & -\mathbf{1}_m \end{pmatrix} \tag{10.5}
$$

is not identical with W. Only ${}^{\mathrm{TTTT}}W$ agrees with W. Besides the supertransposed, one introduces the ordinary transposed

$$
{}^{\mathrm{t}}W = \begin{pmatrix} {}^{\mathrm{t}}a & {}^{\mathrm{t}}\beta \\ {}^{\mathrm{t}}\alpha & {}^{\mathrm{t}}b \end{pmatrix} \in \mathscr{M}(n_2, m_2, n_1, m_1). \tag{10.6}
$$

Thus,

$$
({}^{\mathrm{t}}W)_{ij} = W_{ji}, \qquad ({}^{\mathrm{T}}W)_{ij} = (-)^{\nu_i(1-\nu_j)} W_{ji}. \tag{10.7}
$$

When working with supervectors and supermatrices, the Z_2-grade will often be indicated only by the index as in the equation above. For general matrices, one should indicate whether the Z_2-grade belongs to the first or second index unless it is clear from the context as in the present case. There is no distinction for square matrices as introduced in (10.14). We note that $\nu(W_{ij}) = \nu_i + \nu_j$.

A short but very useful exposé on supermatrices and some supergroups can be found in the article by Rittenberg and Scheunert [224]. The reader may also consult the book due to Berezin [25] and by de Witt [55].

10.2   Chain Rule, Matrix Multiplication

The chain rule for mixed transformations between even and odd variables is given. It defines matrix multiplication and yields the multiplication theorem for supermatrices.

Let the Y_l be differentiable functions of variables X_j with $X_1, \ldots X_{n_3} \in \mathscr{A}_0$, $X_{n_3+1}, \ldots X_{n_3+m_3} \in \mathscr{A}_1$

$$\mathrm{d}\, Y_l = \sum_j V_{lj} \mathrm{d}\, X_j = \sum_j \mathrm{d}\, X_j \,{}^{\mathrm{T}} V_{jl}. \tag{10.8}$$

One obtains

$$\mathrm{d}\, Z_k = \sum_j \Big(\sum_l W_{kl} V_{lj} \Big) \mathrm{d}\, X_j = \sum_j \mathrm{d}\, X_j \Big(\sum_l V_{jl}^T W_{lk}^T \Big). \tag{10.9}$$

This equation includes the chain rule

$$\partial Z_k / \partial X_j = \sum_l (\partial Z_k / \partial Y_l)(\partial Y_l / \partial X_j). \tag{10.10}$$

Equation (10.10) allows the introduction of the matrix product $WV \in \mathscr{M}(n_1, m_1, n_3, m_3)$ for matrices $W \in \mathscr{M}(n_1, m_1, n_2, m_2)$ and $V \in \mathscr{M}(n_2, m_2, n_3, m_3)$

$$(WV)_{kj} = \sum_l W_{kl} V_{lj}. \tag{10.11}$$

One verifies

$$^{\mathrm{T}}(WV) = {}^{\mathrm{T}}V \,{}^{\mathrm{T}}W. \tag{10.12}$$

There is not a similar relation for the ordinary transposed.

The column vectors X, Y and Z are matrices in $\mathscr{M}(n_i, m_i, 1, 0)$, and the Eqs. (10.1), (10.9), (10.3) take the form

$$\mathrm{d}\, Z = W \mathrm{d}\, Y = WV \mathrm{d}\, X, \quad \mathrm{d}\,{}^{\mathrm{T}}Z = \mathrm{d}\,{}^{\mathrm{T}}Y \,{}^{\mathrm{T}}W = \mathrm{d}\,{}^{\mathrm{T}}X \,{}^{\mathrm{T}}V \,{}^{\mathrm{T}}W. \tag{10.13}$$

One verifies that the matrix multiplication satisfies the associative and the distributive law, where matrix addition is defined as usual as the sum of elements of equal indices. The matrices of set $\mathscr{M}(n, m, n, m)$ with fixed n and m constitute an algebra. The matrices in $\mathscr{M}(n, m, n, m)$ are called square matrices. We denote them also by

$$\mathscr{M}(n, m) = \mathscr{M}(n, m, n, m). \tag{10.14}$$

10.3   Berezinian Superdeterminant

The generalization of the Jacobian determinant called Berezinian determinant or superdeterminant for mixed transformations between even and odd elements is derived.

The analog of the Jacobian determinant for the transformation (10.1) has to be determined. It is called Berezinian superdeterminant or, in short, 'Berezinian' or superdeterminant and is denoted by 'sdet'. It is also called the graded determinant and then denoted 'detg'. To derive this Berezinian, one divides the transformation (10.1) into two steps, first the components of Z in $\mathscr{A}_0$, secondly those of Z in $\mathscr{A}_1$ are transformed

$$(\mathrm{d}\,Z_1, \ldots, \mathrm{d}\,Z_n, \mathrm{d}\,Z_{n+1}, \ldots, \mathrm{d}\,Z_{n+m}) \leftarrow \quad (\mathrm{d}\,Z_1, \ldots, \mathrm{d}\,Z_n, \mathrm{d}\,Y_{n+1}, \ldots, \mathrm{d}\,Y_{n+m})$$
$$\leftarrow (\mathrm{d}\,Y_1, \ldots, \mathrm{d}\,Y_n, \mathrm{d}\,Y_{n+1}, \ldots, \mathrm{d}\,Y_{n+m}).$$

$$(10.15)$$

This corresponds to the decomposition of W into

$$W = \begin{pmatrix} a & \alpha \\ \beta & b \end{pmatrix} = \begin{pmatrix} \mathbf{1} & 0 \\ \beta' & b' \end{pmatrix} \begin{pmatrix} a' & \alpha' \\ 0 & \mathbf{1} \end{pmatrix} = \begin{pmatrix} a' & \alpha' \\ \beta'a' & b' + \beta'\alpha' \end{pmatrix}. \qquad (10.16)$$

The solution with respect to the primed quantities yields

$$W = \begin{pmatrix} \mathbf{1} & 0 \\ \beta a^{-1} & b - \beta a^{-1}\alpha \end{pmatrix} \begin{pmatrix} a & \alpha \\ 0 & \mathbf{1} \end{pmatrix}. \qquad (10.17)$$

The first transformation contributes a factor $1/\det(b')$ due to (5.13), the second a factor $\det(a)$. Thus, one obtains

$$\int \prod_k \mathrm{d}\,Z_k f(Z_1, \ldots, Z_{n+m}) = \int \prod_k \mathrm{d}\,Y_k f(Z_1(Y_1, \ldots), \ldots, Z_{n+m}(Y_1, \ldots))\mathrm{sdet}(W) \qquad (10.18)$$

with the superdeterminant

$$\mathrm{sdet}\begin{pmatrix} a & \alpha \\ \beta & b \end{pmatrix} = \frac{\det(a)}{\det(b - \beta a^{-1}\alpha)}. \qquad (10.19)$$

Instead of the transformation (10.15), one may perform the transformation in the sequence

$$(\mathrm{d}\,Z_1, \ldots, \mathrm{d}\,Z_n, \mathrm{d}\,Z_{n+1}, \ldots, \mathrm{d}\,Z_{n+m}) \leftarrow (\mathrm{d}\,Y_1, \ldots, \mathrm{d}\,Y_n, \mathrm{d}\,Z_{n+1}, \ldots, \mathrm{d}\,Z_{n+m})$$
$$\leftarrow (\mathrm{d}\,Y_1, \ldots, \mathrm{d}\,Y_n, \mathrm{d}\,Y_{n+1}, \ldots, \mathrm{d}\,Y_{n+m}).$$

$$(10.20)$$

Then W has to be decomposed

$$W = \begin{pmatrix} a & \alpha \\ \beta & b \end{pmatrix} = \begin{pmatrix} a' & \alpha' \\ 0 & 1 \end{pmatrix} \begin{pmatrix} 1 & 0 \\ \beta' & b' \end{pmatrix} = \begin{pmatrix} a' + \alpha'\beta' & \alpha'b' \\ \beta' & b' \end{pmatrix}, \tag{10.21}$$

which yields

$$W = \begin{pmatrix} a - \alpha b^{-1}\beta & \alpha b^{-1} \\ 0 & 1 \end{pmatrix} \begin{pmatrix} 1 & 0 \\ \beta & b \end{pmatrix} \tag{10.22}$$

and the representation

$$\mathrm{sdet}\begin{pmatrix} a & \alpha \\ \beta & b \end{pmatrix} = \frac{\det(a - \alpha b^{-1}\beta)}{\det(b)} \tag{10.23}$$

for the superdeterminant.

Generalization Transpositions of rows or columns changes the sign of the determinant. This is immediately obvious, if the transpositions are performed within rows or columns of the same Z_2-grade. However, we may, for example, perform permutations, which bring the matrix in the form

$$W = \begin{pmatrix} W^{11} & W^{12} \\ W^{21} & W^{22} \end{pmatrix}, \tag{10.24}$$

where W^{11} and W^{22} are square matrices, $W^{11} \in \mathcal{M}(n_1, m_1)$, $W^{22} \in \mathcal{M}(n_2, m_2)$. Similar arguments yield

$$\mathrm{sdet}(W) = \mathrm{sdet}(W^{11})\mathrm{sdet}(W^{22} - W^{21}\,W^{11\,-1}\,W^{12})$$

$$= \mathrm{sdet}(W^{11} - W^{12}\,W^{22\,-1}\,W^{21})\mathrm{sdet}(W^{22}). \tag{10.25}$$

Equations (10.19) and (10.23) are special cases of (10.25). Equations (10.25) are particularly useful, if $W^{12} = 0$ or $W^{21} = 0$, since then $\mathrm{sdet}(W) = \mathrm{sdet}(W^{11})\mathrm{sdet}(W^{22})$.

Multiplication Theorem One obtains, for the transformation $Z \leftarrow Y \leftarrow X$ considered in (10.10),

$$\int f(Z) \prod \mathrm{d}Z = \int f(Z(Y))\mathrm{sdet}(W) \prod \mathrm{d}Y \tag{10.26}$$

$$= \int f(Z(Y(X)))\mathrm{sdet}(W)\mathrm{sdet}(V) \prod \mathrm{d}X$$

$$= \int f(Z(Y(X)))\mathrm{sdet}(WV) \prod \mathrm{d}X.$$

Since the two expressions in the last line are equal, we find that the multiplication theorem for determinants holds for superdeterminants too.

Theorem 10.1

$$\text{sdet}(WV) = \text{sdet}(W)\text{sdet}(V).$$

Determinants are invariant under transposition of the matrix, $\det(a) = \det({}^{\text{t}}a)$. Using (10.5) and (10.19) or (10.23) one shows

$$\text{sdet}(W) = \text{sdet}({}^{\text{T}}W). \tag{10.27}$$

Later we will also need transformations between matrices $X \to Z$

$$Z = WXV, \quad Z, X \in \mathcal{M}(n_1, m_1, n_2, m_2), \quad W \in \mathcal{M}(n_1, m_1), \quad V \in \mathcal{M}(n_2, m_2). \tag{10.28}$$

We show

Theorem 10.2 *The Berezinian for the transformation $Z = WXV$ reads*

$$[\text{D}\,Z] = \text{sdet}(W)^{n_2 - m_2} \text{sdet}(V)^{n_1 - m_1} [\text{D}\,X],$$

where

$$[\text{D}\,Z] = \prod_{i=1, j=1}^{n_1 + m_1, n_2 + m_2} \text{d}\,Z_{i,j}, \tag{10.29}$$

and similarly for X. Proof: For $\text{d}\,Y = W\text{d}\,X$, one obtains

$$[\text{D}\,Y_{,j}] = \text{sdet}(W)^{\pm 1} [\text{D}\,X_{,j}], \tag{10.30}$$

where the sign in the exponent is positive for $j \le n_2$ and negative for $j > n_2$. Thus,

$$[\text{D}\,Y] = \text{sdet}(W)^{n_2 - m_2} [\text{D}\,X]. \tag{10.31}$$

Similarly, one derives for $\text{d}\,Z = \text{d}\,YV$

$$[\text{D}\,Z] = \text{sdet}(V)^{n_1 - m_1} [\text{D}\,Y], \tag{10.32}$$

which finally yields Theorem 10.2.

10.4   Supertrace and Differential of Superdeterminant

The supertrace, a generalization of the trace is introduced.
   Consider

$$\text{sdet}(\exp(W)) = \lim_{n\to\infty} \text{sdet}[(1 + \frac{W}{n})^n] = \lim_{n\to\infty} [\text{sdet}(1 + \frac{W}{n})]^n$$

$$= \lim_{n\to\infty} [\frac{\det(1 + \frac{a}{n} - \frac{\alpha}{n}(1 + b/n)^{-1}\frac{\beta}{n})}{\det(1 + \frac{b}{n})}]^n$$

$$= \lim_{n\to\infty} (1 + \frac{1}{n}(\text{tr}\,a - \text{tr}\,b))^n = \exp(\text{tr}\,a - \text{tr}\,b). \quad (10.33)$$

One calls

$$\text{str}(W) = \text{str}\begin{pmatrix} a & \alpha \\ \beta & b \end{pmatrix} = \text{tr}\,a - \text{tr}\,b \quad (10.34)$$

the supertrace of W. The supertrace is also called the graded trace and abbreviated 'trg'. Thus, (10.33) reads

$$\text{sdet}(\exp(W)) = \exp(\text{str}(W)). \quad (10.35)$$

Replacing $\exp(W)$ by W reads

$$\text{sdet}(W) = \exp(\text{str}(\ln W)). \quad (10.36)$$

We also use

$$\ln \text{sdet}(W) = \text{str}(\ln W). \quad (10.37)$$

The statements on superdeterminants and supertraces hold only for square matrices. Otherwise they are not defined. For two matrices $W \in \mathcal{M}(n_1, m_1, n_2, m_2)$, $V \in \mathcal{M}(n_2, m_2, n_1, m_1)$ one shows, by elementary multiplication,

$$WV = \begin{pmatrix} a & \alpha \\ \beta & b \end{pmatrix}\begin{pmatrix} c & \gamma \\ \delta & d \end{pmatrix} = \begin{pmatrix} ac + \alpha\delta & \cdots \\ \cdots & bd + \beta\gamma \end{pmatrix}, \quad VW = \begin{pmatrix} ca + \gamma\beta & \cdots \\ \cdots & db + \delta\alpha \end{pmatrix}. \quad (10.38)$$

Since $\text{tr}\,(ac) = \text{tr}\,(ca)$ and $\text{tr}\,(\alpha\delta) = -\,\text{tr}\,(\delta\alpha)$, one obtains the cyclic invariance of the supertrace

$$\text{str}(WV) = \text{str}(VW). \quad (10.39)$$

We determine the differential of the superdeterminant starting with

$$\text{sdet}(W + \mathrm{d}\,W) = \text{sdet}(W)\text{sdet}(1 + W^{-1}\mathrm{d}\,W) = \text{sdet}(W)(1 + \text{str}(W^{-1}\mathrm{d}\,W)),$$
(10.40)

which yields the differential

$$\mathrm{d}\ \text{sdet}(W) = \text{sdet}(W)\ \text{str}(W^{-1}\mathrm{d}\,W).$$
(10.41)

10.5 Parity Transposition

The parity transposition exchanges the bosonic with the fermionic blocks.
The parity transposition exchanges the submatrices of W, yielding $^{\pi}W$,

$$W = \begin{pmatrix} a & \alpha \\ \beta & b \end{pmatrix} \in \mathcal{M}(n_1, m_1, n_2, m_2), \rightarrow\ ^{\pi}W = \begin{pmatrix} b & \beta \\ \alpha & a \end{pmatrix} \in \mathcal{M}(m_1, n_1, m_2, n_2).$$
(10.42)

Thus, the parity transposition exchanges the bosonic block a with the fermionic block b and also the other two blocks, α and β. One easily verifies

$$^{\pi\pi}W = W, \quad ^{\pi}(AB) = {}^{\pi}(A)\,{}^{\pi}(B),$$
(10.43)

$$^{\pi}(A^*) = ({}^{\pi}A)^*, \quad ^{\pi}(A^{\times}) = ({}^{\pi}A)^{\times},$$
(10.44)

$$^{\pi}({}^{t}W1) = {}^{t}({}^{\pi}W), \quad ^{\pi}(\sigma^{(n,m)}) = -\sigma^{(m,n)}, \quad ^{\pi}({}^{T}W) = \sigma\,{}^{T}({}^{\pi}W)\sigma, \quad (10.45)$$

$$\text{str}({}^{\pi}W) = -\text{str}(W), \quad \text{sdet}({}^{\pi}W) = \text{sdet}^{-1}(W).$$
(10.46)

Do not confuse the parity transposition with the parity operator $\mathscr{P}$ introduced in (2.10), (2.11).

Problem

10.1 Check for matrices $\in \mathcal{M}(1, 1)$ that the expressions for the superdeterminants (10.19) and (10.23) agree.

References

[25] F.A. Berezin, *Introduction to Superanalysis* (Reidel, Dordrecht, 1987)
[55] B. DeWitt, *Supermanifolds* (Cambridge University Press, Cambridge, 1984)
[224] V. Rittenberg, M. Scheunert, Elementary construction of graded Lie groups. J. Math. Phys. **19**, 709 (1978)

Chapter 11
Functions of Matrices

Abstract Superdeterminant and supertrace are functions mapping a square super-matrix to an element of the Grassmann algebra. Here mappings of square superma-trices to supermatrices are considered, first the inverse and secondly mappings by general analytic functions.

11.1 The Inverse

The inverse of a square supermatrix is determined.

If the ordinary parts of a and b of the matrix $W = \begin{pmatrix} a & \alpha \\ \beta & b \end{pmatrix} \in \mathcal{M}(n, m)$ are invertible, then W has an inverse. If we decompose

$$W = \begin{pmatrix} a_0 & 0 \\ 0 & b_0 \end{pmatrix} + \text{nil } W, \qquad (11.1)$$

then one obtains, from $WW^{-1} = \mathbf{1}$

$$\begin{pmatrix} a_0 & 0 \\ 0 & b_0 \end{pmatrix} W^{-1} = \mathbf{1} - \text{nil } W \cdot W^{-1} \qquad (11.2)$$

and

$$W^{-1} = \begin{pmatrix} a_0^{-1} & 0 \\ 0 & b_0^{-1} \end{pmatrix} - \begin{pmatrix} a_0^{-1} & 0 \\ 0 & b_0^{-1} \end{pmatrix} \text{nil } W \cdot W^{-1}, \qquad (11.3)$$

which can be iterated and terminates after a finite number of iterations.
If we set

$$W^{-1} = \begin{pmatrix} c & \gamma \\ \delta & d \end{pmatrix}, \qquad (11.4)$$

© Springer-Verlag Berlin Heidelberg 2016
F. Wegner, *Supermathematics and its Applications in Statistical Physics*,
Lecture Notes in Physics 920, DOI 10.1007/978-3-662-49170-6_11

then the solution of the equations

$$WW^{-1} = \begin{pmatrix} ac + \alpha\delta & a\gamma + \alpha d \\ \beta c + b\delta & \beta\gamma + bd \end{pmatrix} = \begin{pmatrix} \mathbf{1} & 0 \\ 0 & \mathbf{1} \end{pmatrix} \qquad (11.5)$$

$$W^{-1}W = \begin{pmatrix} ca + \gamma\beta & c\alpha + \gamma b \\ \delta a + d\beta & \delta\alpha + db \end{pmatrix} = \begin{pmatrix} \mathbf{1} & 0 \\ 0 & \mathbf{1} \end{pmatrix} \qquad (11.6)$$

for the submatrices c, γ, δ, d, yields

$$W^{-1} = \begin{pmatrix} a^{-1} + a^{-1}\alpha b'^{-1}\beta a^{-1} & -a^{-1}\alpha b'^{-1} \\ -b'^{-1}\beta a^{-1} & b'^{-1} \end{pmatrix}$$

$$= \begin{pmatrix} a'^{-1} & -a'^{-1}\alpha b^{-1} \\ -b^{-1}\beta a'^{-1} & b^{-1} + b^{-1}\beta a'^{-1}\alpha b^{-1} \end{pmatrix}. \qquad (11.7)$$

with

$$a' = a - \alpha b^{-1}\beta, \quad b' = b - \beta a^{-1}\alpha. \qquad (11.8)$$

Supergroups Invertible matrices $W \in \mathcal{M}(n, m)$, that is, matrices with

$$\det(\mathrm{ord}\,(a)) \neq 0, \quad \det(\mathrm{ord}\,(b)) \neq 0, \qquad (11.9)$$

form the general linear supergroup GL(n, m) under multiplication. Matrices $W \in \mathcal{M}(n, m)$ with sdet$(W) = 1$ form the special linear supergroup SL(n, m).

Supergroups obey the laws of groups:

Multiplication: The product of any two elements belongs to the elements of the group.
Identity: There is a unit-element $\mathbf{1}$, so that $\mathbf{1}W = W\mathbf{1} = W$.
Inverse: There is an inverse W^{-1} to each element W.
The associative law $(AB)C = A(BC)$ holds.

Due to these group properties the inverse of the product of two invertible matrices $U, V \in \mathcal{M}(n, m)$ obeys

$$(UV)^{-1} = V^{-1}U^{-1}. \qquad (11.10)$$

11.2   Analytic Functions

The mapping of a square supermatrix to a square supermatrix by an analytic function is described.

If a function, f, of a complex variable, x, can be expanded in a Taylor series,

$$f(x) = \sum_k c_k x^k, \qquad (11.11)$$

then also $f(W)$ is defined by its Taylor expansion

$$f(W) = \sum_k c_k W^k. \qquad (11.12)$$

One observes that, term by term,

$$f(W) = Vf(V^{-1}WV)V^{-1} \qquad (11.13)$$

and for a diagonal matrix

$$f(\mathrm{diag}(\lambda_1, \lambda_2, \ldots \lambda_{n+m})) = \mathrm{diag}(f(\lambda_1), f(\lambda_2), \ldots f(\lambda_{n+m})) \qquad (11.14)$$

hold. If W can be brought into a diagonal representation by means of a similarity transformation, then $f(W)$ is defined by means of (11.13) and (11.14). (In Sect. 17.1 we will prove: W can be brought into diagonal form, if all eigenvalues of ord W are different.)

We consider as an example $W \in \mathscr{M}(1,1)$, (10.2). With

$$V = \begin{pmatrix} 1 & \frac{\alpha}{b-a} \\ 0 & 1 \end{pmatrix} \begin{pmatrix} 1 & 0 \\ \frac{\beta}{a-b} & 1 \end{pmatrix}, \quad V^{-1} = \begin{pmatrix} 1 & 0 \\ -\frac{\beta}{a-b} & 1 \end{pmatrix} \begin{pmatrix} 1 & -\frac{\alpha}{b-a} \\ 0 & 1 \end{pmatrix} \qquad (11.15)$$

one obtains

$$V^{-1}WV = \begin{pmatrix} a - \frac{\alpha\beta}{b-a} & 0 \\ 0 & b - \frac{\alpha\beta}{b-a} \end{pmatrix}, \qquad (11.16)$$

from which we deduce

$$f(W) = \begin{pmatrix} f(a) + \alpha\beta\frac{f(b)-f(a)+(a-b)f'(a)}{(a-b)^2} & \frac{\alpha}{a-b}(f(a) - f(b)) \\ \frac{\beta}{a-b}(f(a) - f(b)) & f(b) + \alpha\beta\frac{f(b)-f(a)+(a-b)f'(b)}{(a-b)^2} \end{pmatrix}. \qquad (11.17)$$

If ord $(a-b)$ goes to 0, then one may expand f around a or b, which, with $b = a+n$, yields

$$f\left(\begin{pmatrix} a & \alpha \\ \beta & b \end{pmatrix}\right) = \begin{pmatrix} f(a) + \alpha\beta g_{11} & \alpha g_{12} \\ \beta g_{12} & f(b) - \frac{1}{2}\alpha\beta g_{22} \end{pmatrix} \qquad (11.18)$$

with

$$g_{11} = \sum_{k=0}^{\infty} \frac{n^k}{(k+2)!} f^{(k+2)}(a) = \sum_{k=0}^{\infty} \frac{(k+1)(-n)^k}{(k+2)!} f^{(k+2)}(b), \qquad (11.19)$$

$$g_{12} = \sum_{k=0}^{\infty} \frac{n^k}{(k+1)!} f^{(k+1)}(a) = \sum_{k=0}^{\infty} \frac{(-n)^k}{(k+1)!} f^{(k+1)}(b), \qquad (11.20)$$

$$g_{22} = -\sum_{k=0}^{\infty} \frac{(k+1)n^k}{(k+2)!} f^{(k+2)}(a) = -\sum_{k=0}^{\infty} \frac{(-n)^k}{(k+2)!} f^{(k+2)}(b), \qquad (11.21)$$

where $f^{(k)}$ indicates the kth derivative of f. The function f is not singular as $a - b$ tends to zero despite the denominators $a - b$ and $(a - b)^2$ in (11.17). If $a = b$, then only the terms with $k = 0$ in (11.19)–(11.21) contribute,

$$f\left(\begin{pmatrix} a & \alpha \\ \beta & a \end{pmatrix}\right) = \begin{pmatrix} f(a) + \frac{1}{2}\alpha\beta f''(a) & \alpha f'(a) \\ \beta f'(a) & f(a) - \frac{1}{2}\alpha\beta f''(a) \end{pmatrix}. \qquad (11.22)$$

Problems

11.1 Calculate the inverse of $W \in \mathcal{M}(1, 1)$.

11.2 Another way to determine the inverse of W is to use (11.10) together with the decomposition (10.16) and (10.22), resp. For this purpose, determine the inverse of the matrices

$$\begin{pmatrix} a & \alpha \\ 0 & 1 \end{pmatrix}, \quad \begin{pmatrix} 1 & 0 \\ \beta & b \end{pmatrix}.$$

11.3 Check (11.22) for $f(W) = W^2$.

11.4 Show

$$({}^\mathrm{T}W)^{-1} = {}^\mathrm{T}(W^{-1}). \qquad (11.23)$$

Chapter 12
Supersymmetric Matrices

Abstract The generalization of symmetric and anti-symmetric matrices to super-matrices is introduced. The Gauss integral over both even and odd variables yields the superpfaffian. Orthosymplectic transformations and groups are generalizations of the orthogonal transformations and groups.

12.1 Quadratic Form

A generalization of symmetric and anti-symmetric matrices to supermatrices is introduced.

We introduce the quadratic form of the vector $S \in \mathcal{M}(n, m, 1, 0)$

$$(S, WS) := {}^{\mathrm{T}}SWS \tag{12.1}$$

with a matrix $W \in \mathcal{M}(n, m)$. (S, WS) and its transposed are equal. We obtain, with (10.5) and (10.12),

$$(S, WS) = {}^{\mathrm{T}}(S, WS) = {}^{\mathrm{T}}S\,{}^{\mathrm{T}}W^{\mathrm{TT}}S = {}^{\mathrm{T}}S\,{}^{\mathrm{T}}W\sigma S. \tag{12.2}$$

We may decompose

$$W = W_+ + W_-, \quad W_+ = \tfrac{1}{2}(W + {}^{\mathrm{T}}W\sigma), \quad W_- = \tfrac{1}{2}(W - {}^{\mathrm{T}}W\sigma). \tag{12.3}$$

Then only W_+ contributes to (S, WS), whereas W_- does not,

$$(S, WS) = (S, W_+S), \quad (S, W_-S) = 0, \tag{12.4}$$

since

$$W_+ = {}^{\mathrm{T}}W_+\sigma, \quad W_- = -{}^{\mathrm{T}}W_-\sigma. \tag{12.5}$$

© Springer-Verlag Berlin Heidelberg 2016
F. Wegner, *Supermathematics and its Applications in Statistical Physics*,
Lecture Notes in Physics 920, DOI 10.1007/978-3-662-49170-6_12

Thus, W_+ obeys

$$W_+ = \begin{pmatrix} a & \alpha \\ -\alpha & b \end{pmatrix} \text{ with } {}^{t}a = a, \; {}^{t}b = -b. \tag{12.6}$$

Here, W_+ is called supersymmetric, although this notion is already used otherwise, see part 3. The matrix W_- is called super-skew-symmetric. The bosonic sector a of the supersymmetric matrix W is symmetric, but the fermionic sector is antisymmetric. A super-skew-symmetric matrix W has an antisymmetric bosonic and a symmetric fermionic sector.

12.2 Gauss Integrals IV, Superpfaffian, Expectation Values

The Gauss integral over both even and odd variables yields the superpfaffian.

A supersymmetric matrix in $\mathcal{M}(n, m)$ is symmetric for the special case $m = 0$, and it is antisymmetric for $n = 0$. These cases appeared in the Gauss integrals (5.4) and (5.28). The integral

$$\text{spf}(W) = \int \exp(-\tfrac{1}{2}\, {}^{T}XWX)[DX], \quad [DX] = \prod_{i=1}^{n} \frac{d\,x_i}{\sqrt{2\pi}} d\,\xi_1 d\,\xi_2 \dots d\,\xi_m, \tag{12.7}$$

with $X \in \mathcal{M}(n, m, 1, 0)$ where W is supersymmetric, is called the superpfaffian of W. So that the integral converges, the ordinary part of the bosonic sector has to be positive definite. The substitution $X = VY$ yields

$$\text{spf}(W) = \text{sdet}(V) \int \exp(-\tfrac{1}{2}\, {}^{T}Y\, {}^{T}VWVY)[DY] = \text{sdet}(V)\text{spf}({}^{T}VWV). \tag{12.8}$$

In particular for

$$V = \begin{pmatrix} \mathbf{1} & -a^{-1}\alpha \\ 0 & \mathbf{1} \end{pmatrix} \tag{12.9}$$

one obtains

$$^{T}VWV = \begin{pmatrix} a & 0 \\ 0 & b + {}^{t}\alpha a^{-1}\alpha \end{pmatrix} \tag{12.10}$$

and thus,

$$\text{spf}\begin{pmatrix} a & \alpha \\ -\alpha & b \end{pmatrix} = \frac{\text{pf}(b + {}^{t}\alpha a^{-1}\alpha)}{\sqrt{\det(a)}}. \tag{12.11}$$

On the other hand, the choice

$$V = \begin{pmatrix} \mathbf{1} & 0 \\ b^{-1}\,{}^{t}\alpha & \mathbf{1} \end{pmatrix} \tag{12.12}$$

$$^{t}VWV = \begin{pmatrix} a + \alpha b^{-1}\,{}^{t}\alpha & 0 \\ 0 & b \end{pmatrix} \tag{12.13}$$

yields

$$\mathrm{spf} \begin{pmatrix} a & \alpha \\ -{}^{t}\alpha & b \end{pmatrix} = \frac{\mathrm{pf}(b)}{\sqrt{\det(a + \alpha b^{-1}\,{}^{t}\alpha)}}. \tag{12.14}$$

The superpfaffian is also called the graded pfaffian and denoted 'pfg'. We observe that the superpfaffian of $W \in \mathcal{M}(0, m)$ equals the pfaffian of W. (The determinant of an empty matrix, that is a 0×0-matrix is 1).

Comparing Eqs. (12.11), (12.14), (10.19), (10.23) one sees that the square of the superpfaffian yields the inverse of the superdeterminant

$$\mathrm{spf}^{2}(W) = \mathrm{sdet}^{-1}(W). \tag{12.15}$$

We vary W infinitesimally

$$W + \mathrm{d}\,W = (\mathbf{1} + \tfrac{1}{2}\mathrm{d}\,WW^{-1})W(\mathbf{1} + \tfrac{1}{2}W^{-1}\mathrm{d}\,W), \quad {}^{T}(W^{-1}\mathrm{d}\,W) = \mathrm{d}\,WW^{-1}, \tag{12.16}$$

for supersymmetric W and $\mathrm{d}\,W$, and use (12.8). Then we obtain

$$\mathrm{spf}(W + \mathrm{d}\,W) = \frac{\mathrm{spf}(W)}{\mathrm{sdet}(\mathbf{1} + \tfrac{1}{2}W^{-1}\mathrm{d}\,W)} \tag{12.17}$$

and thus, the differential of the superpfaffian

$$\mathrm{d}\ \mathrm{spf}(W) = -\tfrac{1}{2}\mathrm{spf}(W)\,\mathrm{str}(W^{-1}\mathrm{d}\,W). \tag{12.18}$$

We add the linear term $^{T}AX = {}^{T}X\sigma A$ with $A \in \mathcal{M}(n, m, 1, 0)$ in the exponent of the Gauss integral (12.7). Completion of the square yields

$$\int [\mathrm{D}\,X]\exp(-\tfrac{1}{2}\,{}^{T}XWX + {}^{T}AX) = \mathrm{spf}(W)\exp(\tfrac{1}{2}\,{}^{T}AW^{-1}\sigma A). \tag{12.19}$$

This formula can only be used if W can be inverted, which requires, in particular, an even m.

The Gaussian density

$$d\rho = \mathrm{spf}(W)^{-1} \exp\left(-\tfrac{1}{2}\,{}^{T}XWX\right) [D\,X], \quad X \in \mathcal{M}(m,n,1,0), \quad W = {}^{T}W\sigma. \tag{12.20}$$

yields

$$\langle X\,{}^{T}X\rangle = W^{-1}, \quad \langle \exp({}^{T}AX)\rangle = \exp(\tfrac{1}{2}\,{}^{T}A\langle X\,{}^{T}X\rangle\sigma A). \tag{12.21}$$

12.3   Orthosymplectic Transformation and Group

Orthosymplectic transformation and group are generalizations of the orthogonal transformation and group.

One introduces orthogonal transformations U in real (and also in complex) vector spaces, with the property that they leave the unit matrix invariant, ${}^{t}U\mathbf{1}U = \mathbf{1}$. We introduce something similar for a square supermatrix. We require that a supersymmetric matrix, C, transforms into itself. We cannot do this for the unit matrix, since its fermionic sector is symmetric. Instead we choose a matrix, C, whose fermionic sector is antisymmetric. Moreover we require that C is invertible. This is only possible if m is even, i.e. $m = 2r$. We write

$$C^{(n,2r)} = \begin{pmatrix} \mathbf{1}_n & 0 \\ 0 & \epsilon_r \end{pmatrix} \in \mathcal{M}(n,2r), \quad \epsilon_r = \begin{pmatrix} \epsilon & & 0 \\ & \epsilon & \\ & & \ddots & \\ 0 & & & \epsilon \end{pmatrix}, \quad \epsilon = \begin{pmatrix} 0 & -1 \\ 1 & 0 \end{pmatrix}. \tag{12.22}$$

The scalar product of the zeroth kind is defined as

$$(S,T)_0 = {}^{T}SCT. \tag{12.23}$$

Linear transformations

$$S' = US, \quad T' = UT, \tag{12.24}$$

which leave this scalar product invariant, are called orthosymplectic. They obey the equation

$$ {}^{T}UCU = C. \tag{12.25}$$

These transformations U form the orthosymplectic group $\mathrm{OSp}(n,m)$ under multiplication. C obeys

$$ {}^{T}C = \sigma C = C\sigma, \quad \mathrm{sdet}(C) = 1, \quad C^2 = \sigma. \tag{12.26}$$

From

$$\text{sdet}(^{T}UCU) = (\text{sdet}(U))^2\text{sdet}(C) = \text{sdet}(C),\tag{12.27}$$

one deduces

$$\text{sdet}(U) = \pm 1.\tag{12.28}$$

Problems

12.1 Derive the second Eq. (12.16).

12.2 Calculate the Gauss integral (12.19) if the linear term reads ^{T}ACX.

12.3 Consider an orthosymplectic transformation $U = \mathbf{1} + V$ with infinitesimal

$$V = \begin{pmatrix} a & \alpha \\ \beta & b \end{pmatrix}$$

What are the conditions on the sectors a, α, β, b?

12.4 Show that if $W \in \mathscr{M}(n, m)$ is supersymmetric and $X \in \mathscr{M}(n, m, n', m')$, then also ^{T}XWX is supersymmetric.

12.5 Which of the matrices $\mathbf{1}$, C, σ, and C^{-1} are supersymmetric?

Chapter 13
Adjoint, Scalar Product, Superunitary Groups

Abstract The two types of conjugation allow the introduction of two types of adjoints for matrices, two types of scalar products and corresponding superunitary groups.

13.1 Adjoint

Two types of adjoints of matrices are introduced corresponding to the two types of conjugation.

13.1.1 Adjoint of the First Kind

We define the adjoint of the first kind $W^\dagger$ of $W \in \mathcal{M}(n_1, m_1, n_2, m_2)$ by the requirement

$$\left(\sum_{ij} S_i^* W_{ij} T_j\right)^* = \sum_{ij} T_j^* (W^\dagger)_{ji} S_i \qquad (13.1)$$

for all vectors $S \in \mathcal{M}(n_1, m_1, 1, 0)$, $T \in \mathcal{M}(n_2, m_2, 1, 0)$. The l.h.s. yields $T_j^* W_{ij}^* S_i$. Thus, one obtains

$$(W^\dagger)_{ji} = W_{ij}^*, \text{ i.e. } W^\dagger = {}^t W^*. \qquad (13.2)$$

One sees immediately that

$$W^{\dagger\dagger} = W, \quad (WV)^\dagger = V^\dagger W^\dagger, \quad {}^\pi(W^\dagger) = ({}^\pi W)^\dagger \qquad (13.3)$$

© Springer-Verlag Berlin Heidelberg 2016
F. Wegner, *Supermathematics and its Applications in Statistical Physics*,
Lecture Notes in Physics 920, DOI 10.1007/978-3-662-49170-6_13

and for $W \in \mathscr{M}(n,m)$ of the form (10.2)

$$\mathrm{sdet} W^{\dagger} = \mathrm{sdet} \begin{pmatrix} {}^{\mathfrak{t}}a^* & {}^{\mathfrak{t}}\beta^* \\ {}^{\mathfrak{t}}\alpha^* & {}^{\mathfrak{t}}b^* \end{pmatrix} = \frac{\det({}^{\mathfrak{t}}a^* - {}^{\mathfrak{t}}\beta^*({}^{\mathfrak{t}}b^*)^{-1}{}^{\mathfrak{t}}\alpha^*)}{\det({}^{\mathfrak{t}}b^*)}$$

$$= \frac{\det({}^{\mathfrak{t}}(a - \alpha b^{-1}\beta)^*)}{\det({}^{\mathfrak{t}}b^*)} = (\mathrm{sdet} W)^*. \tag{13.4}$$

A matrix, which obeys $W = W^{\dagger}$, is called hermitian (of the first kind).

13.1.2   Adjoint of the Second Kind

Similarly one defines the adjoint of the second kind $W^{\ddagger}$ of W by the requirement

$$\left(\sum_{ij} S_i^{\times} W_{ij} T_j\right)^{\times} = \sum_{ij} T_j^{\times} (W^{\ddagger})_{ji} S_i. \tag{13.5}$$

The l.h.s. yields $(-)^{v_j(1-v_i)} T_j^{\times} W_{ij}^{\times} S_i$ with $S_i \in \mathscr{A}_{v_i}$, $T_j \in \mathscr{A}_{v_j}$. Thus,

$$(W^{\ddagger})_{ji} = (-)^{v_j(1-v_i)} W_{ij}^{\times}, \text{ i.e. } W^{\ddagger} = {}^{\mathrm{T}}W^{\times} = \begin{pmatrix} {}^{\mathfrak{t}}a^{\times} & {}^{\mathfrak{t}}\beta^{\times} \\ -{}^{\mathfrak{t}}\alpha^{\times} & {}^{\mathfrak{t}}b^{\times} \end{pmatrix} \tag{13.6}$$

starting from (10.2). One shows that

$$W^{\ddagger\ddagger} = W, \quad (WV)^{\ddagger} = V^{\ddagger}W^{\ddagger}, \quad {}^{\pi}(W^{\ddagger}) = \sigma({}^{\pi}W)^{\ddagger}\sigma \tag{13.7}$$

and for square matrices $W \in \mathscr{M}(n,m)$ one obtains

$$\mathrm{sdet} W^{\ddagger} = (\mathrm{sdet} W)^{\times}. \tag{13.8}$$

Matrices $W = W^{\ddagger}$ are called hermitian (of the second kind).

13.1.3   Adjoint and Transposition: Summary

Apart from the relations given in this Section we repeat ${}^{\mathrm{TT}}W = \sigma W \sigma$, (10.5) and note $W^{\times\times} = \sigma W \sigma$. This may be compared with ${}^{\mathrm{tt}}W = W$, $W^{**} = W$. The parity transposed ${}^{\pi}W$ was introduced in Sect. 10.5.

13.2 Scalar Product, Superunitary Group

Scalar products and corresponding superunitary groups are introduced.

13.2.1 First Kind

In (13.1) we have introduced the bilinear form $S^\dagger W T$ with $W = W^\dagger$. If the inverse of W exists, then W can be written as (see Appendix)

$$W = V^\dagger g V, \quad g = \mathrm{diag}(\mathbf{1}_{n'}, -\mathbf{1}_{n''}, \mathbf{1}_{m'}, -\mathbf{1}_{m''}). \tag{13.9}$$

with $n' + n'' = n$, $m' + m'' = m$. Thus,

$$(S, T) = (VS, VT)_1 \tag{13.10}$$

with

$$(S, T)_1 := S^\dagger g T. \tag{13.11}$$

We call (13.11) the scalar product of the first kind. Linear transformations

$$S' = US, \quad T' = UT, \tag{13.12}$$

which leave the scalar product (13.11) invariant, are called pseudounitary (of the first kind). They obey

$$U^\dagger g U = g \tag{13.13}$$

and constitute the pseudounitary group of first kind $\mathrm{UPL}_1(n', n'', m', m'')$. If $n'' = m'' = 0$ then the group is called superunitary and denoted by $\mathrm{UPL}_1(n, m)$.

Besides the vectors $S, T \in \mathcal{M}(n, m, 1, 0)$, one may also consider vectors $S, T \in \mathcal{M}(n, m, 0, 1)$. The scalar product of these vectors is invariant under the transformations (13.12), too. As an example, the matrices $U \in \mathrm{UPL}_1(1, 1)$ are given,

$$U = \begin{pmatrix} e^{ip} & 0 \\ 0 & e^{iq} \end{pmatrix} \begin{pmatrix} 1 - \frac{1}{2}\omega^*\omega & -\omega^* \\ \omega & 1 + \frac{1}{2}\omega^*\omega \end{pmatrix}, \tag{13.14}$$

with $p, q \in \mathscr{A}_0$, $p^* = p$, $q^* = q$, $\omega \in \mathscr{A}_1$.

13.2.1.1　Appendix to Eq. (13.9)

Here we prove Eq. (13.9). The ordinary part of the hermitian matrix W can be diagonalized. This can be written as

$$\text{ord}\,(W)\tilde{U} = \tilde{U}\Lambda, \tag{13.15}$$

where Λ is the diagonal matrix with the eigenvalues of ord (W) and $\tilde{U}$ is the unitary matrix that diagonalizes ord (W). The columns of $\tilde{U}$ are the eigenfunctions of ord (W). We express Λ by diagonal matrices $\tilde{\Lambda}$ and g,

$$\Lambda = \tilde{\Lambda}g\tilde{\Lambda}, \quad \tilde{\Lambda}_i = \sqrt{|\Lambda_i|}, \quad g_i = \text{sign}(\Lambda_i). \tag{13.16}$$

Then

$$\tilde{\Lambda}^{-1}\tilde{U}^\dagger W\tilde{U}\tilde{\Lambda}^{-1} = g + \Delta, \tag{13.17}$$

where Δ is nil-potent and hermitian. We express

$$g + \Delta = (\mathbf{1} + \tilde{R}^\dagger)g(\mathbf{1} + \tilde{R}), \tag{13.18}$$

This equation can be written

$$R^\dagger + R = \Delta - R^\dagger gR, \quad R := g\tilde{R}. \tag{13.19}$$

We may require $R = R^\dagger$ and iterate (13.19), since Δ is nil-potent. Thus, (13.9) is proven with $V = (\mathbf{1} + \tilde{R})\tilde{\Lambda}\tilde{U}^\dagger$.

13.2.2　Second Kind

Similar considerations hold for the bilinear forms

$$(X, Y) = X^\ddagger WY, \quad (X, Y)_2 = X^\ddagger gY, \tag{13.20}$$

which is the scalar product of the second kind. What has been said above, is still valid, if † is replaced by ‡ and the first kind is replaced by the second kind. In particular, the matrices U with

$$U^\ddagger gU = g \tag{13.21}$$

constitute the pseudounitary group of the second kind $\text{UPL}_2(n', n'', m', m'')$. Again for $n'' = m'' = 0$ this is the unitary group of the second kind $\text{UPL}_2(n', m')$. In particular $U \in \mathrm{UPL}_2(1,1)$ has the representation

$$U = \begin{pmatrix} e^{ip} & 0 \\ 0 & e^{iq} \end{pmatrix} \begin{pmatrix} 1 - \frac{1}{2}\omega^\times \omega & -\omega^\times \\ \omega & 1 + \frac{1}{2}\omega^\times \omega \end{pmatrix} \tag{13.22}$$

with $p, q \in \mathscr{A}_0$, $p = p^\times$, $q = q^\times$, $\omega \in \mathscr{A}_1$.

13.3   Gauss Integrals V

We evaluate Gauss integrals over bilinear forms of super-vectors and their hermitian adjoint.

The integral

$$I_+ = \int \exp(-X^\dagger W X)[\mathrm{D}X], \quad [\mathrm{D}X] = \prod_{k=1}^{n} \frac{\mathrm{d}\,\Re x_k \mathrm{d}\,\Im x_k}{\pi} \prod_{k=1}^{m} (\mathrm{d}\,\xi_k^* \mathrm{d}\,\xi_k). \tag{13.23}$$

with $X \in \mathscr{M}(n, m, 1, 0)$, $W \in \mathscr{M}(n, m)$ can be evaluated similarly to the Gauss integral in Sect. 12.2. One performs the transformation

$$\begin{pmatrix} x \\ \xi \end{pmatrix} = V \begin{pmatrix} x \\ \xi' \end{pmatrix}, \quad \begin{pmatrix} x^\dagger & \xi^\dagger \end{pmatrix} = \begin{pmatrix} x^\dagger & \xi''^\dagger \end{pmatrix} V', \tag{13.24}$$

$$V = \begin{pmatrix} \mathbf{1} & 0 \\ -b^{-1}\beta & \mathbf{1} \end{pmatrix}, \quad V' = \begin{pmatrix} \mathbf{1} & -\alpha b^{-1} \\ 0 & \mathbf{1} \end{pmatrix}. \tag{13.25}$$

If W is not hermitian, then $V^\dagger$ will differ from V' and ξ' from ξ''. However, x and $x^\dagger$ are unchanged under this transformation. The components of ξ' and ξ'' need only be independent. Thus, one obtains

$$X^\dagger W X = x^\dagger (a - \alpha b^{-1}\beta) x + \xi''^\dagger b \xi'. \tag{13.26}$$

Since $\mathrm{sdet}\, V = \mathrm{sdet}\, V' = 1$ one obtains (10.19)

$$I_+ = \frac{\det(b)}{\det(a - \alpha b^{-1}\beta)} = \mathrm{sdet}(W)^{-1}. \tag{13.27}$$

Adding a linear term to the bilinear form, one obtains, for $A \in \mathscr{M}(n, m, 1, 0)$,

$$I_+(A) := \int \exp(-X^\dagger W X + A^\dagger X + X^\dagger A)[\mathrm{D}X] = \mathrm{sdet}(W)^{-1} \exp(A^\dagger W^{-1} A). \tag{13.28}$$

Similarly, we consider

$$I_- = \int \exp(-X^\dagger WX)[DX], \quad [DX] = \prod_{k=1}^{m} \frac{d\,\Re x_k\, d\,\Im x_k}{\pi} \prod_{k=1}^{n}(d\,\xi_k^* d\,\xi_k). \quad (13.29)$$

with $X \in \mathcal{M}(n,m,0,1)$, $W \in \mathcal{M}(n,m)$. Now the even and odd elements are exchanged, so that we return to the original integral (13.23), but then a and b as well as α and β are exchanged in W, (10.2). Thus, the result is with (10.23)

$$I_- = \frac{\det(a)}{\det(b - \beta a^{-1}\alpha)} = \text{sdet}\, W \quad (13.30)$$

and for $A \in \mathcal{M}(n,m,0,1)$

$$I_-(A) := \int \exp(-X^\dagger WX + A^\dagger X + X^\dagger A) = \text{sdet}\, W \exp(A^\dagger W^{-1}A). \quad (13.31)$$

The Gaussian densities

$$d\rho_\pm = \text{sdet}(W)^{\pm1} \exp(-X^\dagger WX)[DX], \quad X \in \mathcal{M}(n,m,1,0) \text{ and } X \in \mathcal{M}(n,m,0,1) \quad (13.32)$$

yield the expectation values

$$\langle XX^\dagger \rangle = W^{-1}, \quad \langle \exp(A^\dagger X + X^\dagger A) \rangle = \exp(A^\dagger \langle XX^\dagger \rangle A), \quad (13.33)$$

that is

$$\langle X_i X_j^* \rangle = (W^{-1})_{ij}, \quad (13.34)$$

$$\langle X_i X_j^* X_k X_l^* \rangle = \langle X_i X_j^* \rangle \langle X_k X_l^* \rangle + (-)^{\nu_j\nu_k + \nu_j\nu_l + \nu_k\nu_l} \langle X_i X_l^* \rangle \langle X_k X_j^* \rangle \quad (13.35)$$

Only products with the same number k of factors X^* and factors X yield a non-vanishing expectation value. They are expressed as a sum of $k!$ contributions.

Problems

13.1 Why does $\,{}^t\beta^*(b^*)^{-1}{}^t\alpha^* = {}^t(\alpha b^{-1}\beta)^*$ hold, as used in (13.4)?

13.2 Equation (13.28) can even be generalized to

$$I_+(A,B) := \int \exp(-X^\dagger WX + A^\dagger X + X^\dagger B)[DX] = (\text{sdet}W)^{-1} \exp(A^\dagger W^{-1}B),$$

$B \in \mathcal{M}(1,0,n,m)$.

To prove this, show

$$\int \exp(-(X - \tilde{A})^\dagger W(X - \tilde{B}))[D\,X] = \int \exp(-X^\dagger WX)[D\,X],$$

(Compare Problem 3.6).

Chapter 14
Superreal Matrices, Unitary-Orthosymplectic Groups

Abstract There are no real odd elements under the conjugation of the second kind. However, the introduction of pairs of odd elements (spinors) in matrices allows the definition of superreal supermatrices. The corresponding unitary-orthosymplectic group and its pseudo-form are introduced.

14.1 Matrices and Groups for the Adjoint of Second Kind

Real matrices are generalized to superreal matrices. The elements in the bosonic sector are real. Two-by-two matrices in the fermionic sector are real quaternions. The odd elements form pairs of conjugate complex elements (spinors). The unitary-orthosymplectic group is introduced.

The scalar products

$$(S, T)_0 = {}^\mathrm{T}SCT, \quad (S, T)_2 = S^\ddagger gT \tag{14.1}$$

are invariant under linear transformations $S' = US$, $T' = UT$, which obey both (12.25) and (13.21),

$$^\mathrm{T}UCU = C, \quad U^\ddagger gU = g. \tag{14.2}$$

Solving the equations with respect to U^{-1}, one obtains

$$U^{-1} = C^{-1}\,{}^\mathrm{T}UC = g^{-1}U^\ddagger g. \tag{14.3}$$

We require now the second equation of (14.3) for a class of matrices $W \in \mathcal{M}(n_1, 2r_1, n_2, 2r_2)$. Thus, they obey

$$C^{(2)-1}\,{}^\mathrm{T}WC^{(1)} = g^{(2)-1}W^\ddagger g^{(1)} \tag{14.4}$$

or equivalently

$$W = C^{(1)}g^{(1)}W^\times g^{(2)}C^{(2)-1}, \tag{14.5}$$

© Springer-Verlag Berlin Heidelberg 2016
F. Wegner, *Supermathematics and its Applications in Statistical Physics*,
Lecture Notes in Physics 920, DOI 10.1007/978-3-662-49170-6_14

where $g^{(1)}, C^{(1)} \in \mathcal{M}(n_1, 2r_1)$, $g^{(2)}, C^{(2)} \in \mathcal{M}(n_2, 2r_2)$. C and g are represented by (12.22) and (13.9). $^{\mathrm{T}}W^{\ddagger} = \sigma W^{\times} \sigma$ is used. Iteration and use of $W = \sigma W^{\times\times} \sigma$ yields

$$W = C^{(1)} g^{(1)} C^{(1)-1} g^{(1)} W g^{(2)} C^{(2)} g^{(2)} C^{(2)-1}. \tag{14.6}$$

We require that (14.5) constitutes a restriction on W, but not (14.6). Then the products of matrices before and after W in (14.6) have to be proportional to the unit matrix

$$C^{(i)} g^{(i)} C^{(i)-1} g^{(i)} = c\mathbf{1}. \tag{14.7}$$

This equation is fulfilled in the bosonic sector of the matrix with $c = 1$. The components of g have to be pairwise equal in the fermionic sector. Thus, m' and m'' in (13.9) have to be even. We call a matrix W pseudoreal, if it obeys (14.5) with these C and g. If $g = 1$, i.e. $n'' = m'' = 0$, then we call the matrix superreal.

Let us write $W = \begin{pmatrix} a & \alpha \\ \beta & b \end{pmatrix}$. Then (14.5) requires

$$a_{ij} = g_i^{(1)} a_{ij}^{\times} g_j^{(2)}. \tag{14.8}$$

Thus, $a = a^{\times}$ has to be real for $g_i^{(1)} = g_j^{(2)}$. The components of α, β and b are pairwise related,

$$\alpha_{i,j+1} = g_i^{(1)} \alpha_{ij}^{\times} g_j^{(2)}, \quad j - n_2 \text{ odd}, \tag{14.9}$$

$$\beta_{i+1,j} = g_i^{(1)} \beta_{ij}^{\times} g_j^{(2)}, \quad i - n_1 \text{ odd}, \tag{14.10}$$

$$\left. \begin{array}{l} b_{i+1,j+1} = g_i^{(1)} b_{ij}^{\times} g_j^{(2)}, \\ b_{i+1,j} = -g_i^{(1)} b_{i,j+1}^{\times} g_j^{(2)}, \end{array} \right\} \quad i - n_1, j - n_2 \text{ odd}. \tag{14.11}$$

The four elements (with odd $i - n_1$ and $j - n_2$)

$$\begin{pmatrix} b_{ij} & b_{i,j+1} \\ b_{i+1,j} & b_{i+1,j+1} \end{pmatrix} = \begin{pmatrix} b_{ij} & b_{i,j+1} \\ -b_{i,j+1}^{\times} & b_{ij}^{\times} \end{pmatrix} \tag{14.12}$$

form a real quaternion for $g_i^{(1)} = g_j^{(2)}$. Under the same condition pairs of elements α and β form the spinors

$$\begin{pmatrix} \alpha_{ij} & \alpha_{i,j+1} \end{pmatrix} = \begin{pmatrix} \alpha_{ij} & \alpha_{ij}^{\times} \end{pmatrix}, \quad \begin{pmatrix} \beta_{ij} \\ \beta_{i+1,j} \end{pmatrix} = \begin{pmatrix} \beta_{ij} \\ \beta_{ij}^{\times} \end{pmatrix}. \tag{14.13}$$

One concludes immediately from (14.5), that the sum and the product of two pseudoreal and superreal matrices are again pseudoreal and superreal, respectively. Transformations U, which obey (14.3), are called unitary-orthosymplectic, if $g = 1$, otherwise pseudounitary-orthosymplectic. As intersection of two groups—the orthosymplectic and the (pseudo)unitary group—they constitute a group again, the $\mathrm{UOSp}(n, 2r)$ and $\mathrm{UOSp}_2(n', n'', m', m'')$, respectively.

14.2 Vector Products

Superreal vectors have identical scalar products of zeroth and second kind. For a second kind of vectors both types of products are simply related.

Consider vectors X and Y as matrices in $\mathcal{M}(n, 2r, 1, 0)$, set $g^{(2)} = 1$, and observe $\mathsf{C}^{(2)} = 1_1$. Then (14.4) reads

$$^\mathrm{T}X\mathsf{C} = X^\ddagger g, \tag{14.14}$$

from which one concludes

$$(X, Y)_0 = (X, Y)_2. \tag{14.15}$$

Thus, both scalar products are identical for pseudoreal and superreal vectors. The superreal vectors $(\mathsf{C}^{(1)} = 1)$ are of the form

$$X = \begin{pmatrix} \vdots \\ x_i \\ \vdots \\ \xi_j \\ \xi_j^\times \\ \vdots \end{pmatrix}, \quad Y = \begin{pmatrix} \vdots \\ y_i \\ \vdots \\ \eta_j \\ \eta_j^\times \\ \vdots \end{pmatrix} \in \mathcal{M}(n, 2r, 1, 0). \tag{14.16}$$

With

$$X^\ddagger = \left(\cdots x_i^\times \cdots \xi_j^\times -\xi_j \cdots \right), \quad Y^\ddagger = \left(\cdots y_i^\times \cdots \eta_j^\times -\eta_j \cdots \right) \tag{14.17}$$

one obtains

$$(X, Y)_2 = \sum_i x_i^\times y_i + \sum_j (\xi_j^\times \eta_j - \xi_j \eta_j^\times), \tag{14.18}$$

where actually $x_i = x_i^\times$ and $y_i = y_i^\times$ are real.

A second kind of superreal vectors, which are two-column vectors, can be written

$$X = \begin{pmatrix} \xi & \xi^\times \\ x & \epsilon x^\times \end{pmatrix} \in \mathcal{M}(n, 2r, 0, 2), \quad \xi \in \mathcal{M}(n, 0, 0, 1), \quad x \in \mathcal{M}(0, 2r, 0, 1).$$

(14.19)

It is superreal, $X = CX^\times C^{-1}$. Its hermitian adjoint is

$$X^\ddagger = \begin{pmatrix} -\overline{\xi}^\times & \overline{x}^\times \\ \overline{\xi} & -\overline{x}\epsilon \end{pmatrix} \in \mathcal{M}(0, 2, n, 2r).$$

(14.20)

Introducing a second vector Y of this type by replacing $x \to y, \xi \to \eta$, one obtains

$$(X, Y)_2 = \begin{pmatrix} \overline{x}^\times y - \overline{\xi}^\times \eta & \overline{x}^\times \epsilon y^\times - \overline{\xi}^\times \eta^\times \\ \overline{\xi}\eta - \overline{x}\epsilon y & \overline{\xi}\eta^\times + \overline{x}y^\times \end{pmatrix}.$$

(14.21)

One has to introduce the signs according to (14.9)–(14.11) for the corresponding pseudoreal vectors g. The scalar product (14.21) obeys

$$(X, Y)_0 = C(X, Y)_2.$$

(14.22)

For this more general scalar product, one has, in generalization of (12.23) and (14.15),

$$^T(X, Y)_0 = {}^T(Y, X)_0 \sigma, \quad (X, Y)_2^\ddagger = (Y, X)_2.$$

(14.23)

14.3 Gauss Integrals VI, Superreal Vectors

Gaussian integrals on superreal vectors are determined.
    We consider the integral

$$I_+ = \int \exp(-\tfrac{1}{2}X^\ddagger WX)[DX], \quad [DX] = \prod_k \frac{dx_k}{\sqrt{2\pi}} \prod_k d\xi_k^\times d\xi_k.$$

(14.24)

with superreal $X \in \mathcal{M}(n, 2r, 1, 0)$ and $X^\ddagger \in \mathcal{M}(1, 0, n, 2r)$, (14.16), (14.17). Since X is superreal, it obeys

$$X^\ddagger WX = {}^TXCWX.$$

(14.25)

We require that CW is supersymmetric,

$$CW = {}^T(CW)\sigma \equiv {}^TWC.$$

(14.26)

From (12.7), one obtains

$$I_+ = \int \exp(-\tfrac{1}{2}X^{\ddagger}WX)[DX] = \mathrm{spf}(\mathsf{C}W). \tag{14.27}$$

and for superreal $A \in \mathscr{M}(n, 2r, 1, 0)$

$$I_+(A) = \int \exp(-\tfrac{1}{2}X^{\ddagger}WX + X^{\ddagger}A)[DX] = \mathrm{spf}(\mathsf{C}W)\exp(\tfrac{1}{2}A^{\ddagger}W^{-1}A). \tag{14.28}$$

The Gaussian density

$$\mathrm{d}\rho_+ = \mathrm{spf}(\mathsf{C}W)^{-1}\exp(-\tfrac{1}{2}X^{\ddagger}WX)[DX] \tag{14.29}$$

yields

$$\langle XX^{\ddagger}\rangle = W^{-1}, \quad \langle\exp(X^{\ddagger}A)\rangle = \langle\exp(A^{\ddagger}X)\rangle = \exp(\tfrac{1}{2}A^{\ddagger}\langle XX^{\ddagger}\rangle A). \tag{14.30}$$

Next, we consider the integral

$$I_- = \int \exp(\tfrac{1}{2}\,\mathrm{str}(X^{\ddagger}WX))[DX], \quad [DX] = \prod_{k=1}^{2r}\frac{\mathrm{d}\,\Re x_k \mathrm{d}\,\Im x_k}{\pi}\prod_{k=1}^{n}(\mathrm{d}\,\xi_k^{\times}\mathrm{d}\,\xi_k) \tag{14.31}$$

with the superreal vectors X and $X^{\ddagger}$, Eqs. (14.19), (14.20). Provided the condition $W = \mathsf{C}^{-1}\,{}^{\mathsf{T}}W\mathsf{C}$ (14.26) is fulfilled,

$$X^{\ddagger}WX = \left(-\xi^{\times}\ x^{\times}\right)W\begin{pmatrix}\xi\\x\end{pmatrix} \tag{14.32}$$

and, comparing with (13.30), yields

$$I_- = \mathrm{sdet}(W). \tag{14.33}$$

Further, we obtain with superreal $A \in \mathscr{M}(n, 2r, 0, 2)$

$$\int [DX]\exp\,\mathrm{str}(\tfrac{1}{2}X^{\ddagger}WX + X^{\ddagger}A) = \mathrm{sdet}(W)\exp(-\tfrac{1}{2}\,\mathrm{str}(A^{\ddagger}W^{-1}A)). \tag{14.34}$$

The Gaussian density

$$\mathrm{d}\rho_- = \mathrm{sdet}(W)^{-1}\exp(\tfrac{1}{2}\,\mathrm{str}(X^{\ddagger}WX))[DX] \tag{14.35}$$

yields

$$\langle X_{ij} X_{kl}^{\ddagger} \rangle = \delta_{jk} (W^{-1})_{il}, \tag{14.36}$$

$$\langle \exp(\operatorname{str}(X^{\ddagger} A)) \rangle = \langle \exp(\operatorname{str}(A^{\ddagger} X)) \rangle = \exp(-\tfrac{1}{2} \operatorname{str}(A^{\ddagger} \langle XX^{\ddagger} \rangle A)). \tag{14.37}$$

In deriving this result, one uses

$$(X^{\ddagger} WX)_{kj} = -\tfrac{1}{2} \delta_{kj} \operatorname{str}(X^{\ddagger} WX). \tag{14.38}$$

One can obtain this result by considering

$$^{\mathrm{T}}(X^{\ddagger} WX) = {}^{\mathrm{T}}X \, {}^{\mathrm{T}}W \, {}^{\mathrm{T}}X^{\ddagger}. \tag{14.39}$$

Insertion of

$$^{\mathrm{T}}X = CX^{\ddagger} C^{-1}, \qquad {}^{\mathrm{T}}X^{\ddagger} = \sigma X^{\times} \sigma = \sigma C^{-1} XC\sigma \tag{14.40}$$

and (14.26) yields

$$^{\mathrm{T}}(X^{\ddagger} WX) = C(X^{\ddagger} WX)C^{-1} \tag{14.41}$$

and thus (14.36).

Expectation values of products of an odd number of factors X vanish. For four factors X one obtains three contributions

$$\langle X_i X_j X_k X_l \rangle = \langle X_i X_j \rangle \langle X_k X_l \rangle + (-)^{\nu_j \nu_k} \langle X_i X_k \rangle \langle X_j X_l \rangle + (-)^{\nu_l (\nu_j + \nu_k)} \langle X_i X_l \rangle \langle X_j X_k \rangle. \tag{14.42}$$

In general a product of $2k$ factors X yields a sum of $(2k-1)!!$ terms.

Summary of Gauss Integrals We have considered Gaussian integrals and expectation values several times. We give a summary:

Integrals of $\exp(-x^{\dagger} ax)$ and $\exp(-\xi^{\dagger} a\xi)$ in Sect. 3.3, their supersymmetric generalization $\exp(-X^{\dagger} WX)$ in Sect. 13.3, integrals of $\exp(-\tfrac{1}{2} {}^{\mathrm{t}}xax)$ and $\exp(\tfrac{1}{2} {}^{\mathrm{t}}\xi a\xi)$ in Sects. 5.1 and 5.3 and their supersymmetric generalization $\exp(-\tfrac{1}{2} {}^{\mathrm{T}}XWX)$ in Sect. 12.2, finally integrals of $\exp(\pm \tfrac{1}{2} X^{\ddagger} WX)$ are determined in this Sect. 14.3.

Problem

14.1 Show that, if W obeys two of the following three conditions, then also the third one holds: (i) W is hermitian, $W = W^{\ddagger}$, (ii) W is superreal $W = CW^{\times}C^{-1}$, (iii) CW is supersymmetric, $CW = {}^{\mathrm{T}}(CW)\sigma$.

14.2 Show for superreal $A \in \mathcal{M}(n_1, 2r_1, n_3, 2r_3)$ and $B \in \mathcal{M}(n_2, 2r_2, n_3, 2r_3)$

$$^{\mathrm{T}}(BA^{\ddagger}) = \mathsf{C}AB^{\ddagger}\mathsf{C}^{-1} \tag{14.43}$$

and similarly $^{\mathrm{T}}(A^{\ddagger}B) = \mathsf{C}B^{\ddagger}A$ for superreal $A \in \mathcal{M}(n_3, 2r_3, n_2, 2r_2)$ and $B \in \mathcal{M}(n_3, 2r_3, n_1, 2r_1)$.

Chapter 15
Integral Theorems for the Unitary Group

Abstract The integral over a function of supervectors and supermatrices invariant under superunitary transformations can be reduced by cancelling equal numbers of even and odd components.

15.1 Integral Theorem for Functions of Vectors Invariant Under Superunitary Groups

The integral over a function of supervectors invariant under superunitary transformations can be reduced by cancelling equal numbers of even and odd components. After an introduction we present the theorem. Then we show it for the simplest case and finally we generalize it to several vectors and several components.

15.1.1 Introduction

Parisi and Sourlas [204] were probably the first to give an integral theorem for functions of vectors invariant under superunitary operations as presented in this section, in the simplest case given by (15.15).

The conjecture that there should be theorems for integrals over functions of vectors and matrices invariant under superunitary or unitary-orthosymplectic groups came from the formulation of random-matrix theory (Chap. 21) and of the diffusive models (Chap. 22). If one expresses the average over the product of two Green's functions at different energies, then the governing action contains fields for both energies. If one determines the averaged one-particle Green's function, then the integrals over the fields of the other energy have to yield the action for the fields of the energy of interest. Thus, one may expect that the symmetry of the fields to be integrated over is sufficient to reduce the integral to the action for one field only. Indeed this can be shown. Generally we expand the functions in Grassmann variables and determine the relation between the expansion coefficients. Then the integral over the coefficient multiplied by the product of all odd variables is performed. This is possible, since it can be expressed by derivatives of the body of the function.

© Springer-Verlag Berlin Heidelberg 2016
F. Wegner, *Supermathematics and its Applications in Statistical Physics*,
Lecture Notes in Physics 920, DOI 10.1007/978-3-662-49170-6_15

Integral theorems have been derived by Efetov [65, 66], Rothstein [225], Constantinescu and de Groote [52], Wegner [274]. A survey is given by Kieburg, Kohler, and Guhr [143].

15.1.2   Theorem for Superunitary Vectors of First Kind

We show that if a function, f, of supervectors is invariant under arbitrary super-unitary transformations of these vectors, then the integral over all these vectors is unchanged, if one even and one odd component of the vectors is set to zero and only integration over the remaining components is performed.

Consider a function, f, of N vectors $S(x)$, $x = 1, 2, \ldots N$ with $S(x) \in \mathscr{M}(n, m, 1, 0) \cup \mathscr{M}(n, m, 0, 1)$ which is invariant against arbitrary superunitary transformations,

$$f\{S\} = f\{US\}, \quad U \in UPL_1(n, m). \tag{15.1}$$

Examples of such functions are functions which depend only on scalar products $(S(x), S(y))_1$.

Provided the function f is sufficiently often differentiable, and f as well as its derivatives decay sufficiently rapidly for large arguments of the ordinary parts of S, then

$$\int [D\,S] f\{S\} = \int [D\,S'] f\{S'\} \tag{15.2}$$

holds, where

$$[D\,S] = \prod_{x=1}^{N} \prod_{i=1}^{n+m} D\,S_i(x), \ [D\,S'] = \prod_{x=1}^{N} \prod_{i=2}^{n+m-1} D\,S_i(x), \tag{15.3}$$

$$D\,S_i = \begin{cases} \frac{1}{\pi}\mathrm{d}\,\Re S_i\mathrm{d}\,\Im S_i, & S_i \in \mathscr{A}_0, \\ \mathrm{d}\,S_i^*\mathrm{d}\,S_i & S_i \in \mathscr{A}_1. \end{cases} \tag{15.4}$$

$$S_i'(x) = \begin{cases} S_i(x) & 1 < i < n + m, \\ 0 & i = 1 \text{ or } i = n + m. \end{cases} \tag{15.5}$$

To show this for general functions f we fix the components with indices $1 < i < n + m$ and use the invariance of f under superunitary transformations U, which leave these components unchanged, but transform between the components S_1 and S_{n+m}. Then it is sufficient to show that for $n = m = 1$

$$\int [D\,S] f\{S\} = f\{0\} \tag{15.6}$$

holds. We proceed in two steps: In Sect. 15.1.3 we show that (15.6) holds for $N = 1$, and in Sect. 15.1.4 we show by complete induction that (15.6) holds for natural N.

15.1.3   Proof of the Theorem for $N = 1$

The theorem is first shown for one vector only. We write without loss of generality

$$S(1) = (S, \theta) \in \mathcal{M}(1, 1, 1, 0), \quad S = R + iI, \quad R, I \text{ real.} \tag{15.7}$$

We expand in the Grassmann variables,

$$f = f_0(R, I) + \theta f_1(R, I) + \theta^* f_2(R, I) + \theta^* \theta f_3(R, I). \tag{15.8}$$

Under the transformation (13.14), for infinitesimal p, q, ω, ω^*, one obtains

$$\delta f = p h_0 + i q h_1 + \omega h_2 + \omega^* h_3 = 0 \tag{15.9}$$

with

$$h_1 = \theta f_1 - \theta^* f_2 = 0, \quad \text{i.e.} f_1 = f_2 = 0. \tag{15.10}$$

Further, one has

$$h_0 = -I \frac{\partial f_0}{\partial R} + R \frac{\partial f_0}{\partial I} + \theta^* \theta (-I \frac{\partial f_3}{\partial R} + R \frac{\partial f_3}{\partial I}) = 0, \tag{15.11}$$

which yields

$$f_0 = f_0(R^2 + I^2) = f_0(S^* S), \quad f_3 = f_3(S^* S). \tag{15.12}$$

Finally one concludes from

$$h_2 = \theta^* (\frac{\partial f_0}{\partial S^*} - S f_3) = 0, \quad h_3 = \theta(-\frac{\partial f_0}{\partial S} + S^* f_3) = 0 \tag{15.13}$$

that

$$f = f_0 + \theta^* \theta f_0'(S^* S) = f_0(S^* S + \theta^* \theta). \tag{15.14}$$

Thus, f is a function of the scalar product $(S(1), S(1))_1$. Application of this argument on several components of a vector $S \in \mathcal{M}(n, m, 1, 0)$ yields generally $f = f(S^\dagger S)$.

The integration yields

$$\int \mathrm{D}\, SD\, \theta f(S^*S + \theta^*\theta) = -\frac{1}{\pi}\int \mathrm{d}\, R \mathrm{d}\, I f'(S^*S)$$

$$= -\int \mathrm{d}\,(S^*S)f'(S^*S) = f(0) - f(\infty). \quad (15.15)$$

Instead of considering $S \in \mathcal{M}(1,1,1,0)$, the same argument can be applied to the components S_1 and S_{n+m} for $S \in \mathcal{M}(n,m,1,0)$ and yields (15.6) for $N = 1$, where $f(S^*S \to \infty) \to 0$ is assumed.

Instead of $S \in \mathcal{M}(1,1,1,0)$ we can apply the same argument for $S = (\theta, S) \in \mathcal{M}(1,1,0,1)$.

Here, we assumed that R and I are real. Later we will consider the case, where $S = R + iI$, $S^* = R - iI$, but allow R and I to be complex. For this case the proof also holds.

15.1.4 Generalization to Natural N

The argument is generalized to an arbitrary number of vectors. Let us introduce

$$I(S(N)) = \int \mathrm{D}\, S(1)\mathrm{D}\, S(2)\cdots\mathrm{D}\, S(N-1)f\{S\}. \quad (15.16)$$

One obtains, with $S'(x) = US(x)$,

$$\mathrm{D}\, S'(x) = \frac{1}{\pi}\mathrm{d}\,\Re S'\mathrm{d}\,\Im S'\mathrm{d}\,\theta'^*\mathrm{d}\,\theta' = \frac{1}{2\pi\mathrm{i}}\mathrm{d}\, S'^*\mathrm{d}\, S'\mathrm{d}\,\theta'^*\mathrm{d}\,\theta'$$

$$= \frac{1}{2\pi\mathrm{i}}\mathrm{sdet}U^\dagger\mathrm{sdet}\,^\mathrm{T}U\mathrm{d}\, S^*\mathrm{d}\, S\mathrm{d}\,\theta^*\mathrm{d}\,\theta = \mathrm{D}\, S(x) \quad (15.17)$$

and the invariance of f

$$I(S(N)) = \int \mathrm{D}\, S'(1)\mathrm{D}\, S'(2)\cdots\mathrm{D}\, S'(N-1) \quad (15.18)$$

$$\times f(S'(1), S'(2),\dots S'(N-1), US(N)) = I(US(N)),$$

i.e. $I(S(N))$ is invariant under superunitary transformations of $S(N)$. Thus, one concludes

$$I = \int \mathrm{D}\, S(N)I(S(N)) = I(S(N)) = 0. \quad (15.19)$$

In order to evaluate I, one sets $S(N)$ equal to zero and is now only left with the integrations over $S(1), \cdots S(N - 1)$. Thus, if (15.6) holds for $N - 1$, then it holds for N, too. Since it has been shown in Sect. 15.1.3 for $N = 1$, it holds also for all natural N.

15.1.5 Consequences

Due to the theorem, the integral depends only on $m - n$ components. We return to the original integral, (15.2), with general n and m. In the special case $n = m$, the integral yields $f\{0\}$, otherwise one is left with the integral

$$\int [D\,S]'f\{S'\} \tag{15.20}$$

with vectors $S' \in \mathcal{M}(n - m, 0, 1, 0) \cup \mathcal{M}(n - m, 0, 0, 1)$, if $n > m$, otherwise with vectors $S' \in \mathcal{M}(0, m - n, 1, 0) \cup \mathcal{M}(0, m - n, 0, 1)$.

If, in particular, f is explicitly defined as a function of the scalar products $(S(x), S(y))_1$, then the integral depends only on f and $n - m$.

The integral $\int [D\,S]f\{S\}$ is for complex vectors with a components, $S(x) \in \mathcal{M}(a, 0, 1, 0)$ only defined for $a > 0$. The observation that the integral for $S \in \mathcal{M}(n, m, 1, 0) \cup \mathcal{M}(n, m, 0, 1)$ depends only on $n - m$, allows one to define it also for integer $a \leq 0$. One takes $n - m = a, n \geq 0, m \geq 0$.

One obtains the same theorem and proof for functions of S, which are invariant under superunitary transformations of the second kind.

15.2 Integral Theorem for Quasihermitian Matrices: Superunitary Group

Integrals over a function of quasihermitian matrices $\in \mathcal{M}(n, m)$ invariant under superunitary transformations can be reduced to the same integrals over matrices $\in \mathcal{M}(n-a, m-a)$, where the cancelled components are set to zero with the exception of the diagonal matrix elements, which have to agree in the bosonic and fermionic sector, but are otherwise arbitrary.

15.2.1 Introduction and Theorem, 'Quasihermitian'

Here we formulate the theorem which reduces the integral over the matrices from $\mathcal{M}(n, m)$ to $\mathcal{M}(n - 1, m - 1)$.

We will now derive an integral theorem for matrices analogous to the one for vectors in the preceding section. One may ask for integrals over functions $f\{Q(x)\}$ of hermitian matrices $Q(x)$, which are invariant under unitary transformations,

$$f\{UQ(x)U^\dagger\} = f\{Q(x)\}. \tag{15.21}$$

However, the requirement of hermiticity leads to a problem of convergence. We will see in Sect. 15.2.2 that a function $f(Q)$ of a matrix $Q \in \mathcal{M}(1,1)$, which is invariant under superunitary transformations and differentiable, assumes for

$$Q = \begin{pmatrix} \lambda & 0 \\ 0 & \lambda \end{pmatrix} \tag{15.22}$$

a value independent of λ. This is surprising, since Q, (15.22), is invariant against superunitary transformations and does not transform into one with a different λ. The elements $Q_{\alpha\beta} \in \mathcal{A}_0$ of a hermitian matrix $Q \in \mathcal{M}(n,m)$ can be given by

$$Q_{\alpha\beta} = R_{\alpha\beta} + iI_{\alpha\beta}, \quad R_{\alpha\beta} = R_{\beta\alpha}, \quad I_{\alpha\beta} = -I_{\beta\alpha}, \tag{15.23}$$

where R and I are real.

If an integral over a real variable is given, then we usually shift the path of integration into the complex plane. Here we will do the same separately for the variables $R_{\alpha\beta}$ and $I_{\alpha\beta}$, keeping (15.23) and call the matrix Q quasihermitian.

The Integral Theorem If f is sufficiently often differentiable and decays for large arguments $|R|$, $|I|$ sufficiently rapidly to 0, then

$$\int [\mathrm{D}\,Q] f\{Q\} = (\mathrm{is})^{(2m-1)N} \int [\mathrm{D}\,Q'] f\{Q'\} \tag{15.24}$$

with

$$[\mathrm{D}\,Q] = \prod_x \left\{ \prod_{\alpha > \beta, Q_{\alpha\beta} \in \mathcal{A}_0} \left(\frac{1}{\pi} \mathrm{d}\,R_{\alpha\beta}(x)\mathrm{d}\,I_{\alpha\beta}(x) \right) \prod_\alpha \left(\frac{1}{\sqrt{2\pi}} \mathrm{d}\,R_{\alpha\alpha}(x) \right) \right.$$

$$\left. \times \prod_{\alpha \le n} \prod_{\beta > n} (\mathrm{d}\,Q_{\alpha\beta}(x)\mathrm{d}\,Q_{\beta\alpha}(x)) \right\}. \tag{15.25}$$

The integral over Q' contains only the integration of the components $1 < \alpha < n+m$, $1 < \beta < n+m$. The other components $Q'_{\alpha\beta}(x)$ are set to $\lambda(x)\delta_{\alpha\beta}$. N is the number of matrices Q, s is a sign depending on the directions of integration of R and I. First we show that the theorem holds for $N = 1$, $Q \in \mathcal{M}(1,1)$. Then the proof is generalized to an arbitrary number N of matrices $Q \in \mathcal{M}(n,m)$ with natural n and m.

15.2.2   Integral Theorem for One Matrix $\in \mathcal{M}(1,1)$

Here we show the theorem for one matrix $\in \mathcal{M}(1,1)$. The result is given by the function of the matrix $\lambda \mathbf{1}$ with arbitrary λ.

First, the most general differentiable function f of

$$Q = \begin{pmatrix} a & \alpha^* \\ \alpha & b \end{pmatrix} \in \mathcal{M}(1,1), \tag{15.26}$$

is determined, which fulfils (15.21). We expand

$$f = g(a,b) + \alpha g_1(a,b) + \alpha^* g_2(a,b) + \alpha\alpha^* h(a,b). \tag{15.27}$$

With

$$U = \begin{pmatrix} e^{ip} & 0 \\ 0 & e^{iq} \end{pmatrix} \tag{15.28}$$

one obtains

$$UQU^\dagger = \begin{pmatrix} a & e^{i(p-q)}\alpha^* \\ e^{i(q-p)}\alpha & b \end{pmatrix} \tag{15.29}$$

and thus, from Eq. (15.21), $g_1 = g_2 = 0$.

The infinitesimal transformation (13.14)

$$U = \begin{pmatrix} 1 & -\omega^* \\ \omega & 1 \end{pmatrix}, \quad U^\dagger = \begin{pmatrix} 1 & \omega^* \\ -\omega & 1 \end{pmatrix} \tag{15.30}$$

yields

$$UQU^\dagger = \begin{pmatrix} a - \omega^*\alpha + \omega\alpha^* & \alpha^* - \omega^*(b-a) \\ \alpha - \omega(b-a) & b - \omega^*\alpha + \omega\alpha^* \end{pmatrix} \tag{15.31}$$

and thus,

$$f(UQU^\dagger) = f(Q) + (-\omega^*\alpha + \omega\alpha^*)\chi \tag{15.32}$$

with

$$\chi = \frac{\partial g}{\partial a} + \frac{\partial g}{\partial b} + (a-b)h = 0, \tag{15.33}$$

which, due to (15.21), has to vanish. If we insert $a = b = \lambda$ in (15.33), then it follows that $g(\lambda, \lambda)$ does not depend on λ, i.e. $f(\lambda \mathbf{1})$ does not depend on λ, as claimed before.

We have to determine

$$\int D\, Qf(Q) = \frac{1}{2\pi} \int d\, a d\, b d\, \alpha^* d\, \alpha (g + \alpha \alpha^* h)$$

$$= \frac{1}{2\pi} \int d\, a d\, b h = \frac{1}{2\pi} \int d\, a d\, b \frac{\partial_a g + \partial_b g}{b - a}. \qquad (15.34)$$

First Evaluation   We express a and b as functions of ρ and ϕ,

$$a = \lambda + e^{i\gamma_+} \rho \cos(\phi - \gamma_+),$$
$$b = \lambda + e^{i\gamma_-} \rho \cos(\phi - \gamma_-) \qquad (15.35)$$

and integrate in the positive ρ and ϕ direction

$$d\, a d\, b = e^{i(\gamma_+ + \gamma_-)} \rho |\sin(\gamma_+ - \gamma_-)| d\, \rho d\, \phi,$$
$$b - a = \rho i e^{i(\gamma_+ + \gamma_- - \phi)} \sin(\gamma_- - \gamma_+) \qquad (15.36)$$
$$\frac{\partial g}{\partial a} + \frac{\partial g}{\partial b} = e^{-i\phi} \left(\frac{\partial}{\partial \rho} - \frac{i}{\rho} \frac{\partial}{\partial \phi} \right) g.$$

Hence

$$\int D\, Qf(Q) = -\frac{is}{2\pi} \int d\, \rho d\, \phi \left(\frac{\partial g}{\partial \rho} - \frac{i}{\rho} \frac{\partial g}{\partial \phi} \right). \qquad (15.37)$$

The ϕ-integral of the second term vanishes. Integration of the first term yields

$$\int D\, Qf(Q) = -\frac{is}{2\pi} \int d\, \phi (g(\rho = \infty) - g(\rho = 0)) = +isg(\rho = 0) = +isf(\lambda \mathbf{1}). \qquad (15.38)$$

Thus, (15.24) is proven for $N = n = m = 1$.

We observe that the sign s of the integral depends on how the paths of integration for a and b cross. If b approaches from the right, then s is positive, if it comes from the left, then it is negative.

Second Evaluation   Another evaluation without restriction to the paths (15.35) can be obtained by starting from the following integral over the region R indicated in

Fig. 15.1 Region R to be integrated over. a and b are parametrized $a = a(x, y)$, $b = b(x, y)$, $x, y \in \mathbb{R}$

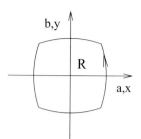

Fig. 15.1 of x, y or equivalently a, b. Stokes' theorem yields

$$\oint_{\partial R} \left(db \frac{g}{b-a} - da \frac{g}{b-a} \right) = \int_R da\, db \left(\partial_a \frac{g}{b-a} + \partial_b \frac{g}{b-a} \right) \tag{15.39}$$

$$= \int_R da\, db \left(\frac{\partial_a g + \partial_b g}{b-a} + g \left(\partial_a \frac{1}{b-a} + \partial_b \frac{1}{b-a} \right) \right).$$

For $b \neq a$, one has $\partial_a \frac{1}{b-a} + \partial_b \frac{1}{b-a} = 0$. However, there is a contribution at $b = a$, which can be easily seen by performing the integral for $g = 1$,

$$\oint_{\partial R} \frac{d(b-a)}{b-a} = \ln(b-a), \tag{15.40}$$

which, after circling around $b = a$, yields $\pm 2\pi i$. The sign is given by $s = \text{sign}(\Im \frac{\partial(a,b^*)}{\partial(x,y)})$. Thus we obtain

$$\int_R da\, db \frac{\partial_a g + \partial_b g}{b-a} = \oint_{\partial R} \left(db \frac{g}{b-a} - da \frac{g}{b-a} \right) \tag{15.41}$$

$$+ 2\pi i \int_R da\, db\, g(a,a)s\delta(\Re(b-a))\delta(\Im(b-a)).$$

Provided $g(a, b)$ decays sufficiently rapidly as R increases in all directions the integral (15.34) yields

$$\int DQ f(Q) = i s f(\lambda \mathbf{1}). \tag{15.42}$$

Again (15.24) is proven for $N = n = m = 1$. As before the sign s of the integral depends on how the paths of integration for a and b cross.

15.2.3 Integral Theorem for N Matrices $Q \in \mathcal{M}(1,1)$

We generalize the theorem to integrals over N matrices. We show first that

$$I(Q(x_N)) = \int \mathrm{D}\, Q(x_1) \cdots \mathrm{D}\, Q(x_{N-1}) f\{Q\} \qquad (15.43)$$

is invariant under unitary transformations of $Q(x_N)$,

$$I(UQ(x_N)U^\dagger)$$

$$= \int \mathrm{D}\, Q(x_1) \cdots \mathrm{D}\, Q(x_{N-1}) f(Q(x_1), \cdots Q(x_{N-1}), UQ(x_N)U^\dagger)$$

$$= \int \mathrm{D}\, Q(x_1) \cdots \mathrm{D}\, Q(x_{N-1}) f(U^\dagger Q(x_1)U, \cdots U^\dagger Q(x_{N-1})U, Q(x_N))$$

$$(15.44)$$

With

$$Q'(x_i) = U^\dagger Q(x_i)U, \quad \mathrm{D}\, Q'(x_i) = \mathrm{sdet}^0(U^\dagger)\mathrm{sdet}^0(U)\mathrm{D}\, Q(x_i) = \mathrm{D}\, Q(x_i),$$
$$(15.45)$$

where Theorem 10.2 is used, we obtain

$$I(UQ(x_N)U^\dagger) = \int \mathrm{D}\, Q'(x_1) \ldots \mathrm{D}\, Q'(x_{N-1}) f(Q'(x_1), \cdots Q(x'_{N-1}), Q(x_N))$$

$$= I(Q(x,N)). \qquad (15.46)$$

This property yields

$$\int \mathrm{D}\, Q(x_N) I(Q(x_N)) = \mathrm{is} I(\lambda(x_N)\mathbf{1})$$

$$= \mathrm{is} \int \mathrm{D}\, Q(x_1) \cdots \mathrm{D}\, Q(x_{N-1}) f(Q(x_1), \cdots Q(x_{N-1}), \lambda(x_N)\mathbf{1})$$

$$(15.47)$$

according to Sect. 15.2.2. The integrand in (15.47) is also invariant under superunitary transformations. Thus, the integral over one $Q(x_i)$ after the other can be replaced by setting $Q(x_i) = \lambda(x_i)\mathbf{1}$ in the function. This proves the integral theorem for N matrices $Q \in \mathcal{M}(1,1)$,

$$\int [\mathrm{D}\, Q] f(Q(x_1), \ldots Q(x_N)) = (\mathrm{is})^N f(\lambda_1 \mathbf{1}, \ldots m\lambda_N \mathbf{1}). \qquad (15.48)$$

15.2.4   *Integral Theorem for N Matrices $Q \in \mathcal{M}(n, m)$*

Finally the theorem is generalized to N matrices $\in \mathcal{M}(n, m)$. We decompose $Q(x)$ into the matrix

$$\bar{Q}(x) = \begin{pmatrix} Q_{11} & Q_{1,n+m} \\ Q_{n+m,1} & Q_{n+m,n+m} \end{pmatrix}, \tag{15.49}$$

into vectors

$$P_\alpha(x) = \begin{pmatrix} Q_{1\alpha} \\ Q_{n+m,\alpha} \end{pmatrix}, \quad \bar{P}_\alpha(x) = \begin{pmatrix} Q_{\alpha 1} & Q_{\alpha,n+m} \end{pmatrix}, \quad 1 < \alpha < n + m \tag{15.50}$$

and into the Matrix Q', which consists of the remaining components of Q. We write

$$f\{Q\} = \bar{f}\{\bar{Q}, P, \bar{P}, Q'\}. \tag{15.51}$$

From (15.21) one obtains with superunitary transformations, which transform only the components with indices $\alpha = 1$ and $n + m$,

$$\bar{f}\{U\bar{Q}U^\dagger, UP, \bar{P}U^\dagger, Q'\} = \bar{f}\{\bar{Q}, P, \bar{P}, Q'\}. \tag{15.52}$$

Equation (15.25) yields

$$\int \mathrm{D}Q = \int \mathrm{D}\bar{Q} \int \mathrm{D}P \int \mathrm{D}Q' \tag{15.53}$$

with

$$\mathrm{D}P = \prod_{x=1}^{N} \prod_{\alpha=2}^{n+m-1} \mathrm{D}P_\alpha(x), \tag{15.54}$$

$$\mathrm{D}P_\alpha(x) = \begin{cases} \frac{1}{\pi}\mathrm{d}R_{1\alpha}(x)\mathrm{d}I_{1\alpha}(x)\mathrm{d}Q_{\alpha,n+m}(x)\mathrm{d}Q_{n+m,\alpha}(x) & 1 < \alpha \leq n \\ \frac{1}{\pi}\mathrm{d}R_{n+m,\alpha}(x)\mathrm{d}I_{n+m,\alpha}(x)\mathrm{d}Q_{1\alpha}(x)\mathrm{d}Q_{\alpha 1}(x) & n < \alpha < n + m \end{cases}$$

$$\tag{15.55}$$

We put

$$I\{P, \bar{P}, Q'\} = \int \mathrm{D}\bar{Q}\bar{\bar{f}}\{\bar{Q}, P, \bar{P}, Q'\}, \tag{15.56}$$

then

$$I\{UP, \bar{P}U^{\dagger}, Q'\} = \int D\bar{Q}\bar{f}\{\bar{Q}, UP, \bar{P}U^{\dagger}, Q'\}$$

$$= \int D\bar{Q}\bar{f}\{U^{\dagger}\bar{Q}U, P, \bar{P}, Q'\} = I\{P, \bar{P}, Q'\}, \qquad (15.57)$$

where the considerations of Sect. 15.2.3 are used in the last step. The invariance (15.57) yields according to the integral theorem (15.6)

$$\int D\,PI\{P, \bar{P}, Q'\} = (-)^{(m-1)N}I\{0, 0, Q'\}. \qquad (15.58)$$

The sign arises from an interchange of the differentials $\mathrm{d}\,Q_{1\alpha}\mathrm{d}\,Q_{\alpha 1}$ in (15.55) in comparison to $\mathrm{d}\,S_i^*\mathrm{d}\,S_i$ in (15.4). Thus one obtains

$$\int D\,Qf\{Q\} = \int D\bar{Q}\int D\,P\int D\,Q'\bar{f}\{\bar{Q}, P, \bar{P}, Q'\}$$

$$= (-)^{(m-1)N}\int D\bar{Q}\int D\,Q'\bar{f}\{\bar{Q}, 0, 0, Q'\}$$

$$= (-)^{(m-1)N}(\mathrm{i}s)^N\int \mathrm{d}\,Q'\bar{f}\{\lambda\mathbf{1}, 0, 0, Q'\}$$

$$= (\mathrm{i}s)^{(2m-1)N}\int D\,Q'f\{Q'\}, \qquad (15.59)$$

where the invariance

$$\bar{f}\{\bar{Q}, 0, 0, Q'\} = \bar{f}\{U\bar{Q}U^{\dagger}, 0, 0, Q'\} \qquad (15.60)$$

and (15.48) are used. This proves the integral theorem (15.24).

15.2.5   Final Remarks

The integral theorem may be applied several times, which reduces the integrals over $Q \in \mathcal{M}(n, m)$ to those of $Q \in \mathcal{M}(n - a, m - a)$. For antihermitian matrices one obtains the same results as for hermitian ones, since the matrices Q have only to be replaced by $\mathrm{i}Q$.

Multiple application of the integral theorem (15.24) yields, in the special case $n = m$,

$$\int D\,Qf\{Q\} = (\mathrm{i}s)^{Nm^2}f\{\lambda(x)\mathbf{1}\}. \qquad (15.61)$$

The integration of functions of hermitian matrices for $n \neq m$ reduces to the integral over functions of hermitian matrices with $|n-m|$ rows and columns. If, in particular, f is a function of supertraces $\mathrm{str}(Q(x_1)Q(x_2) \cdots)$, then the integral depends only on f and $n - m$. This observation allows one to define the integral for matrices of negative dimensions a. Put $a = n - m$, $n \geq 0$, $m \geq 0$. With this generalization of the definition of integrals over functions of hermitian c-number matrices with $a \times a$ elements, the integral of such a function for negative dimension a is the integral over a function of hermitian c-number matrices with $|a| \times |a|$ elements, where the function is again invariant under unitary transformations of the matrix. Roughly speaking there is an equivalence $\mathrm{U}(a) \sim \mathrm{UPL}(m + a, m) \sim \mathrm{UPL}(n, n - a) \sim \mathrm{U}(-a)$.

These considerations were performed for $\mathrm{UPL}_1(n, m)$. They hold analogously for $\mathrm{UPL}_2(n, m)$. One may consider quasiantihermitian matrices instead of quasihermitian matrices. Then one has to put

$$R_{\alpha\beta} = -R_{\beta\alpha}, \quad I_{\alpha\beta} = I_{\beta\alpha} \tag{15.62}$$

in (15.23), which corresponds to an exchange of R and I and thus, yields the same result.

15.3   Matrix as a Set of Vectors

Functions of $W \in \mathcal{M}(1, 1)$ invariant under independent unitary transformations on both sides of W are expressed by the functions of their body.
   We consider a matrix

$$W = \begin{pmatrix} a & \alpha \\ \beta & b \end{pmatrix} \in \mathcal{M}(1, 1) \tag{15.63}$$

and look for the general form of a function $F(a, a^*, b, b^*, \alpha, \alpha^*, \beta, \beta^*)$ which is invariant under independent (pseudo)unitary transformations from both sides of W,

$$W' = UWV, \quad U^\dagger g^U U = g^U, \quad g^U = \mathrm{diag}(g_1^U, g_2^U), \tag{15.64}$$

and similarly for V.
   We may consider W consisting of two row-vectors. Thus, the function should be a function of the scalar products obtained from

$$W \begin{pmatrix} 1 & 0 \\ 0 & t \end{pmatrix} W^\dagger = \begin{pmatrix} aa^* + t\alpha\alpha^* & a\beta^* + tb^*\alpha \\ a^*\beta + tb\alpha^* & tbb^* + \beta\beta^* \end{pmatrix}, \quad t = \frac{g_2^V}{g_1^V} = \pm 1. \tag{15.65}$$

F can be expressed as

$$F = F_0(aa^* + t\alpha\alpha^*, bb^* + t\beta\beta^*) \tag{15.66}$$
$$+(a\beta^* + tb^*\alpha)(a^*\beta + tb\alpha^*)tF_1(aa^* + t\alpha\alpha^*, bb^* + t\beta\beta^*).$$

Since F is a function of two vectors invariant under (pseudo)unitary transformations, its integral is given by

$$\int [DW]F(W) = F(0) = F_0(0), \quad [DW] = \pi^{-2}d\,\Re a\,d\,\Im a\,d\,\Re b\,d\,\Im b\,d\alpha^*\,d\alpha\,d\beta^*\,d\beta. \tag{15.67}$$

W may also be considered consisting of two column vectors. Thus, it is a function of the scalar products obtained from

$$W^\dagger \begin{pmatrix} 1 & 0 \\ 0 & u \end{pmatrix} W = \begin{pmatrix} a^*a + u\beta^*\beta & a^*\alpha + ub\beta^* \\ a\alpha^* + ub^*\beta & ub^*b + \alpha^*\alpha \end{pmatrix}, \quad u = \frac{g_2^U}{g_1^U} = \pm 1. \tag{15.68}$$

Then F can be expressed as

$$F = F_0(a^*a + u\beta^*\beta, b^*b + u\alpha^*\alpha) \tag{15.69}$$
$$+(a^*\alpha + ub\beta^*)(a\alpha^* + ub^*\beta)uF_2(a^*a + u\beta^*\beta, b^*b + u\alpha^*\alpha).$$

The ordinary parts of both expressions (15.66) and (15.69) agree since, in both cases, we have chosen the same function F_0. Expanding F in the odd elements and setting $A = a^*a, B = b^*b$, we obtain

$$F = F_0 - t\alpha^*\alpha(F'_{0A} + BF_1) + t\beta^*\beta(-F'_{0B} + AF_1) - \alpha^*\beta^*abF_1 + \alpha\beta a^*b^*F_1$$
$$+\alpha^*\alpha\beta^*\beta(F''_{0AB} - AF'_{1A} + BF'_{1B})$$
$$= F_0 + u\alpha^*\alpha(F'_{0B} - AF_2) + u\beta^*\beta(F'_{0A} + BF_2) - \alpha^*\beta^*abF_2 + \alpha\beta a^*b^*F_2$$
$$+\alpha^*\alpha\beta^*\beta(F''_{0AB} - AF'_{2A} + BF'_{2B}). \tag{15.70}$$

Comparison of the $\alpha^*\beta^*$- and $\alpha\beta$-terms yield $F_2 = F_1$. Then also the $\alpha^*\alpha\beta^*\beta$-terms agree. The $\alpha^*\alpha$- and $\beta^*\beta$-terms yield

$$t(F'_{0A} + BF_1) = u(-F'_{0B} + AF_1). \tag{15.71}$$

which allows F_1 to be expressed in terms of the function F_0. F reads, in terms of F_0,

$$F = F_0 + (-tu\alpha^*\alpha + \beta^*\beta)\frac{AF'_{0A} + BF'_{0B}}{uA - tB} + (ab\beta^*\alpha^* + a^*b^*\alpha\beta)\frac{tF'_{0A} + uF'_{0B}}{uA - tB}$$
$$+\alpha^*\alpha\beta^*\beta[\frac{uBF''_{0BB} - tAF''_{0AA}}{uA - tB} + (uA + tB)\frac{tF'_{0A} + uF'_{0B}}{(uA - tB)^2}]. \tag{15.72}$$

The integral over F can be easily obtained from the coefficient of $\alpha^*\alpha\beta^*\beta$ in (15.70) with

$$\int \pi^{-2} d\,\Re a\; d\,\Im a\; d\,\Re b\; d\,\Im b = \int_0^\infty d\,A d\,B. \tag{15.73}$$

The terms depending on F_1 can be rewritten

$$-AF'_{1A} + BF'_{1B} = -\partial_A(AF_1) + \partial_B(BF_1) \tag{15.74}$$

and vanish, since the integrands vanish at 0 and ∞,

$$-\int d\,B(AF_1)|_0^\infty + \int d\,A(BF_1)|_0^\infty = 0 \tag{15.75}$$

Thus, only

$$\int_0^\infty d\,A \int_0^\infty d\,B F''_{0AB} = F_0(0) \tag{15.76}$$

remains, which had been claimed in (15.67). We have assumed in (15.65) and (15.68) that products of W and $W^\dagger$ are scalar products, and F depends only on them. One can without this assumption using the invariance (15.64) show that F has the same form as that in (15.72) and (15.67) holds.

Problem

15.1 Calculate $\int [D\,W] F(W)$ from the $\alpha^*\alpha\beta^*\beta$-term in (15.72).

References

[52] F. Constantinescu, H.F. de Groote, The integral theorem for supersymmetric invariants. J. Math. Phys. **30**, 981 (1989)
[65] K.B. Efetov, Supersymmetry and theory of disordered metals. Adv. Phys. **32**, 53 (1983)
[66] K.B. Efetov, *Supersymmetry in Disorder and Chaos* (Cambridge University Press, Cambridge, 1997)
[143] M. Kieburg, H. Kohler, T. Guhr, Integration of Grassmann variables over invariant functions in flat superspaces. J. Math. Phys. **50**, 013528 (2009)
[204] G. Parisi, N. Sourlas, Random magnetic fields, supersymmetry, and negative dimensions. Phys. Rev. Lett. **43**, 744 (1979)

[225] M.J. Rothstein, Integration on noncompact supermanifolds. Trans. Am. Math. Soc. **299**, 387 (1987)

[261] J.J.M. Verbaarschot, H.A. Weidenmüller, M.R. Zirnbauer, Grassmann integration in stochastic quantum physics: the case of compound-nucleus scattering. Phys. Rep. **129**, 367 (1985)

[274] F. Wegner, unpublished notes (1983/84), compare acknowledgment in [52], ref. [5] in [143], ref. [17] in [261]

Chapter 16
Integral Theorems
for the (Unitary-)Orthosymplectic Group

Abstract In this chapter it is also shown that integrals over functions of vectors and matrices invariant under the (unitary-)orthosymplectic group can be reduced by cancelling equal numbers of even and odd components. We have to determine the form of the functions invariant under these groups for the smallest non-trivial cases and their integral. Generalization to several vectors and matrices and to those with more components is performed as for the superunitary case.

An introduction to the integral theorems and references were given in Sect. 15.1.1.

16.1 Integral Theorem for Vectors

There are similar theorems for functions of vectors invariant under orthosymplectic and unitary-orthosymplectic transformations as in Sect. 15.1. Since the number of components $m = 2r$ is even, the number of components n and m has to be reduced by two.

16.1.1 Invariance Under the Orthosymplectic Group

If $f\{S\} = f\{US\}$ for $S(x) \in \mathcal{M}(n, 2r, 1, 0)$, $x = 1, 2, \ldots N$, $U \in OSp(n, 2r)$, f sufficiently often differentiable and sufficiently rapidly approaching zero for $|S| \to \infty$, then

$$\int [D\,S]f\{S\} = \int [D\,S']f\{S'\} \tag{16.1}$$

holds with

$$[D\,S] = \prod_{x=1}^{N} \prod_{i=1}^{n} \frac{d\,S_i(x)}{\sqrt{2\pi}} \prod_{i=n+1}^{n+2r} d\,S_i(x), \tag{16.2}$$

© Springer-Verlag Berlin Heidelberg 2016
F. Wegner, *Supermathematics and its Applications in Statistical Physics*,
Lecture Notes in Physics 920, DOI 10.1007/978-3-662-49170-6_16

correspondingly for [D S'], where now i runs only from 3 to $n + 2r - 2$,

$$S_i'(x) = \begin{cases} S_i(x) & 3 \le i \le n + 2r - 2 \\ 0 & i = 1, 2, n + 2r - 1, n + 2r. \end{cases} \tag{16.3}$$

The proof runs analogously to the superunitary case. One needs to show only that the integral for $N = 1$, $n = m = 2$ yields $f\{0\}$. An infinitesimal orthosymplectic transformation $U = 1 + \delta U$ obeys $C\delta U +^{\mathrm{T}} \delta UC = 0$, hence

$$\delta U = \begin{pmatrix} 0 & p_1 & \omega_1 & \omega_2 \\ -p_1 & 0 & \omega_3 & \omega_4 \\ -\omega_2 & -\omega_4 & p_2 & p_3 \\ \omega_1 & \omega_3 & p_4 & -p_2 \end{pmatrix} \tag{16.4}$$

with infinitesimal ps and ωs. The components of the vector are $S_1, S_2, \theta_1, \theta_2$. $\partial f / \partial p_1 = 0$ yields $f = f(S_1^2 + S_2^2, \theta_1\theta_2)$. $\partial f / \partial p_2 = 0$ yields

$$f = f_0(S_1^2 + S_2^2) + \theta_1\theta_2 f_3(S_1^2 + S_2^2). \tag{16.5}$$

Finally the variation with respect to the ωs yields

$$\delta f = (\omega_1\theta_1 S_1 + \omega_2\theta_2 S_1 + \omega_3\theta_1 S_2 + \omega_4\theta_2 S_2)(2\frac{\partial f_0}{\partial q} - f_3(q)), \tag{16.6}$$

with $q = S_1^2 + S_2^2$. Thus, one obtains

$$f = f(S_1^2 + S_2^2 + 2\theta_1\theta_2). \tag{16.7}$$

f depends only on the scalar product

$$(S, S)_0 = {}^{\mathrm{T}}SCS = S_1^2 + S_2^2 + 2\theta_1\theta_2. \tag{16.8}$$

The proof continues as in the superunitary case ($S_1 = R$, $S_2 = I$) yielding

$$\int [\mathrm{D}\, S] f(S) = f(0). \tag{16.9}$$

For more components and N vectors one argues in similar fashion to the paragraph after (15.15) and Sects. 15.1.4 and 15.1.5.

16.1.2   Invariance Under the Unitary-Orthosymplectic Group

Suppose $f\{S\} = f\{US\}$ for all $U \in \mathrm{UOSp}(n, 2r)$ and $S(x) \in \mathcal{M}(n, 2r, 1, 0)$ for $x = 1, 2, \ldots N$, $S(x) \in \mathcal{M}(n, 2r, 0, 2)$ for $x = N + 1, \ldots N + R$, and f sufficiently often differentiable and sufficiently rapidly vanishing for $|\mathrm{ord}\, S| \to \infty$. Then

$$\int [\mathrm{D}\, S] f\{S\} = \int [\mathrm{D}\, S'] f\{S'\} \tag{16.10}$$

with

$$[\mathrm{D}\, S] = \prod_{n=1}^{N} \left(\prod_{i=1}^{n} \frac{\mathrm{d}\, S_i(x)}{\sqrt{2\pi}} \prod_{i=n+1}^{n+r} \mathrm{d}\,\theta_i(x)^\times \mathrm{d}\,\theta_i(x) \right)$$
$$\times \prod_{x=N+1}^{N+R} \left(\prod_{i=1}^{n} (\mathrm{d}\,\theta_i^\times(x) \mathrm{d}\,\theta_i(x)) \prod_{i=n+1}^{n+2r} (\frac{1}{\pi} \mathrm{d}\,\Re S_i(x) \mathrm{d}\,\Im S_i(x)) \right). \tag{16.11}$$

For $[\mathrm{D}\, S']$, i runs from 3 to $n + 2r - 2$. All other components are set to zero. Also here it is sufficient to consider first the cases of one vector $N = 1, R = 0$ and $N = 0, R = 1$, respectively.

The transformation matrix $U = \mathbf{1} + \Delta$ obeys $U^{\ddagger} = \mathbf{1} + \Delta^{\ddagger} = \mathbf{1} - \Delta$ for infinitesimal Δ. Hence, it can be parametrized

$$\Delta = \begin{pmatrix} 0 & q & \omega^\times & \omega \\ -q & 0 & \eta^\times & \eta \\ -\omega & -\eta & \mathrm{i}p & s \\ \omega^\times & \eta^\times & -s^\times & -\mathrm{i}p \end{pmatrix}, \quad q, p \in \mathbb{R}. \tag{16.12}$$

For

$$S = \begin{pmatrix} S_1 \\ S_2 \\ \theta \\ -\theta^\times \end{pmatrix} \in \mathcal{M}(2, 2, 1, 0) \tag{16.13}$$

one obtains, in analogy to the orthosymplectic case, that f depends only on

$$(S, S)_2 = S_1^2 + S_2^2 + 2\theta^\times \theta. \tag{16.14}$$

The proof continues as before.

The variation $\delta S = \Delta S$ for a vector

$$
S = \begin{pmatrix} \theta_1^\times & \theta_1 \\ \theta_2^\times & \theta_2 \\ S_2^\times & S_1 \\ -S_1^\times & S_2 \end{pmatrix} \in \mathcal{M}(2,2,0,2) \tag{16.15}
$$

is given by

$$
\begin{array}{llll}
\delta S_1 = & ipS_1 & +sS_2 & -\omega\theta_1 & -\eta\theta_2 \\
\delta S_1^\times = & -ipS_1^\times & +s^\times S_2^\times & -\omega^\times\theta_1^\times & -\eta^\times\theta_2^\times \\
\delta S_2 = & -ipS_2 & -s^\times S_1 & +\omega^\times\theta_1 & +\eta^\times\theta_2 \\
\delta S_2^\times = & ipS_2^\times -sS_1^\times & & -\omega\theta_1^\times & -\eta\theta_2^\times \\
\delta\theta_1 = q\theta_2 & & & +\omega S_2 +\omega^\times S_1 \\
\delta\theta_1^\times = q\theta_2^\times & & & -\omega S_1^\times +\omega^\times S_2^\times \\
\delta\theta_2 = -q\theta_1 & & & +\eta S_2 +\eta^\times S_1 \\
\delta\theta_2^\times = -q\theta_1^\times & & & -\eta S_1^\times +\eta^\times S_2^\times
\end{array} \tag{16.16}
$$

Let us for the moment write

$$
f = f(S_1, k_0, k_1, k_2), \quad k_0 = S_1 S_2, \quad k_1 = S_1^\times S_1, \quad k_2 = S_2^\times S_2. \tag{16.17}
$$

All even components of S can be expressed by S_1 and the ks. Then the variation with respect to p yields

$$
\frac{\partial f}{\partial p} = iS_1 \frac{\partial f}{\partial S_1}. \tag{16.18}
$$

Thus, f does not depend on S_1. Further variations yield

$$
\frac{\partial f}{\partial s} = S_2^2 \frac{\partial f}{\partial k_0} + S_2 S_1^\times \left(\frac{\partial f}{\partial k_1} - \frac{\partial f}{\partial k_2} \right), \quad \frac{\partial f}{\partial s^\times} = -S_1^2 \frac{\partial f}{\partial k_0} - S_1 S_2^\times \left(\frac{\partial f}{\partial k_1} - \frac{\partial f}{\partial k_2} \right). \tag{16.19}
$$

As a consequence

$$
\frac{\partial f}{\partial k_0} = 0, \quad \frac{\partial f}{\partial k_1} = \frac{\partial f}{\partial k_2}. \tag{16.20}
$$

Thus, f does not depend on k_0 and depends on the components $S_i \in \mathscr{A}_0$ only by

$$
k = k_1 + k_2 = S_1^\times S_1 + S_2^\times S_2, \quad f = f(k, \theta_1, \theta_1^\times, \theta_2, \theta_2^\times). \tag{16.21}
$$

Next q is varied. Then only the following combinations of the θs are allowed

$$
\begin{aligned}
f = {} & f_0(k) + \theta_1\theta_2 f_1(k) + (\theta_1^\times\theta_1 + \theta_2^\times\theta_2)f_2(k) + (\theta_1^\times\theta_2 - \theta_2^\times\theta_1)f_3(k) \\
& + \theta_1^\times\theta_2^\times f_4(k) + \theta_1^\times\theta_2^\times\theta_1\theta_2 f_5(k).
\end{aligned}
\tag{16.22}
$$

By varying ω, $\omega^\times$, η, and $\eta^\times$ one obtains

$$
\begin{aligned}
\delta k = {} & \delta(\theta_1^\times\theta_1 + \theta_2^\times\theta_2) = \omega^\times(-\theta_1^\times S_1 + \theta_1 S_2^\times) - \omega(\theta_1 S_1^\times + \theta_1^\times S_2) \\
& + \eta^\times(-\theta_2^\times S_1 + \theta_2 S_2^\times) - \eta(\theta_2 S_1^\times + \theta_2^\times S_2)
\end{aligned}
\tag{16.23}
$$

and

$$
\begin{aligned}
\delta f = {} & \delta k(f_0' + f_2) + \delta k(\theta_1^\times\theta_1 + \theta_2^\times\theta_2)(f_2' - f_5) \tag{16.24} \\
& + (\omega^\times\theta_2 - \eta^\times\theta_1)(S_1 f_1 + S_2^\times f_3) + (-\omega\theta_2 + \eta\theta_1)(-S_2 f_1 + S_1^\times f_3) \\
& + (-\omega\theta_2^\times + \eta\theta_1^\times)(-S_2 f_3 + S_1^\times f_4) + (\omega^\times\theta_2^\times - \eta^\times\theta_1^\times)(S_1 f_3 + S_2^\times f_4) \\
& + (\omega^\times\theta_1^\times + \eta^\times\theta_2^\times)\theta_1\theta_2(-S_1 f_1' - S_2^\times f_3') + (-\omega\theta_1^\times - \eta\theta_2^\times)\theta_1\theta_2(S_2 f_1' - S_1^\times f_3') \\
& + (\omega^\times\theta_1 + \eta^\times\theta_2)\theta_1^\times\theta_2^\times(S_1 f_3' + S_2^\times f_4') + (-\omega\theta_1 - \eta\theta_2)\theta_1^\times\theta_2^\times(-S_2 f_3' + S_1^\times f_4').
\end{aligned}
$$

One concludes

$$
f_2 = -\frac{\partial f_0}{\partial k}, \quad f_5 = \frac{\partial f_2}{\partial k}, \quad f_1 = f_3 = f_4 = 0,
\tag{16.25}
$$

i.e.

$$
f = f(S_1^\times S_1 + S_2^\times S_2 + \theta_1\theta_1^\times + \theta_2\theta_2^\times).
\tag{16.26}
$$

Thus, also here, the function f is only a function of $(S^\ddagger, S)_0 = (S, S)_2$. The proof continues as in the superunitary case ($n = m = 2$).

16.2   Integral Theorem for Quasihermitian and Quasireal Matrices: Invariance Under UOSp

Integrals over a function of quasihermitian and quasireal matrices $\in \mathcal{M}(n, 2r)$ invariant under unitary-orthosymplectic transformations can be reduced to the same integrals over matrices $\in \mathcal{M}(n - 2a, 2r - 2a)$, where the cancelled components are set to zero with the exception of the diagonal matrix elements, which have to agree in the bosonic and fermionic sector, but are otherwise arbitrary.

16.2.1   *Theorem*

An integral theorem for functions $f\{Q\}$, $Q(x) \in \mathcal{M}(n, 2r)$, which obey

$$f\{UQ(x)U^{\ddagger}\} = f\{Q(x)\}, \tag{16.27}$$

will be derived. The elements $Q_{\alpha\beta} \in \mathcal{A}_0$ of a hermitian superreal matrix can be written

$$Q_{\alpha\beta} = Q_{\beta\alpha} \qquad 1 \le \alpha, \beta \le n \tag{16.28}$$

$$Q_{n+2\alpha-1,n+2\beta-1} = R_{\alpha\beta} + iI^{(3)}_{\alpha\beta},$$

$$Q_{n+2\beta,n+2\alpha} = R_{\alpha\beta} - iI^{(3)}_{\alpha\beta},$$

$$Q_{n+2\alpha-1,n+2\beta} = I^{(2)}_{\alpha\beta} + iI^{(1)}_{\alpha\beta} \quad 1 \le \alpha, \beta \le r \tag{16.29}$$

$$Q_{n+2\alpha,n+2\beta-1} = -I^{(2)}_{\alpha\beta} + iI^{(1)}_{\alpha\beta},$$

$$R_{\alpha\beta} = R_{\beta\alpha}, \qquad I^{(k)}_{\alpha\beta} = -I^{(k)}_{\beta\alpha} \tag{16.30}$$

with real elements R, I and Q, the latter for $1 \le \alpha, \beta \le n$. The odd components of Q obey

$$Q_{n+2\beta-1,\alpha} = Q_{\alpha,n+2\beta}, \quad Q_{n+2\beta,\alpha} = -Q_{\alpha,n+2\beta-1}, \quad 1 \le \alpha \le n, \quad 1 \le \beta \le r. \tag{16.31}$$

Now we set with real q, r and j

$$\begin{aligned}
Q_{\alpha\beta}(x) &= \lambda(x)\delta_{\alpha\beta} + q_{\alpha\beta}(x)e^{i\gamma+} & 1 \le \alpha, \beta \le n \\
R_{\alpha\beta}(x) &= \lambda(x)\delta_{\alpha\beta} + r_{\alpha\beta}(x)e^{i\gamma-} & 1 \le \alpha, \beta \le r \\
I^{(k)}_{\alpha\beta}(x) &= \qquad\qquad j^{(k)}_{\alpha\beta}(x)e^{i\gamma-} & 1 \le \alpha, \beta \le r
\end{aligned} \tag{16.32}$$

and it will be shown that, for these quasihermitian, quasireal matrices Q, provided f is sufficiently often differentiable and falls off for large $|q|$, $|r|$, $|j|$ sufficiently rapidly, the following integral theorem holds:

$$\int \mathrm{D}\, Qf\{Q\} = (is)^N \int \mathrm{D}\, Q'f\{Q'\} \tag{16.33}$$

with

$$\mathrm{D}\, Q = \prod_{x=1}^{N}\left\{ \prod_{n \ge \alpha > \beta \ge 1}\left(\frac{1}{\sqrt{2\pi}}\mathrm{d}\, Q_{\alpha\beta}\right) \prod_{\alpha=1}^{n}\left(\frac{1}{\sqrt{4\pi}}\mathrm{d}\, Q_{\alpha\alpha}\right)\right.$$

$$\times \prod_{r\geq\alpha>\beta\geq1} (\frac{1}{\pi^2}d\,R_{\alpha\beta} \prod_{i=1}^{3} d\,I^{(i)}_{\alpha\beta}) \prod_{\alpha=1}^{r}(\frac{1}{\sqrt{2\pi}}d\,R_{\alpha\alpha})$$

$$\times \prod_{\alpha=1}^{n}\prod_{\beta=1}^{r}(d\,Q_{n+2\beta-1,\alpha}d\,Q_{n+2\beta,\alpha})\Bigg\}. \tag{16.34}$$

The integral over Q' contains only the integration of the components $3 \leq \alpha, \beta \leq n + 2r - 2$. The other components $Q'_{\alpha\beta}$ are set equal $\lambda(x)\delta_{\alpha\beta}$. s is given by

$$s = \text{sign} \sin(\gamma_- - \gamma_+) \tag{16.35}$$

provided $\gamma_- - \gamma_+$ is not an integer multiple of π. If it is an integer multiple, then no convergence is expected. Firstly the proof is given for $N = 1$, $Q \in \mathcal{M}(2,2)$.

16.2.2   Invariant Function $f(Q)$, $Q \in \mathcal{M}(2,2)$

The superreal hermitian matrix $Q \in \mathcal{M}(2,2)$ may be written

$$Q = \begin{pmatrix} a & b & \alpha^\times & \alpha \\ b & c & \beta^\times & \beta \\ \alpha & \beta & e & 0 \\ -\alpha^\times & -\beta^\times & 0 & e \end{pmatrix}, \quad a,b,c,e \in \mathbb{R}. \tag{16.36}$$

Similar to the derivation in Sect. 15.2.2 one derives the general differentiable function invariant under UOSp$(2,2)$. Then for the infinitesimal transformation (16.12), one obtains

$$\delta Q = UQU^\ddagger - Q = \Delta Q - Q\Delta \tag{16.37}$$

with

$$\begin{aligned}
\delta a &= 2qb & -2\omega\alpha^\times & +2\omega^\times\alpha \\
\delta b &= q(c-a) & -\omega\beta^\times & +\omega^\times\beta & -\eta\alpha^\times & +\eta^\times\alpha \\
\delta c &= -2qb & & & -2\eta\beta^\times & +2\eta^\times\beta \\
\delta e &= & -\omega\alpha^\times & +\omega^\times\alpha & -\eta\beta^\times & +\eta^\times\beta \\
\delta\alpha &= q\beta +ip\alpha -s\alpha^\times & +\omega(e-a) & & -\eta b \\
\delta\alpha^\times &= q\beta^\times -ip\alpha^\times +s^\times\alpha & & +\omega^\times(e-a) & & -\eta^\times b \\
\delta\beta &= -q\alpha +ip\beta -s\beta^\times & -\omega b & & +\eta(e-c) \\
\delta\beta^\times &= -q\alpha^\times -ip\beta^\times +s^\times\beta & & -\omega^\times b & & +\eta^\times(e-c) \\
\end{aligned}$$
$$\tag{16.38}$$

We expand f in the Grassmann variables α, β, $\alpha^\times$, $\beta^\times$. Out of 16 terms only six remain, which are invariant under variation of p,

$$f(Q) = F + \alpha\alpha^\times g + \beta\beta^\times h + \beta\alpha^\times i + \alpha\beta^\times j + \alpha\alpha^\times\beta\beta^\times K, \qquad (16.39)$$

where F, g, h, i, j, K depend on a, b, c, e. Next we vary s and $s^\times$,

$$\delta f = s\alpha^\times\beta^\times(i - j) + s^\times\alpha\beta(i - j), \qquad (16.40)$$

hence $j = i$ and

$$f(Q) = F + \alpha\alpha^\times g + \beta\beta^\times h + (\beta\alpha^\times + \alpha\beta^\times)i + \alpha\alpha^\times\beta\beta^\times K. \qquad (16.41)$$

The variation of f with respect to q, ω, $\omega^\times$, η, and $\eta^\times$ yields

$$\begin{aligned}
\delta f = {}& q\phi_1 + q\alpha\alpha^\times\phi_2 + q\beta\beta^\times\phi_3 + q(\alpha\beta^\times + \beta\alpha^\times)\phi_4 + q\alpha\alpha^\times\beta\beta^\times\phi_5 \\
&+(\omega\alpha^\times - \omega^\times\alpha)(\phi_6 + \beta\beta^\times\phi_8) + (\omega\beta^\times - \omega^\times\beta)(\phi_7 + \alpha\alpha^\times\phi_9) \\
&+(\eta\beta^\times - \eta^\times\beta)(\phi_{10} + \alpha\alpha^\times\phi_{12}) + (\eta\alpha^\times - \eta^\times\alpha)(\phi_{11} + \beta\beta^\times\phi_{13})
\end{aligned}$$
$$(16.42)$$

with

$$\phi_1 = \Box F, \quad \phi_2 = \Box g - 2i, \quad \phi_3 = \Box h + 2i, \quad \phi_4 = \Box i + g - h, \quad \phi_5 = \Box K,$$
$$\phi_6 = -2\partial_a F - \partial_e F + (e - a)g - bi,$$
$$\phi_7 = -\partial_b F - bh + (e - a)i,$$
$$\phi_8 = -2\partial_a h - \partial_e h + \partial_b i + (e - a)K, \qquad (16.43)$$
$$\phi_9 = -\partial_b g + 2\partial_a i + \partial_e i - bK,$$
$$\phi_{10} = -2\partial_c F - \partial_e F + (e - c)h - bi,$$
$$\phi_{11} = -\partial_b F - bg + (e - c)i,$$
$$\phi_{12} = -2\partial_c g - \partial_e g + \partial_b i + (e - c)K,$$
$$\phi_{13} = -\partial_b h + 2\partial_c i + \partial_e i - bK.$$

and the operator

$$\Box = 2b(\partial_a - \partial_c) + (c - a)\partial_b. \qquad (16.44)$$

The invariance requires that all ϕ_i vanish. $\phi_7 = \phi_{11}$ yields

$$b(h - g) + (a - c)i = 0, \qquad (16.45)$$

which is solved by

$$g - h = 2(a - c)H, \quad i = 2bH. \tag{16.46}$$

With

$$2G := g + h \tag{16.47}$$

one obtains

$$g = G + (a - c)H, \quad h = G + (c - a)H. \tag{16.48}$$

We express g, h, and i by G and H. Then (16.41) reads

$$f(Q) = F + (\alpha\alpha^\times + \beta\beta^\times)G \tag{16.49}$$
$$+ [\alpha\alpha^\times(a - c) + \beta\beta^\times(c - a) + 2(\alpha\beta^\times + \beta\alpha^\times)b]H + \alpha\alpha^\times\beta\beta^\times K.$$

$\phi_i = 0$ for $i = 1 \ldots 5$ yields

$$\Box F = \Box G = \Box H = \Box K = 0. \tag{16.50}$$

Equation (16.50) implies that these functions depend only on

$$t = \tfrac{1}{2}(\lambda_1 + \lambda_2) = \tfrac{1}{2}(a + c), \quad k = \tfrac{1}{4}(\lambda_1 - \lambda_2)^2 = \tfrac{1}{4}(a - c)^2 + b^2, \tag{16.51}$$

and on e, where λ_1 and λ_2 are the eigenvalues of the matrix $\begin{pmatrix} a & b \\ b & c \end{pmatrix}$. This allows one to express the derivatives ∂_a, ∂_b, ∂_c by ∂_t, and ∂_k, as

$$\partial_a = \tfrac{1}{2}\partial_t + \tfrac{1}{2}(a - c)\partial_k,$$
$$\partial_b = b\partial_k, \tag{16.52}$$
$$\partial_c = \tfrac{1}{2}\partial_t + \tfrac{1}{2}(c - a)\partial_k.$$

Then $\phi_6 = \phi_{10} = 0$ yield

$$-\partial_t F - \partial_e F + (e - t)G - 2kH = 0, \tag{16.53}$$
$$\partial_k F + (t - e)H + \tfrac{1}{2}G = 0. \tag{16.54}$$

Next $\phi_7 = \phi_{11} = 0$ are satisfied by (16.54). $\phi_8 = \phi_{12} = 0$ yield

$$-\partial_t G + 4k\partial_k H + 4H - \partial_e G + (e - t)K = 0 \tag{16.55}$$
$$\partial_k G - \partial_t H - \partial_e H + \tfrac{1}{2}K = 0. \tag{16.56}$$

$\phi_9 = \phi_{13} = 0$ are satisfied by (16.56). Equation (16.55) follows from the three other equations, since

$$(16.55) = -2\partial_k \, (16.53) - 2(\partial_t + \partial_e) \, (16.54) + 2(e - t)(16.56). \tag{16.57}$$

From (16.56), (16.53)+2(t-e)(16.54) and (t-e)(16.53)+2k(16.54) one obtains

$$K = -2\partial_k G + 2\partial_t H + 2\partial_e H, \tag{16.58}$$

$$\mathcal{N}H = \tfrac{1}{2}\partial_t F + \tfrac{1}{2}\partial_e F + (e - t)\partial_k F, \tag{16.59}$$

$$\mathcal{N}G = (e - t)(\partial_t F + \partial_e F) + 2k\partial_k F, \tag{16.60}$$

$$\mathcal{N} := (t - e)^2 - k = (\lambda_1 - e)(\lambda_2 - e). \tag{16.61}$$

Thus, $f(Q)$ is given by (16.49), where G, H, and K are given by F (16.58)–(16.60). F, G, H, and K depend only on t, k, and e given by (16.51).

For $\lambda_1 = \lambda_2 = e$, one obtains $k = 0$, $t = e$ and from (16.59)

$$\partial_t F + \partial_e F = 0. \tag{16.62}$$

Thus, $f(\lambda \mathbf{1})$ does not depend on λ.

16.2.3   The Integral for $N = 1$, $Q \in \mathcal{M}(2,2)$

With

$$\int DQ = \frac{1}{8\pi^2} \int da \, db \, dc \, de \, d\alpha \, d\alpha^\times \, d\beta \, d\beta^\times, \tag{16.63}$$

one obtains

$$\int DQf = \frac{1}{8\pi^2} \int da \, db \, dc \, de \, K = \frac{1}{4\pi} \int dk \, dt \, de \, K(k, t, e). \tag{16.64}$$

The k-integral runs from 0 to $e^{2i\gamma} + \infty$, whereas t and e are integrated from $\lambda - e^{i\gamma} \pm \infty$ to $\lambda + e^{i\gamma} \pm \infty$. The integrals over $\partial_t H$ and $\partial_e H$ vanish, since H vanishes for fixed e as $t \to \pm\infty$ and for fixed t as $e \to \pm\infty$. The $\partial_k G$ term in (16.58) yields

$$\int DQf = \frac{1}{2\pi} \int dt \, de \, G(k = 0, t, e), \tag{16.65}$$

since G vanishes for large arguments. Equation (16.60) yields

$$\int DQf = \frac{1}{2\pi} \int dt \, de \, \left. \frac{\partial_t F + \partial_e F}{e - t} \right|_{k=0}, \tag{16.66}$$

which gives

$$\int DQf = isF(e = t = \lambda, k = 0) = isf(\lambda \mathbf{1}) \tag{16.67}$$

in analogy to (15.41) and thus, proves (16.33) for this special case.

16.2.4   The General Case

The generalization to N matrices $Q \in \mathcal{M}(2,2)$ and $\mathcal{M}(n,2r)$ runs in analogy to the considerations in Sects. 15.2.3 and 15.2.4. It is interesting that a system of unitarysymplectic symmetry $USp(2a)$ corresponds to a system with orthogonal symmetry $O(-2a)$, roughly

$$USp(2a) \sim UOSp(n, n + 2a) \sim UOSp(2r - 2a, 2r) \sim O(-2a) \tag{16.68}$$

with $n + 2a = 2r$.

16.3   Integral Theorem for Quasiantihermitian Quasireal Matrices

Integrals over a function of quasiantihermitian quasireal matrices$\in \mathcal{M}(n,2r)$ invariant under unitary-orthosymplectic transformations can be reduced to the same integrals over matrices $\in \mathcal{M}(n-2a, 2r-2a)$, where the cancelled components are set to zero.

16.3.1   The Theorem

Superreal antihermitian matrices $Q(x) \in \mathcal{M}(n,2r)$ have a structure different from hermitian matrices. Multiplication of a hermitian matrix with i makes it antihermitian, but does not leave it superreal. The elements $Q_{\alpha\beta} \in \mathcal{A}_0$ of an antihermitian superreal matrix obey

$$Q_{\alpha\beta} = -Q_{\beta\alpha}, \quad 1 \le \alpha, \beta \le n \tag{16.69}$$

and (16.29) with

$$R_{\alpha\beta} = -R_{\beta\alpha}, \quad I_{\alpha\beta}^{(k)} = I_{\beta\alpha}^{(k)} \tag{16.70}$$

with real R, I and Q, the latter for $1 \leq \alpha, \beta \leq n$. With (16.32) and $\lambda(x) = 0$ we show for quasireal quasiantihermitian matrices Q: If $f\{Q\}$ is invariant under UOSp transformations (16.27) and decays for large $|r|$, $|q|$, $|j|$ sufficiently rapidly along the path of integration, then

$$\int D\, Qf\{Q\} = (-\mathrm{i}s)^N \int D\, Q' f\{Q'\} \tag{16.71}$$

holds with

$$DQ = \prod_x \left\{ \prod_{1 \leq \alpha < \beta \leq n} \left(\frac{1}{\sqrt{2\pi}} \mathrm{d}\, Q_{\alpha\beta} \right) \prod_{1 \leq \alpha < \beta \leq r} \left(\frac{1}{\pi^2} \mathrm{d}\, R_{\alpha\beta} \prod_i \mathrm{d}\, I_{\alpha\beta}^{(i)} \right) \right.$$

$$\left. \prod_{\alpha i} \left(\frac{1}{\sqrt{2\pi}} \mathrm{d}\, I_{\alpha\alpha}^{(i)} \right) \prod_{\alpha=1}^{n} \prod_{\beta=1}^{r} \left(\mathrm{d}\, Q_{n+2\beta-1,\alpha} \mathrm{d}\, Q_{n+2\beta,\alpha} \right) \right\}, \tag{16.72}$$

where $\int D\, Q'$ denotes the integral over $Q_{\alpha\beta}$ with $3 \leq \alpha, \beta \leq n + 2r - 2$. The other components are set to zero. Again we begin with the proof for $N = 1$, $Q \in \mathcal{M}(2,2)$.

16.3.2 Invariant Function $f(Q)$, $Q \in \mathcal{M}(2,2)$

We first determine the form $f(Q)$ of a superreal antihermitian matrix Q

$$Q = \begin{pmatrix} 0 & a & \alpha^\times & \alpha \\ -a & 0 & \beta^\times & \beta \\ -\alpha & -\beta & \mathrm{i}b & c \\ \alpha^\times & \beta^\times & -c^\times & -\mathrm{i}b \end{pmatrix}, \quad a, b \text{ real.} \tag{16.73}$$

invariant under unitary orthosymplectic transformations. Variation of Q with (16.12), (16.37) yields

$$
\begin{aligned}
\delta a &= & & \omega\beta^\times & -\omega^\times\beta & -\eta\alpha^\times & +\eta^\times\alpha \\
\delta b &= & \mathrm{i}s c^\times & -\mathrm{i}s^\times c & +\mathrm{i}\omega\alpha^\times & +\mathrm{i}\omega^\times\alpha & +\mathrm{i}\eta\beta^\times & +\mathrm{i}\eta^\times\beta \\
\delta c &= & 2\mathrm{i}pc & -2\mathrm{i}sb & -2\omega\alpha & & -2\eta\beta & \\
\delta c^\times &= & -2\mathrm{i}pc^\times & +2\mathrm{i}s^\times b & -2\omega^\times\alpha^\times & & -2\eta^\times\beta^\times & \\
\delta\alpha &= q\beta & +\mathrm{i}p\alpha & -s\alpha^\times & & -\mathrm{i}\omega b & +\omega^\times c & -\eta a \\
\delta\alpha^\times &= q\beta^\times & -\mathrm{i}p\alpha^\times & & +s^\times\alpha & -\omega c^\times & +\mathrm{i}\omega^\times b & -\eta^\times a \\
\delta\beta &= -q\alpha & +\mathrm{i}p\beta & -s\beta^\times & & +\omega a & -\mathrm{i}\eta b & +\eta^\times c \\
\delta\beta^\times &= -q\alpha^\times & -\mathrm{i}p\beta^\times & & +s^\times\beta & +\omega^\times a & -\eta c^\times & +\mathrm{i}\eta^\times b
\end{aligned}
\tag{16.74}
$$

Again, f is expanded in the Grassmann variables α, β, $\alpha^\times$, $\beta^\times$. Eight terms remain after variation of p,

$$f(Q) = F + \alpha\alpha^\times G + \beta\beta^\times h + \beta\alpha^\times i + \alpha\beta^\times j + \alpha^\times\beta^\times cK + \alpha\beta c^\times l + \alpha\alpha^\times\beta\beta^\times M, \tag{16.75}$$

where the functions F, G, h, i, j, K, l, M depend on a, b, and $c^\times c$. Variation of q yields

$$\delta f = q(\beta\alpha^\times + \alpha\beta^\times)(G - h) + q(\alpha\alpha^\times - \beta\beta^\times)(-i - j), \tag{16.76}$$

hence, $h = G$, $i = -j$ and thus,

$$f(Q) = F + (\alpha\alpha^\times + \beta\beta^\times)G + (-\beta\alpha^\times + \alpha\beta^\times)j + \alpha^\times\beta^\times cK + \alpha\beta c^\times l + \alpha\alpha^\times\beta\beta^\times M. \tag{16.77}$$

Variation of s and $s^\times$ yields

$$\begin{aligned}
\delta f =\ & (ic^\times s - ics^\times)\phi_1 + (ic^\times s - ics^\times)(\alpha\alpha^\times + \beta\beta^\times)\phi_2 \\
& + (ic^\times s - ics^\times)\alpha\alpha^\times\beta\beta^\times\phi_3 + (\alpha\beta^\times - \beta\alpha^\times)(ic^\times s\phi_6 - ics^\times\phi_7) \\
& + \alpha^\times\beta^\times(s\phi_8 - ic^2 s^\times\phi_4) + \alpha\beta(ic^{\times 2}s\phi_5 + s^\times\phi_9),
\end{aligned} \tag{16.78}$$

$$\phi_1 = \Box F, \quad \phi_2 = \Box G, \quad \phi_3 = \Box M, \quad \phi_4 = \Box K, \quad \phi_5 = \Box l,$$

$$\phi_6 = \Box j + il, \quad \phi_7 = \Box j + iK, \tag{16.79}$$

$$\phi_8 = icc^\times\Box K - 2ibK - 2j, \quad \phi_9 = -icc^\times\Box l + 2ibl + 2j,$$

$$\Box := \frac{\partial}{\partial b} - 2b\frac{\partial}{\partial(cc^\times)}. \tag{16.80}$$

Due to the invariance, all ϕ_i have to vanish. Since $\phi_1 = \ldots = \phi_5 = 0$, the functions F, G, K, M, and l depend only on a and $r^2 = b^2 + cc^\times$. From $\phi_8 = \phi_9 = 0$, one obtains

$$j = -ibK, \quad l = K. \tag{16.81}$$

Also $\phi_6 = \phi_7 = 0$ are satisfied. This yields inserted in (16.77)

$$f(Q) = F + (\alpha\alpha^\times + \beta\beta^\times)G + (i\beta\alpha^\times b - i\alpha\beta^\times b + \alpha\beta c^\times + \alpha^\times\beta^\times c)K + \alpha\alpha^\times\beta\beta^\times M. \tag{16.82}$$

It remains to vary f with respect to ω, $\omega^\times$, η, $\eta^\times$. One obtains

$$\begin{aligned}
\delta f =\ & (-\omega^\times\beta + \omega\beta^\times + \eta^\times\alpha - \eta\alpha^\times)\phi_{10} \\
& + (-i\omega^\times\alpha b - i\omega\alpha^\times b + \omega^\times\alpha^\times c + \omega\alpha c^\times)(\phi_{11} + \beta\beta^\times\phi_{12})
\end{aligned}$$

$$+(-\omega\beta^\times\alpha\alpha^\times + \omega^\times\beta\alpha\alpha^\times + \eta^\times\alpha\beta\beta^\times + \eta\alpha^\times\beta\beta^\times)\phi_{13}$$

$$+(-i\eta^\times\beta b - i\eta\beta^\times b + \eta^\times\beta^\times c + \eta\beta c^\times)(\phi_{11} + \alpha\alpha^\times\phi_{12}) \qquad (16.83)$$

with

$$\phi_{10} = \partial_a F + aG - r^2 K,$$

$$\phi_{11} = -2\partial_{r^2} F + G - aK,$$

$$\phi_{12} = -2\partial_{r^2} G + \partial_a K + M,$$

$$\phi_{13} = \partial_a G - 3K - 2r^2\partial_{r^2} K + aM. \qquad (16.84)$$

$\phi_{10} = \phi_{11} = \phi_{12} = 0$ yield

$$(r^2 - a^2)K = \partial_a F + 2a\partial_{r^2} F, \qquad (16.85)$$

$$(r^2 - a^2)G = a\partial_a F + 2r^2\partial_{r^2} F \qquad (16.86)$$

$$M = 2\partial_{r^2} G - \partial_a K. \qquad (16.87)$$

Equation $\phi_{13} = 0$ is fulfilled, since

$$\phi_{13} = 2\partial_{r^2}\phi_{10} + \partial_a\phi_{11} + a\phi_{12}. \qquad (16.88)$$

Thus, $f(Q)$ is given by (16.82). Equations (16.85)–(16.87) express G, K, and M in terms of the function F.

Inserting $r^2 = a^2$ in (16.85) shows that $F(a, a^2)$ does not depend on a.

16.3.3   The Integral for $N = 1$, $Q \in \mathcal{M}(2, 2)$

The integral yields with (16.82)

$$\int DQ f(Q) = \frac{1}{4\pi^2}\int da\, db\, dc_1\, dc_2\, d\alpha^\times\, d\alpha\, d\beta^\times\, d\beta\, f$$

$$= \frac{1}{4\pi^2}\int da\, db\, dc_1\, dc_2\, M(a, b^2 + c_1^2 + c_2^2) \qquad (16.89)$$

where $c = c_1 + ic_2$, $c^\times = c_1 - ic_2$. The a-integral over the K-term in (16.87) vanishes. Thus, we are left with the G-term. Transforming from the 'Cartesian coordinates' b, c_1, c_2 to the radius r, we use $db\, dc_1\, dc_2 = 4\pi r^2 dr$. Then we obtain

$$\int DQ f(Q) = -\frac{1}{\pi}\int da\, dr\, rG(a, r^2) = -\frac{1}{\pi}\int da\, dr\, r\frac{a\partial_a F + r\partial_r F}{r^2 - a^2}, \qquad (16.90)$$

where r runs from 0 to $e^{i\gamma-}\infty$. We decompose

$$\frac{a\partial_a F + r\partial_r F}{r^2 - a^2} = \frac{\partial_a F + \partial_r F}{2(r-a)} + \frac{\partial_a F - \partial_r F}{2(-r-a)}, \tag{16.91}$$

and obtain two integrals. In the second integral, we change the sign of r and use $F(a, (-r)^2) = F(a, r^2)$ to obtain

$$\int D\,Qf(Q) = -\frac{1}{2\pi} \int d\,a d\,r \frac{\partial_a F + \partial_r F}{r - a}, \tag{16.92}$$

where r now runs from $-e^{i\gamma-}\infty$ to $+e^{i\gamma-}\infty$. As for the evaluation of (15.34), we obtain

$$\int D\,Qf(Q) = -isF(0,0) = -isf(0), \tag{16.93}$$

which shows (16.71) holds for this special case.

16.3.4   The General Case

The generalization to N matrices, $Q \in \mathcal{M}(n, 2r)$, runs analogously to the considerations in Sects. 15.2.3 and 15.2.4. The statements on the general case in Sect. 16.2.4 hold similarly.

16.3.5   Matrix as a Set of Vectors

The integral over a function of a superreal matrix, $W \in \mathcal{M}(2,2)$, invariant under independent unitary-orthosymplectic transformations on both sides of W, is proportional to the function at $W = 0$, provided the function decays sufficiently rapidly for large arguments along the paths of integration.

The arguments run similarly as for the function of $W \in \mathcal{M}(1,1)$ invariant under independent unitary transformations on both sides, as given in Sect. 15.3. However, the derivation of the explicit form of the function is lengthy due to the large number (16) of independent variables.

Chapter 17
More on Matrices

Abstract In this chapter the following topics are considered: (1) the eigenvalue problem; (2) a functional equation for square matrices, and (3) the Berezinian for matrices with linearly dependent matrix elements. We find (1) If the ordinary part of a supermatrix is not degenerate, then it can be diagonalized by a similarity transformation. If it is degenerate, then even if it is hermitian, it can normally not be diagonalized. However, superreal hermitian matrices have two-fold degenerate eigenvalues in the fermionic sector. If this is the only degeneracy, then diagonalization is possible. (2) A differentiable function F of a square matrix obeying the functional equation $F(A)F(B) = F(AB)$ vanishes identically, or is a power of the superdeterminant. (3) The Berezinian of matrices with linearly dependent matrix elements is determined.

17.1 Eigenvalue Problem

The eigenvalue problem is considered. If the body of a matrix $\in \mathcal{M}(n, m)$ with $n > 0$, $m > 0$ has degenerate eigenvalues, then normally the matrix cannot be diagonalized by a similarity transformation, even if the matrix is hermitian.

The eigenvalue problem for a matrix $H \in \mathcal{M}(n, m)$ is formulated as a solution of the equation

$$HV = VD, \quad V \in \mathcal{M}(n, m), \quad D = \mathrm{diag}(\lambda_1, \ldots, \lambda_n, \mu_1, \ldots, \mu_m). \qquad (17.1)$$

If such a pair of V and D with nonsingular V, can be found, then D contains the eigenvalues of H. The columns of V are the right eigenvectors. Equation (17.1) can then be rewritten

$$V^{-1}H = DV^{-1}. \qquad (17.2)$$

Thus, the rows of V^{-1} are the left-eigenvectors.

We approach the solution in two steps. In the first step, we consider the ordinary parts of H, V, D.

© Springer-Verlag Berlin Heidelberg 2016
F. Wegner, *Supermathematics and its Applications in Statistical Physics*,
Lecture Notes in Physics 920, DOI 10.1007/978-3-662-49170-6_17

Definition 17.1 A matrix $H \in \mathcal{M}(n, m)$ is called non-degenerate, if all eigenvalues of ord (H) differ.

The ordinary parts a_i and b_i of the eigenvalues of the bosonic sector H_b and of the fermionic sector H_f, respectively, are obtained from the characteristic polynomials

$$\det(e\mathbf{1} - \text{ord}(H_b)) = \prod_{i=1}^{n}(e - a_i), \quad \det(e\mathbf{1} - \text{ord}(H_f)) = \prod_{j=1}^{m}(e - b_j). \quad (17.3)$$

Non-degenerate means that all a_i and b_j are different.

Theorem 17.1 *If $H \in \mathcal{M}(n, m)$ is non-degenerate, then it can be diagonalized by means of a similarity transformation.*

This can be seen in the following way: Since $\det(a_i\mathbf{1} - \text{ord}(H_b))$ vanishes, it indicates that there is a solution, $V_b^{(i)}$, to the homogeneous equation ord $(H_b)V_b^{(i)} = a_i V_b^{(i)}$, similarly for the fermionic sector ord $(H_f)V_f^{(i)} = b_i V_f^{(i)}$. This can be rewritten as

$$\text{ord}(H)V^{(0)} = V^{(0)}\begin{pmatrix} \text{diag}(a) & 0 \\ 0 & \text{diag}(b) \end{pmatrix}, \quad V^{(0)} := \begin{pmatrix} V_b & 0 \\ 0 & V_f \end{pmatrix}, \quad (17.4)$$

where the $V^{(i)}$s are the columns of the Vs.

Thus, we obtain

$$V^{(0)-1}HV^{(0)} = W = \text{diag}(a, b) + \begin{pmatrix} \kappa^2 A & \kappa\alpha \\ \kappa\beta & \kappa^2 B \end{pmatrix}, \quad (17.5)$$

where $\kappa \in \mathcal{A}_0$ has been introduced as an expansion parameter, and the ordinary parts of $A, B \in \mathcal{A}_0$ vanish, and $\alpha, \beta \in \mathcal{A}_1$. We consider the first eigenvector (the other eigenvectors can be considered similarly) of W,

$$\begin{pmatrix} x \\ \xi \end{pmatrix}, \quad x_i = \begin{cases} 1 & i = 1, \\ \sum_{l>0}\kappa^{2l}x_i^{(l)} & i > 1 \end{cases}, \quad \xi_i = \sum_{l>0}\kappa^{2l-1}\xi_i^{(l)} \quad (17.6)$$

with eigenvalue

$$\lambda_1 = a_1 + \sum_{l>0}\kappa^{2l}\Delta^{(l)}. \quad (17.7)$$

One can easily expand in powers of κ, since the eigenvalue equations read

$$(b_i - a_1)\xi_i^{(l)} = \sum_k \Delta^{(k)}\xi_i^{(l-k)} - B_{ij}\xi_j^{(l-1)} - \beta_{ij}x_j^{(l-1)} - \beta_{i,1}\delta_{l,1}, \quad (17.8)$$

$$\Delta^{(l)} = A_{11}\delta_{l,1} + A_{1,j}x_j^{(l-1)} + \alpha_{1j}\xi_j^{(l)},\tag{17.9}$$

$$(a_i - a_1)x_i^{(l)} = \sum_k \Delta^{(k)}x_i^{(l-k)} - A_{ij}x_j^{(l-1)} - \alpha_{ij}\xi_j^{(l)} - A_{i1}\delta_{l,1}, \quad i \neq 1.\tag{17.10}$$

There is a solution if a_1 differs from all other a_i, b_i.

Thus, we obtain the eigenvectors with eigenvalue λ and similarly with eigenvalues μ,

$$W\begin{pmatrix} x \\ \xi \end{pmatrix} = \lambda \begin{pmatrix} x \\ \xi \end{pmatrix}, \quad W\begin{pmatrix} \eta \\ y \end{pmatrix} = \mu \begin{pmatrix} \eta \\ y \end{pmatrix}.\tag{17.11}$$

Putting the column vectors together, one obtains

$$W\begin{pmatrix} x & \eta \\ \xi & y \end{pmatrix} = \begin{pmatrix} x & \eta \\ \xi & y \end{pmatrix}\begin{pmatrix} \mathrm{diag}(\lambda) & 0 \\ 0 & \mathrm{diag}(\mu) \end{pmatrix}\tag{17.12}$$

and

$$HV^{(0)}\begin{pmatrix} x & \eta \\ \xi & y \end{pmatrix} = V^{(0)}\begin{pmatrix} x & \eta \\ \xi & y \end{pmatrix}\begin{pmatrix} \mathrm{diag}(\lambda) & 0 \\ 0 & \mathrm{diag}(\mu) \end{pmatrix}.\tag{17.13}$$

The multiplication theorem yields

$$\mathrm{sdet}(H) = \mathrm{sdet}(W) = \mathrm{sdet}\begin{pmatrix} \mathrm{diag}(\lambda) & 0 \\ 0 & \mathrm{diag}(\mu) \end{pmatrix} = \frac{\prod_{i=1}^n \lambda_i}{\prod_{j=1}^m \mu_j}.\tag{17.14}$$

If the a_i, b_j are all different, then there exist left- and right-eigenvectors. If, however, two such eigenvalues coincide, then, in general, there are no such eigenvectors.

Example Consider

$$W = \begin{pmatrix} a & \alpha & \beta \\ \gamma & b & 0 \\ \delta & 0 & c \end{pmatrix} \in \mathcal{M}(1,2).\tag{17.15}$$

Eigenvalues and eigenvectors of

$$WX^{(i)} = \lambda^{(i)}X^{(i)}, \quad WY^{(i)} = \mu^{(i)}Y^{(i)}\tag{17.16}$$

are

$$\lambda^{(1)} = a + \frac{\alpha\gamma}{a-b} + \frac{\beta\delta}{a-c} + \frac{(2a-b-c)\alpha\beta\gamma\delta}{(a-b)^2(a-c)^2},$$

$$X^{(1)} = \begin{pmatrix} 1 \\ \frac{\gamma}{a-b}(1 - \frac{\beta\delta}{(a-b)(a-c)}) \\ \frac{\delta}{a-c}(1 - \frac{\gamma\alpha}{(a-b)(a-c)}) \end{pmatrix}, \qquad (17.17)$$

$$\mu^{(1)} = b - \frac{\alpha\gamma}{b-a} + \frac{\alpha\beta\gamma\delta}{(b-a)^2(b-c)}, \qquad Y^{(1)} = \begin{pmatrix} \frac{\alpha}{b-a}(1 + \frac{\beta\delta}{(b-a)(b-c)}) \\ 1 \\ -\frac{\alpha\delta}{(b-a)(b-c)} \end{pmatrix},$$

$$ (17.18) $$

$$\mu^{(2)} = c - \frac{\beta\delta}{c-a} + \frac{\alpha\beta\gamma\delta}{(c-a)^2(c-b)}, \qquad Y^{(2)} = \begin{pmatrix} \frac{\beta}{c-a}(1 + \frac{\alpha\gamma}{(c-a)(c-b)}) \\ -\frac{\beta\gamma}{(c-a)(c-b)} \\ 1 \end{pmatrix}.$$

$$ (17.19) $$

Hermitian Matrices It is apparent from the above expressions (17.17)–(17.19) that even hermitian matrices, whose bodies have degenerate eigenvalues, will, in general, not have corresponding eigenvalues and eigenfunctions.

If W is hermitian $W^\dagger = W$, then

$$WS^{(i)} = \zeta^{(i)}S^{(i)} \rightarrow S^{(i)\dagger}W = \zeta^{(i)*}S^{(i)\dagger}, \qquad (17.20)$$

where S stands for both X and Y, and ζ for λ and μ, respectively. Thus, evaluating $S^{(i)\dagger}WS^{(j)}$ with both expressions (17.20) one obtains

$$(S^{(i)\dagger}S^{(j)})(\zeta^{(i)*} - \zeta^{(j)}) = 0. \qquad (17.21)$$

Thus, for $i = j$, one finds that the eigenvalues $\zeta^{(i)}$ are real, $\zeta^{(i)*} = \zeta^{(i)}$, since ord $(S^{(i)})$ does not vanish identically. If ord $\lambda^{(i)} \neq$ ord $\lambda^{(j)}$, then $S^{(i)}$ and $S^{(j)}$ are orthogonal, $S^{(i)\dagger}S^{(j)} = 0$.

17.2   Diagonalization of Superreal Hermitian Matrices

Hermitian supereal matrices have a twofold degeneracy in the fermionic sector. This allows diagonalization if no other degeneracies appear.

Much of the analysis for hermitian matrices of the first kind given above hold for superreal hermitian matrices too. One has only to replace $*$ by $^\times$ and † by ‡.

Since W is superreal (14.5), $W = \mathbf{C} W^{\times} \mathbf{C}^{-1}$ holds. If S is an eigenvector with eigenvalue ζ, $WS = \zeta S$, then $\mathbf{C} W^{\times} \mathbf{C}^{-1} S = \zeta S$ holds. The complex conjugate yields $\mathbf{C} W^{\times \times} \mathbf{C}^{-1} S^{\times} = \zeta^{\times} S^{\times}$ and finally (note $\mathbf{C} W^{\times \times} \mathbf{C}^{-1} = \mathbf{C}^{-1} W \mathbf{C}$)

$$WS = \zeta S \;\succ\; W \mathbf{C} S^{\times} = \zeta^{\times} \mathbf{C} S^{\times}. \tag{17.22}$$

Thus, $\mathbf{C} S^{\times}$ is an eigenfunction with eigenvalue $\zeta^{\times}$. If W is hermitian, then ζ is superreal and consequently both S and $\mathbf{C} S^{\times}$ are eigenvectors with the same eigenvalue.

For $X \in \mathcal{M}(n, 2r, 1, 0)$ one obtains $\mathbf{C} X^{\times}$

$$X = \begin{pmatrix} x \\ \xi \end{pmatrix} \;\succ\; \mathbf{C} X^{\times} = \begin{pmatrix} x^{\times} \\ \epsilon_r \xi^{\times} \end{pmatrix}. \tag{17.23}$$

The linear combinations $X_+ = X + \mathbf{C} X^{\times}$ and $X_- = \mathrm{i}(X - \mathbf{C} X^{\times})$ obey $\mathbf{C} X_{\pm}^{\times} = X_{\pm}$ and are superreal eigenfunctions. Generically they will be proportional to each other and thus constitute only one independent eigenfunction.

However, for $Y \in \mathcal{M}(n, 2r, 0, 1)$

$$Y = \begin{pmatrix} \eta \\ y \end{pmatrix} \;\succ\; \mathbf{C} Y^{\times} = \begin{pmatrix} \eta^{\times} \\ \epsilon_r y^{\times} \end{pmatrix}. \tag{17.24}$$

one obtains

$$(\mathbf{C} Y^{\times})^{\ddagger} Y = {}^{\mathrm{t}}\eta \eta - {}^{\mathrm{t}}y \epsilon_r y = 0. \tag{17.25}$$

Thus, $\mathbf{C} Y^{\times}$ and Y are orthogonal, which implies that the eigenfunctions appear pairwise with the same eigenvalues. This corresponds to the Kramers degeneracy. $\left(Y \; \mathbf{C} Y^{\times} \right) \in \mathcal{M}(n, 2r, 0, 2)$ is a superreal double-vector.

Theorem 17.2 *Superreal hermitian matrices have pairwise equal eigenvalues μ in the fermionic sector.*

We show explicitly that this degeneracy does not forbid the diagonalization. Let us expand Y

$$\eta_i = \sum_{l>0} \kappa^{2l-1} \eta_i^{(l)}, \quad y_i = \begin{cases} \mathbf{1}_2 & i = 1, \\ \sum_{l>0} \kappa^{2l} y_i^{(l)}, & i > 1 \end{cases} \quad \mu_1 = b_1 + \sum_{l>0} \kappa^{2l} \Delta^{(l)}, \tag{17.26}$$

where $\eta_i^{(l)}$ are real row-spinors and $y_i^{(l)}$ real quaternions. We use W, (17.5) with superreal elements $A_{..} \in \mathcal{M}(1, 0)$, $\alpha_{..} \in \mathcal{M}(1, 0, 0, 2)$, $\beta_{..} \in \mathcal{M}(0, 2, 1, 0)$, $B_{..} \in$

$\mathscr{M}(0,2)$. Then one may iterate starting with $l = 1$

$$(b_1 - a_i)\eta_i^{(l)} = -\sum_k \Delta^{(k)}\eta_i^{(l-k)} + A_{ij}\eta_j^{(l-1)} + \alpha_{ij}y_j^{(l-1)} + \alpha_{i1}\delta_{l,1}, \quad (17.27)$$

$$\Delta^{(l)}1_2 = B_{11}\delta_{l,1} + B_{1j}y_j^{(l-1)} + \beta_{1j}\eta_j^{(l)}, \quad (17.28)$$

$$(b_1 - b_i)y_i^{(l)} = -\sum_k \Delta^{(k)}y_i^{(l-k)} + B_{i1}\delta_{l,1} + B_{ij}y_j^{(l-1)} + \beta_{ij}\eta_j^{(l)}, i > 1.$$

$$(17.29)$$

Example In the case

$$W = \begin{pmatrix} a & \alpha & \alpha^\times \\ -\alpha^\times & b & 0 \\ \alpha & 0 & b \end{pmatrix}, \quad (17.30)$$

the eigenvalues and eigenvectors read

$$\lambda^{(1)} = a + \frac{2}{a-b}\alpha^\times\alpha, \ X^{(1)} = \begin{pmatrix} 1 \\ -\frac{1}{a-b}\alpha^\times \\ \frac{1}{a-b}\alpha \end{pmatrix}, \quad (17.31)$$

$$\mu^{(1)} = b + \frac{1}{a-b}\alpha^\times\alpha, \ Y^{(1)} = \begin{pmatrix} -\frac{\alpha}{a-b} & -\frac{\alpha^\times}{a-b} \\ 1 & 0 \\ 0 & 1 \end{pmatrix}. \quad (17.32)$$

17.3   Functional Equation for Matrices

The multiplication theorem for superdeterminants was given in (10.1). Here we derive an inversion of this theorem.

Theorem 17.3 *Let $F(A) \in \mathscr{A}_0$, $A \in \mathscr{M}(n,m)$ be a holomorphic function of the matrix elements of A.*
  If the functional equation $F(AB) = F(A)F(B)$ holds for all A and B, then

$$F(A) = \text{sdet}(A)^k, \quad (17.33)$$

unless F vanishes identically.

We sketch the steps of the proof:

1. $1A = A$ yields $F(1) = 1$,
2. $F(V)F(V^{-1}) = F(1) = 1$ yields $F(V^{-1}) = F(V)^{-1}$.

3. Nearly all A (see Theorem 17.1) can be diagonalized by a similarity transformation, $A = V\Lambda V^{-1}$. Thus, $F(A) = F(\Lambda)$, with $\Lambda = \mathrm{diag}(\lambda_1, \lambda_2, \ldots \lambda_n, \mu_1, \mu_2, \ldots \mu_m)$.

4. $F(\Lambda)$ is invariant under transpositions of the λs and transpositions of the μs. The matrix representation of the transposition $T^{(kl)}$ of k and l can be written as

$$T_{ij}^{(kl)} = \delta_{ij} - (\delta_{ik} - \delta_{il})(\delta_{jk} - \delta_{jl}), \quad k \neq l. \tag{17.34}$$

Then $T^{(kl)}AT^{(kl)}$ exchanges columns k and l and rows k and l of A. Note that $T^{(kl)2} = \mathbf{1}$.

5. Thus, $f(x) := F(\mathrm{diag}(x, x^{-1}, 1, \ldots)) = F(\mathrm{diag}(x^{-1}, x, 1, \ldots))$. Since the product of these matrices equals $\mathbf{1}$, $f(x) = \pm 1$, since f is holomorphic in x, $f(x) = f(1) = 1$.

6. Thus, $F(\mathrm{diag}(\lambda_1, \lambda_2, \ldots)) = F(\mathrm{diag}(x\lambda_1, x^{-1}\lambda_2, \ldots))$, which, with $x = \lambda_2$, yields $F(\lambda_1\lambda_2, 1, \ldots)$. Thus, F depends only on the products $P_\lambda = \prod_i \lambda_i$ and $P_\mu = \prod_i \mu_i$, $F(A) = f(P_\lambda^{(A)}, P_\mu^{(A)})$.

7. For two matrices A and B we have $f(x, y)f(u, v) = f(xu, yv)$ with $x = P_\lambda^{(A)}$, $y = P_\mu^{(A)}$, $u = P_\lambda^{(B)}$, $v = P_\mu^{(B)}$. Taking the derivative with respect to x yields $kf(u, v) = u\partial f(u, v)/\partial u$ with $k = \partial f(x, 1)/\partial x|_{x=1}$ with the solution $f(u, v) = C(v)u^k$. The analogous argument for v yields $f(u, v) = Cu^k v^{k'}$. The unit matrix $u = 1, v = 1$ yields $f = 1$. Thus $C = 1$.

8. Finally the relation between k and k' has to be found. We consider

$$AB = \begin{pmatrix} a & \alpha \\ \beta & b \end{pmatrix} \begin{pmatrix} c & 0 \\ 0 & d \end{pmatrix} = \begin{pmatrix} ac & d\alpha \\ c\beta & bd \end{pmatrix} \in \mathcal{M}(1, 1). \tag{17.35}$$

For larger matrices these entries are in the upper left corners of the corresponding sectors. One adds nonvanishing entries in the diagonal of the bosonic and fermionic sectors. They have no effect to our considerations. The eigenvalues of these matrices are

$$\lambda^{(A)} = a + \frac{\alpha\beta}{a - b}, \quad \mu^{(A)} = b + \frac{\alpha\beta}{a - b}, \quad \lambda^{(B)} = c, \quad \mu^{(B)} = d, \tag{17.36}$$

$$\lambda^{(AB)} = ac + cd\frac{\alpha\beta}{ac - bd}, \quad \mu^{(AB)} = bd + cd\frac{\alpha\beta}{ac - bd}.$$

Thus $F(A)F(B) = F(AB)$ requires

$$(ac)^k[1 + \frac{k}{a(a - b)}\alpha\beta](bd)^{k'}[1 + \frac{k'}{b(a - b)}\alpha\beta]$$

$$= (ac)^k[1 + \frac{kd}{a(ac - bd)}\alpha\beta](bd)^{k'}[1 + \frac{k'c}{b(ac - bd)}\alpha\beta]. \tag{17.37}$$

Comparing the coefficients of $\alpha\beta$, one observes that this relation is identically fulfilled only for $k' = -k$. Since $P_\lambda^{(A)}/P_\mu^{(A)} = \mathrm{sdet}(A)$ and for all matrices $\mathrm{sdet}(A)\mathrm{sdet}(B) = \mathrm{sdet}(AB)$ holds, the above theorem is proven.

17.4 Berezinian for Transformation of Matrices with Linearly Dependent Matrix Elements

The Berezinian for the transformation of supersymmetric, super-skew-symmetric, and for superreal hermitian and antihermitian matrices is given.

Theorem 10.2 gives the Berezinian for the transformation of matrices whose matrix elements are linearly independent. However, this derivation cannot be applied for matrices whose matrix elements are linearly dependent as for supersymmetric and super-skew-symmetric matrices. These will be considered here.

The transformation

$$W' = {}^{\mathrm{T}}VWV, \quad W, W', V \in \mathcal{M}(n, m) \tag{17.38}$$

transforms a supersymmmetric/super-skew-symmetric matrix W again into a supersymmetric/super-skew-symmetric matrix W',

$$W = \pm {}^{\mathrm{T}}W\sigma \rightarrow W' = \pm {}^{\mathrm{T}}W'\sigma \tag{17.39}$$

Theorem 17.4 *The Berezinian for the transformation (17.38) $W \rightarrow W'$ with (17.39) reads*

$$[\mathrm{D}\,W'] = \mathrm{sdet}(V)^{n-m\pm 1}[\mathrm{D}\,W].$$

This can be shown by means of the functional theorem 17.3. If we choose

$$W'' = {}^{\mathrm{T}}V'WV', \quad W' = {}^{\mathrm{T}}V''W''V'', \tag{17.40}$$

then $V = V'V''$. Denote the Berezinian by F,

$$[\mathrm{D}\,W''] = F(V')[\mathrm{D}\,W], \quad [\mathrm{D}\,W'] = F(V'')[\mathrm{D}\,W''], \tag{17.41}$$

then

$$[\mathrm{D}\,W'] = F(V'')F(V')[\mathrm{D}\,W] = F(V'V'')[\mathrm{D}\,W]. \tag{17.42}$$

Thus due to the inversion theorem 17.3 $F(V) = \mathrm{sdet}(V)^k$. The exponent k can be determined by choosing V diagonal with matrix elements λ in the diagonal of the bosonic sector and μ in the fermionic sector. Then the various sectors contribute

$$[\text{D } W'] = \lambda^{n(n\pm1)} \cdot (\lambda\mu)^{-nm} \cdot \mu^{m(m\mp1)}[\text{D } W]$$
$$= (\lambda^n \mu^{-m})^{n-m\pm1}[\text{D } W] = \text{sdet}(V)^{n-m\pm1}[\text{D } W]. \qquad (17.43)$$

Similarly one derives, for superreal hermitian and antihermitian matrices,

$$W' = V^{\ddagger}WV, \quad W = \pm W^{\ddagger}, \quad W, W', V \in \mathscr{M}(n, 2r). \qquad (17.44)$$

Theorem 17.5 *The Berezinian for superreal hermitian and antihermitian matrices reads*

$$[\text{D } W'] = (\text{sdet}V)^{n-2r\pm1}[\text{D } W].$$

The derivation runs parallel to that of super-(skew)-symmetric matrices given above $(m = 2r)$.

Part III
Supersymmetry in Statistical Physics

In this last part we consider several applications of supersymmetry in statistical physics. Supersymmetry is used with various meanings. We distinguish between these as follows:

1. The notion of supersymmetry was first introduced in high energy physics as a symmetry of spacetime. Two pairs of anticommuting space components are added to the conventional four-dimensional space. The supersymmetric theory predicts bosonic and fermionic particles with degenerate masses. As of yet, they have not been observed. As to whether supersymmetry does not exist or symmetry breaking has prevented its observation is an open question. Very similar techniques can be used in systems described by stochastic time-dependent equations. We consider both in Chap. 19.

2. In a number of cases the action or Hamiltonian is given as a product of two operators, often denoted by Q and $Q^\dagger$. They yield systems with the same spectrum. Or they have the property $Q^2 = 0$ and yield pairs of eigenenergies E and $-E$. The notion of supersymmetric quantum mechanics is used in such cases. We consider this in Chap. 18.

3. Disorder in a number of models in d dimensions yield a formulation, which allows the reduction to the pure model in $d - 2$ dimensions. This formulation is considered in Chap. 20.

4. Supersymmetry in target space appears for particles in random one-particle potentials, which we briefly considered in Chap. 4. The mapping of random matrix models on nonlinear sigma-models is considered in Chap. 21. Diffusive models, that is tight-binding models with random on-site and hopping matrix elements, can be mapped on models of interacting matrices. This is derived in Chap. 22. Finally, in Chap. 23 we consider the Anderson transition, in particular the scaling theory of conductivity and multifractality close to the mobility edge. A few more aspects are addressed in this chapter: Besides the three

Wigner-Dyson classes of disordered systems there are the chiral classes and the Bogolubov-de Gennes classes. All of them are related to the ten symmetric spaces. A short account of the physics of two-dimensional disordered systems, which is particularly rich, is given. Finally the concept of superbosonization is mentioned.

The last chapter summarizes the book. A few related subjects and relevant papers are mentioned for the interested reader.

Chapter 18
Supersymmetric Models

Abstract Supersymmetry in high-energy physics predicts bosonic and fermionic states with equal energies. Certain general ideas of supersymmetry were adopted to models in solid-state physics. They are often described by pairs of operators Q and $Q^\dagger$, which allow for pairs of Hamiltonians $Q^\dagger Q$ and $QQ^\dagger$ with the same spectrum. An example is the hydrogen atom. A class of models has the property $Q^2 = 0$. They yield either chiral models with pairs of energy levels E and $-E$ if the Hamiltonian is linear in Q and $Q^\dagger$, or models with pairs of states with the same energy if the Hamiltonian is bilinear in Q and $Q^\dagger$.

18.1 Supersymmetric Quantum Mechanics

In this section we consider models with supersymmetric partner Hamiltonians, which have the property that both Hamiltonians have the same spectrum.

Supersymmetric quantum mechanics deals with Hamiltonians, which have the same spectrum of non-zero energy states. Formally they may be considered to belong to Hilbert spaces with different number of fermions. Some examples can be found in the textbook by Schwabl [236].

18.1.1 Supersymmetric Partners

Supersymmetric Hamiltonians are often defined [289] in terms of the *charges* Q_i with the property

$$\{Q_i, Q_j\} = \delta_{ij} H, \quad [Q_i, H] = 0, \tag{18.1}$$

where the second set of equations follows from the first one. For two charges one may consider

$$Q_1 = \tfrac{1}{2}(\sigma_1 p_x + \sigma_2 W(x)), \quad Q_2 = \tfrac{1}{2}(\sigma_2 p_x - \sigma_1 W(x)), \tag{18.2}$$

© Springer-Verlag Berlin Heidelberg 2016
F. Wegner, *Supermathematics and its Applications in Statistical Physics*,
Lecture Notes in Physics 920, DOI 10.1007/978-3-662-49170-6_18

with momentum operator $p_x = \frac{\mathrm{d}}{\mathrm{i}\,\mathrm{d}x}$, Pauli operators σ_i, and function $W(x)$, which yields

$$H = \tfrac{1}{2}(p_x^2 + W^2(x) + \sigma_3 \frac{\mathrm{d}\,W(x)}{\mathrm{d}\,x}) = \begin{pmatrix} H_+ & 0 \\ 0 & H_- \end{pmatrix}. \tag{18.3}$$

One obtains the same $H_\pm$ from

$$Q^\dagger = p_x + \mathrm{i}W(x), \quad Q = p_x - \mathrm{i}W(x), \tag{18.4}$$

$$H_+ = \tfrac{1}{2}QQ^\dagger, \quad H_- = \tfrac{1}{2}Q^\dagger Q. \tag{18.5}$$

The Hamiltonian (18.3) can be obtained from

$$H = \tfrac{1}{2}(f^\dagger Q + Q^\dagger f)^2 = f^\dagger f H_+ + f f^\dagger H_-, \tag{18.6}$$

with fermion creation and annihilation operators $f^\dagger$ and f, respectively. These operators commute with Q and $Q^\dagger$. The connection between the fermion operators and the spin operators lies in the fact that the operators σ^+ and σ^- have the same anticommutation relations as $f^\dagger$ and f. The Hamiltonian H_+ acts in the Hilbert space with occupied fermion, and H_- in that without fermion. Correspondingly H_- is called the bosonic sector and H_+ the fermionic sector in (18.3). Supersymmetry (see Sect. 19.2) is characterized by operators, which connect states with different fermion number, but equal eigenenergy. Here these operators are $f^\dagger Q$ and $Q^\dagger f$ in (18.6).

Equation (18.5) yields

$$H_- Q^\dagger = Q^\dagger H_+, \quad H_+ Q = Q H_-. \tag{18.7}$$

Hence, eigenstates of H_+ and H_-

$$H_+ \psi_+ = E_+ \psi_+, \quad H_- \psi_- = E_- \psi_- \tag{18.8}$$

can easily be transformed into eigenstates of the other Hamiltonian,

$$H_-(Q^\dagger \psi_+) = E_+(Q^\dagger \psi_+), \quad H_+(Q\psi_-) = E_-(Q\psi_-). \tag{18.9}$$

Thus, H_+ and H_- have the same eigenvalues, and the eigenvectors are connected by $Q^\dagger$ and Q. These Hamiltonians are called supersymmetric partners. Only if $Q^\dagger \psi_+ = 0$ or $Q\psi_- = 0$, are there no such pairs. Thus, the number of eigenstates with energy zero may differ. The difference of the number of states at zero energy is called the Witten index [290].

Ground State The condition $Q\psi_- = 0$ yields the ground state of H_-,

$$\psi_-(x) = \exp\left(-\int^x \mathrm{d}x' W(x')\right), \tag{18.10}$$

provided it is normalizable. Similarly $Q^\dagger \psi_+ = 0$ yields

$$\psi_+(x) = \exp\left(+\int^x dx' W(x')\right). \tag{18.11}$$

Usually only one of the functions $\psi_\pm$ is normalizable. If $W(x)$ approaches zero in the limit $x \to \pm\infty$ sufficiently rapidly, then both states are continuum states and both H_+ and H_- have ground states of energy 0.

Thus, the Hamiltonians

$$H_\pm = \tfrac{1}{2}p_x^2 + V_\pm(x), \quad V_\pm(x) = \tfrac{1}{2}(W^2(x) \pm W'(x)), \tag{18.12}$$

describe two systems with potential V_+ and V_-, which have the same spectrum. Only the number of states with zero energy may differ. More on this subject can be found in the article by Cooper et al. [51]. Below, we give some examples.

18.1.2 Harmonic Oscillator

The operators Q and $Q^\dagger$ are known as creation and annihilation operators for the harmonic oscillator

$$W(x) = x, \quad V_\pm = \tfrac{1}{2}(x^2 \pm 1), \tag{18.13}$$

where in our dimensionless units $\hbar\omega = 1$.

18.1.3 The $\cosh^{-2}$-Potential

As a second example we choose

$$W(x) = c\tanh(x), \quad V_\pm = \frac{1}{2}\left(c^2 - \frac{c(c \mp 1)}{\cosh^2(x)}\right). \tag{18.14}$$

and define the Hamiltonians

$$H_c = \tfrac{1}{2}p_x^2 - \frac{c(c+1)}{2\cosh^2(x)} \tag{18.15}$$

Then $H_c = H_- - \tfrac{1}{2}c^2$ and $H_{c-1} = H_+ - \tfrac{1}{2}c^2$ have the same spectrum. In particular H_0 and H_1 have the same spectrum. The eigenfunctions of H_0 are the plane waves,

ψ_+. Application of $Q^\dagger$ yields the eigenfunctions ψ_- of H_1,

$$\psi_+ = \mathrm{e}^{\mathrm{i}kx}, \quad \psi_- = (k + \mathrm{i}\tanh(x))\mathrm{e}^{\mathrm{i}kx}, \quad E_\pm - \tfrac{1}{2} = \tfrac{1}{2}k^2. \tag{18.16}$$

This potential is reflection free. One can continue in this way and find that all hamiltonians H_c with integer c are reflection free.

Bound States The ground state of H_c is obtained from $Q\psi_- = 0$. One finds the eigenfunction and the binding energy are

$$\psi_c^{(0)}(x) \propto \cosh^{-c}(x), \quad E_c^{(0)} = -\tfrac{1}{2}c^2. \tag{18.17}$$

Using that H_c and H_{c+1} have the same spectrum, one obtains the spectrum of the bound states

$$E_c^{(k)} = -\tfrac{1}{2}(c - k)^2 \tag{18.18}$$

for integer $k < c$. The eigenfunction can be calculated recursively by application of $Q^\dagger$,

$$\psi_c^{(k)}(x) \propto (-\frac{\mathrm{d}}{\mathrm{d}x} + c\tanh x)\psi_{c-1}^{(k-1)}(x) \tag{18.19}$$

starting from $\psi_{c-k}^{(0)}$.

18.1.4 Supersymmetric δ-Potential

Another example is

$$W(x) = \omega\delta(x), \quad V_s = \tfrac{1}{2}(\omega^2\delta^2(x) + s\omega\delta'(x)), \quad s = \pm 1. \tag{18.20}$$

The extremely strong $\delta^2(x)$-term of the potential prevents any transmission across the potential as long as $|s| \neq 1$,

$$\psi(-0) = \psi(0) = 0, \quad \psi'(-0) \text{ and } \psi'(+0) \text{ arbitrary}. \tag{18.21}$$

Surprisingly, the potential becomes transmitting in the supersymmetric cases $s = \pm 1$,

$$\psi(+0) = \mathrm{e}^{\pm\omega}\psi(-0), \quad \psi'(+0) = \mathrm{e}^{\mp\omega}\psi'(-0). \tag{18.22}$$

See [50] and Problem 18.3.

18.1.5 Hydrogen Spectrum

The fact that the Hamiltonians $Q^\dagger Q$ and $QQ^\dagger$ have the same spectrum, was used by Schrödinger[234, 235] in 1940 and 1941 for the determination of the hydrogen spectrum. The radial wave function $u(r)$ for the electron, $\psi(\mathbf{r}) = \frac{u(r)}{r} Y_{l,m}(\theta, \phi)$, in the Coulomb potential of the proton obeys

$$H_l u_{n,l}(r) = E_{n,l} u_{n,l}(r), \quad H_l = -\frac{1}{2} \frac{\partial^2}{\partial r^2} - \frac{1}{r} + \frac{l(l+1)}{2r^2}, \tag{18.23}$$

where the Rydberg constant is set to one half and the Bohr radius to unity. Using

$$Q_l = \frac{1}{i} \left(\frac{\partial}{\partial r} - \frac{l}{r} + \frac{1}{l} \right) \tag{18.24}$$

one obtains

$$2H_l = Q_l Q_l^\dagger - \frac{1}{l^2}, \quad 2H_{l-1} = Q_l^\dagger Q_l - \frac{1}{l^2}. \tag{18.25}$$

Setting $Q_n u_{n,n-1} = 0$ yields the eigenstate

$$u_{n,n-1} \sim r^n e^{-r/n}, \quad E_{n,n-1} = -\frac{1}{2n^2}. \tag{18.26}$$

Then, from (18.25), one obtains

$$Q_l^\dagger H_l = H_{l-1} Q_l^\dagger. \tag{18.27}$$

Thus,

$$H_l Q_{l+1}^\dagger Q_{l+2}^\dagger \cdots Q_{n-2}^\dagger Q_{n-1}^\dagger = Q_{l+1}^\dagger Q_{l+2}^\dagger \cdots Q_{n-2}^\dagger Q_{n-1}^\dagger H_{n-1}, \tag{18.28}$$

from which one concludes that

$$u_{n,l}(r) = Q_{l+1}^\dagger Q_{l+2}^\dagger \cdots Q_{n-2}^\dagger Q_{n-1}^\dagger u_{n,n-1}(r) \tag{18.29}$$

obeys Eq. (18.23) with $E_{n,l} = -1/(2n^2)$. This shows that the energies $E_{n,l}$ depend only on the main quantum number n.

18.2   Chiral and Supersymmetric Models with $Q^2 = 0$

We consider models with $Q^2 = 0$. Among these models are chiral models with H linear in Q and $Q^\dagger$ with pairs of states with energies $\pm E$, and models with H bilinear in $Q, Q^\dagger$, with pairs of states with the same energy.

In this section we consider a class of systems characterized by the condition

$$Q^2 = 0, \quad Q^{\dagger 2} = 0. \tag{18.30}$$

Let $\mathscr{H}$ be the Hilbert space, in which Q acts. Let us denote the Hilbert space spanned by $Q|\psi\rangle$ with $|\psi\rangle \in \mathscr{H}$ by $\mathscr{H}_-$, and the Hilbert space spanned by $Q^\dagger|\psi\rangle$ with $|\psi\rangle \in \mathscr{H}$ by $\mathscr{H}_+$. Then, due to (18.30), the states in $\mathscr{H}_+$ and $\mathscr{H}_-$ are orthogonal to each other, i.e.

$$(Q^\dagger|\psi'\rangle)^\dagger Q|\psi\rangle = \langle\psi'|Q^2|\psi\rangle = 0. \tag{18.31}$$

States $|\psi\rangle$ orthogonal to both the states in $\mathscr{H}_+$ and $\mathscr{H}_-$ span the Hilbert space $\mathscr{H}_0$. They apparently obey

$$Q|\psi\rangle = Q^\dagger|\psi\rangle = 0 \tag{18.32}$$

and the total Hilbert space is $\mathscr{H} = \mathscr{H}_+ \oplus \mathscr{H}_- \oplus \mathscr{H}_0$.

18.2.1   Chiral Models

The Hamiltonian can be expressed as

$$H = \begin{pmatrix} 0 & \hat{Q}^\dagger \\ \hat{Q} & 0 \end{pmatrix} = Q + Q^\dagger \quad \text{with} \quad Q = \begin{pmatrix} 0 & 0 \\ \hat{Q} & 0 \end{pmatrix}, \quad Q^\dagger = \begin{pmatrix} 0 & \hat{Q}^\dagger \\ 0 & 0 \end{pmatrix}. \tag{18.33}$$

The matrix $\hat{Q}$ need not be quadratic. If its dimension is $n_1 \times n_2$, then at least $|n_1 - n_2|$ eigenvalues of H vanish. This chiral Hamiltonian has the property $Q^2 = 0$ and

$$H\sigma = -\sigma H, \quad \sigma := \begin{pmatrix} 1 & 0 \\ 0 & -1 \end{pmatrix}. \tag{18.34}$$

Thus, for an eigenstate $|\psi\rangle$ with energy E there is an eigenstate $\sigma|\psi\rangle$ with energy $-E$, since

$$H\sigma|\psi\rangle = -\sigma H|\psi\rangle = -E\sigma|\psi\rangle. \tag{18.35}$$

Thus, energies E and $-E$ occur pairwise with the exception of states with $E = 0$. Such models are called chiral. Free electrons on bipartite lattices, which are allowed to hop only from one sublattice to the other, have this property.

18.2.2   Fermions on a Lattice

Fendley et al. [77], Fendley and Schoutens [76] (see also in [117]) consider systems of spinless fermions on a lattice with sites r,

$$Q = \sum_r c_r f_r\{n\}, \quad Q^\dagger = \sum_r c_r^\dagger f_r^*\{n\}, \tag{18.36}$$

$$f_r\{n\} := f(n_{r+\delta_1}, n_{r+\delta_2}, \ldots), \quad n_{r+\delta} := c_{r+\delta}^\dagger c_{r+\delta}, \tag{18.37}$$

where f_r does not depend on n_r. Q decreases the number of fermions by one, $Q^\dagger$ increases it by one. The authors require $Q^2 = Q^{\dagger 2} = 0$. Since

$$2Q^2 = \sum_{rr'} c_r c_{r'} (f_{r,r'}^{(0)} f_{r',r}^{(1)} - f_{r,r'}^{(1)} f_{r',r}^{(0)}) \tag{18.38}$$

with

$$f_{r,r'}^{(\hat{n})} = f_r\{n\}|_{n_{r'}=\hat{n}}. \tag{18.39}$$

one requires

$$\frac{f_{r,r'}^{(1)}}{f_{r,r'}^{(0)}} = \frac{f_{r',r}^{(1)}}{f_{r',r}^{(0)}}, \tag{18.40}$$

which yields

$$f_r\{n\} = \prod_\delta (u_\delta + v_\delta n_{r+\delta}), \quad u_{-\delta} = u_\delta, \quad v_{-\delta} = v_\delta. \tag{18.41}$$

The Hamiltonian reads

$$H = \{Q^\dagger, Q\} = T + V, \quad T = \sum_{r',r} T_{r',r} c_{r'}^\dagger c_r, \quad V = \sum_r f_r^*\{n\} f_r\{n\}, \tag{18.42}$$

with

$$T_{r',r} = f_{r',r}^{(0)*} f_{r,r'}^{(0)} - f_{r',r}^{(1)*} f_{r,r'}^{(1)}. \tag{18.43}$$

The potential V is a function of the occupation numbers n, the kinetic energy T allows particles to hop between lattice sites, depending, however, on the occupation numbers n of some neighbors.

The fermion number is conserved. The operators Q and $Q^\dagger$ commute with H. Thus, for positive energies, there are always two states, a bosonic and a fermionic state, that are degenerate if we call states with an even (odd) number of fermions bosonic (fermionic).

The number of bosonic states minus the number of fermionic states is called the Witten index[290]. It can be written

$$W = \operatorname{tr}\left((-)^F e^{-\beta H}\right), \tag{18.44}$$

where F is the number of fermions. The factor $e^{-\beta H}$ serves for convergence, but only states with zero energy contribute.

Problems

18.1 Supersymmetric box Put the supersymmetric system in a box ranging from $x = -L$ to $x = +L$ by choosing $W = W_0(x) + h\Theta(x - L) + h\Theta(-x - L)$.

1. Determine the boundary conditions at $x = \pm L$ assuming that $W_0(x)$ varies only smoothly at $x = \pm L$. Take the limit $h \to \infty$.
2. What are the solutions for constant $W_0(x)$?

18.2 Given

$$Q = c_1(1 + vc_2^\dagger c_2) + c_2(1 + vc_1^\dagger c_1).$$

Determine $H = Q^\dagger Q + QQ^\dagger$ and the eigenstates and energies.

18.3 In order to derive (18.21), (18.22) put $W = w$ in the interval $x = (0, a)$ and take the limit $a \to 0$ for fixed $wa = \omega$.

References

[50] A. Comtet, C. Texier, Y. Tourigny, Product of random matrices and generalized quantum point scatterers. J. Stat. Phys. **140**, 427 (2010)
[51] F. Cooper, A. Khare, U. Sukhatme, Supersymmetry and quantum mechanics. Phys. Rep. **251** 267 (1995)
[76] P. Fendley, K. Schoutens, Exact results for strongly-correlated fermions in 2+1 dimensions. Phys. Rev. Lett. **95**, 046403 (2005)

[77] P. Fendley, K. Schoutens, J. de Boer, Lattice models with $N = 2$ supersymmetry. Phys. Rev. Lett. **90**, 120402 (2003)

[117] L. Hujse, N. Moran, J. Vala, K. Schoutens, Exact ground state of a staggered supersymmetric model for lattice fermions. Phys. Rev. B **84**, 115124 (2011)

[234] E. Schrödinger, A method of determining quantum-mechanical eigenvalues and eigenfunctions. Proc. R. Ir. Acad. A **46**, 9 (1940)

[235] E. Schrödinger, Further studies on solving eigenvalue problems by factorization. Proc. R. Ir. Acad. A **46**, 183 (1940)

[236] F. Schwabl, *Quantenmechanik*, 2nd edn. (Springer, Berlin, 1990)

[289] E. Witten, Dynamical breaking of supersymmetry. Nucl. Phys. B **188**, 513 (1981)

[290] E. Witten, Constraints on supersymmetry breaking. Nucl. Phys. B **202**, 253 (1982)

Chapter 19
Supersymmetry in Stochastic Field Equations and in High Energy Physics

Abstract We start from the purely dissipative Langevin equation and the corresponding Fokker-Planck equation. The corresponding field theory is supersymmetric. This symmetry yields the fluctuation-dissipation theorem. The formulation has much in common with supersymmetry in high energy physics. In both theories, anticommuting coordinates are introduced in addition to the commuting coordinates.

19.1 Stochastic Time-Dependent Equations

We introduce the purely dissipative Langevin equation and derive the corresponding Fokker-Planck equation. The fields of the corresponding theory are supersymmetric fields. They depend in addition to the time on a pair of anticommuting coordinates. The supersymmetry yields the fluctuation-dissipation theorem.

19.1.1 Langevin and Fokker-Planck Equation

We start with a set of equations of motion for classical particles with coordinates $q_i(t)$, called Langevin equations

$$m_i \ddot{q}_i(t) = -\frac{\partial V}{\partial q_i} - \gamma_i \dot{q}_i(t) + \eta_i(t). \qquad (19.1)$$

Here $-\gamma \dot{q}_i(t)$ is a friction term and $\eta_i(t)$ a stochastic force with Gaussian distribution

$$\langle \eta_i(t) \rangle = 0, \quad \langle \eta_i(t)\eta_j(t') \rangle = \kappa_i \delta_{ij}\delta(t - t'), \qquad (19.2)$$

which, due to the δ-function in time, is called white-noise, since it extends over all frequencies. If the motion is strongly damped, then $m_i \ddot{q}_i(t)$ becomes negligible and

© Springer-Verlag Berlin Heidelberg 2016
F. Wegner, *Supermathematics and its Applications in Statistical Physics*,
Lecture Notes in Physics 920, DOI 10.1007/978-3-662-49170-6_19

Eq. (19.1) reduces to

$$\dot{q}_i(t) = -f_i(q(t)) + \eta_i(t), \quad f_i(q(t)) = \frac{1}{\gamma_i}\frac{\partial V}{\partial q_i}, \quad \langle \eta_i(t)\eta_j(t')\rangle = \frac{\kappa_i}{\gamma_i^2}\delta_{ij}\delta(t-t').$$

$$(19.3)$$

Due to this noise the system does not move in a deterministic way, but its development has to be described by a probability distribution, $P(q, t)$. Suppose we consider the average of a function, $F(q)$, as a function of time. We expand for a small time interval, δt,

$$\langle F(q(t+\delta t))\rangle = \langle F(q(t))\rangle + \langle \frac{\partial F}{\partial q_i}(q_i(t+\delta t) - q_i(t))\rangle + \frac{1}{2}\langle \frac{\partial^2 F}{\partial q_i^2}(q_i(t+\delta t) - q_i(t))^2\rangle.$$

$$(19.4)$$

The term with the first derivative of F contains $-f\delta t$, while in the term with the second derivative

$$\langle (q_i(t+\delta t) - q_i(t))^2\rangle = \int_t^{t+\delta t} d t_1 \int_t^{t+\delta t} d t_2 \langle \eta_i(t)\eta_i(t')\rangle = \frac{\kappa_i}{\gamma_i^2}\delta t \qquad (19.5)$$

enters. Thus, we obtain, from Eq. (19.4),

$$\frac{\partial}{\partial t}\langle F(q,t)\rangle = -\langle f_i\frac{\partial F}{\partial q_i}\rangle + \Omega_i\langle \frac{\partial^2 F}{\partial q_i^2}\rangle, \quad \Omega_i = \frac{\kappa_i}{2\gamma_i^2}. \qquad (19.6)$$

Using

$$\langle F(q)\rangle = \int d q F(q)P(q) \qquad (19.7)$$

yields, after partial integrations,

$$\int d q F(q)\dot{P}(q) = \int d q F(q)\left(\frac{\partial}{\partial q_i}(f_i(q)P(q)) + \Omega_i\frac{\partial^2 P}{\partial q_i^2}\right), \qquad (19.8)$$

which holds for arbitrary F and yields the Fokker-Planck equation, which describes the evolution of the probability distribution P,

$$\dot{P}(q,t) = H_{\text{F.P.}}P(q,t), \quad H_{\text{F.P.}} = \frac{\partial}{\partial q_i}(f_i(q) + \Omega_i\frac{\partial}{\partial q_i}). \qquad (19.9)$$

Assuming $f_i = \Omega_i\frac{\partial E}{\partial q_i}$ this 'Fokker-Planck' hamiltonian can be transformed by

$$e^{E(q)/2}H_{\text{F.P.}}e^{-E(q)/2} = -\Omega_i(-\frac{\partial}{\partial q_i} + \frac{1}{2}\frac{\partial E}{\partial q_i})(\frac{\partial}{\partial q_i} + \frac{1}{2}\frac{\partial E}{\partial q_i}) = -\Omega_i A_i^\dagger A_i \qquad (19.10)$$

into an hermitian one, which is negative-semi-definite. The transformed Hamiltonian applied on $\exp(-E/2)$ vanishes and $H_{\text{F.P.}}\exp(-E) = 0$. If $\exp(-E)$ is normalizable, then it is the equilibrium distribution, given by $\exp(-V/(k_BT))$. The condition $f_i = \Omega_i\frac{\partial E}{\partial q_i}$ shows that κ and γ are connected by $\kappa_i = 2k_BT\gamma_i$.

19.1.2   Time-Dependent Correlation Functions

In this section we express the time-dependent correlation functions as expectation values of a theory governed by an action, S. In order to facilitate the calculation, we rescale the variables q by introducing

$$\tilde{q}_i = \sqrt{\gamma_i}q_i, \quad \tilde{\eta}_i = \sqrt{\gamma_i}\eta_i, \tag{19.11}$$

which yields

$$\dot{\tilde{q}}_i = -\tilde{f}_i + \tilde{\eta}_i \tag{19.12}$$

with

$$\tilde{f}_i = \frac{\partial\tilde{V}}{\partial\tilde{q}}, \quad \tilde{V}(\tilde{q}) := V(q), \quad \langle\tilde{\eta}_i(t)\tilde{\eta}_j(t')\rangle = 2k_BT\delta_{ij}\delta(t-t'). \tag{19.13}$$

Henceforth, we drop the $\tilde{\ }$. The probability distribution of the noise η (19.3) is then proportional to

$$\exp\left(-\int dt\sum_i\frac{1}{4k_BT}\eta_i^2(t)\right)[D\,\eta]$$

$$= \int[D\,\eta][D\,\lambda]\exp\left(\int dt\sum_i(k_BT\lambda_i^2(t) + \lambda_i(t)\eta_i(t))\right) \tag{19.14}$$

with λ integrated along the imaginary axis. The coordinates q_i depend on the fluctuating forces $\eta_i(t)$, thus, $q_i(t) = q_i(t,\{\eta\})$. Thus, the expectation value of a function, F, of the coordinates can be written

$$\langle F\{q(t)\}\rangle = N^{-1}\int[D\,\lambda][D\,\eta]F\{q,\eta\}e^{\sum_i(k_BT\lambda_i^2(t)-\lambda_i(t)\eta_i(t))} \tag{19.15}$$

with some normalization N. Instead of integrating over η we prefer to integrate over q. Thus, we introduce the Jacobian [we express η in terms of q using (19.3)],

$$\frac{\partial(\eta)}{\partial(q)} = \det_{ij}(\delta_{ij}\frac{\partial}{\partial t} + \frac{\partial f_i}{\partial q_j}). \tag{19.16}$$

This determinant can be expressed by the Grassmann variables ψ and $\psi^\times$,

$$\int [\mathrm{D}\,\psi^\times \mathrm{D}\,\psi]\exp(\psi_i^\times(\delta_{ij}\partial_t + \frac{\partial f_i}{\partial q_j})\psi_j). \qquad (19.17)$$

Then one obtains

$$F = \int [\mathrm{D}\,q][\mathrm{D}\,\psi^\times][\mathrm{D}\,\psi][\mathrm{D}\,\lambda]F\{q(t)\}\exp(-S) \qquad (19.18)$$

with

$$S(q,\psi^\times,\psi,\lambda) = \int \mathrm{d}\,t \sum_i(-k_\mathrm{B}T\lambda_i^2 + \lambda_i(\dot{q}_i + f_i(q)) - \psi_i^\times\dot{\psi}_i - \psi_i^\times\sum_j\frac{\partial f_i}{\partial q_j}\psi_j) \qquad (19.19)$$

where $q,\psi^\times,\psi,\lambda$ depend on time t. We introduce two additional Grassmann variables, $\theta^\times$ and θ, and, with the time t, may define the supercoordinates

$$Q_i(t) := q_i(t) + \psi_i^\times(t)\theta + \theta^\times\psi_i(t) + \lambda_i(t)\theta^\times\theta. \qquad (19.20)$$

The action may then be written as

$$S\{Q\} = \int \mathrm{d}\,t\mathrm{d}\,\theta\mathrm{d}\,\theta^\times(k_\mathrm{B}T\sum_i\bar{\mathscr{D}}Q_i\mathscr{D}Q_i - V\{Q\}) \qquad (19.21)$$

with

$$\bar{\mathscr{D}} = \frac{\partial}{\partial\theta}, \quad \mathscr{D} = \frac{\partial}{\partial\theta^\times} - \frac{1}{k_\mathrm{B}T}\theta\frac{\partial}{\partial t}. \qquad (19.22)$$

19.1.3   Supersymmetry and Fluctuation-Dissipation Theorem

Starting from the action S, we determine the invariance of the correlation functions and the response function, and their relation: the fluctuation-dissipation theorem.

In addition to the operators $\mathscr{D}$ and $\bar{\mathscr{D}}$, we introduce

$$\mathscr{D}' = \frac{\partial}{\partial\theta^\times}, \quad \bar{\mathscr{D}}' = \frac{\partial}{\partial\theta} + \frac{1}{k_\mathrm{B}T}\theta^\times\frac{\partial}{\partial t}. \qquad (19.23)$$

We then obtain the anticommutators

$$\mathscr{D}^2 = \bar{\mathscr{D}}^2 = 0, \quad \{\mathscr{D},\bar{\mathscr{D}}\} = -\frac{1}{k_\mathrm{B}T}\frac{\partial}{\partial t}, \qquad (19.24)$$

$$\mathscr{D}'^2 = \bar{\mathscr{D}}'^2 = 0, \quad \{\mathscr{D}', \bar{\mathscr{D}}'\} = \frac{1}{k_B T} \frac{\partial}{\partial t}, \tag{19.25}$$

$$\{\mathscr{D}, \mathscr{D}'\} = 0 = \{\bar{\mathscr{D}}, \bar{\mathscr{D}}'\}, \tag{19.26}$$

$$\{\mathscr{D}, \bar{\mathscr{D}}'\} = 0 = \{\bar{\mathscr{D}}, \mathscr{D}'\}. \tag{19.27}$$

The action is invariant under transformations by $\mathscr{D}'$ and $\bar{\mathscr{D}}'$. Let $\epsilon \in \mathscr{A}_1$. Then a shift of Q, i.e.

$$\delta Q = \epsilon \mathscr{D}' Q, \tag{19.28}$$

yields

$$\begin{aligned}
\delta(\bar{\mathscr{D}} Q \mathscr{D} Q) &= \bar{\mathscr{D}}(\delta Q)\mathscr{D} Q + \bar{\mathscr{D}} Q \mathscr{D}(\delta Q) = \bar{\mathscr{D}}(\epsilon \mathscr{D}' Q)\mathscr{D} Q + \bar{\mathscr{D}} Q \mathscr{D}(\epsilon \mathscr{D}' Q) \\
&= -\epsilon\{\bar{\mathscr{D}}, \mathscr{D}'\}\mathscr{D} Q + \epsilon \mathscr{D}'(\bar{\mathscr{D}} Q)\mathscr{D} Q - \bar{\mathscr{D}} Q \epsilon\{\mathscr{D}, \mathscr{D}'\}Q \\
&\quad -\epsilon \bar{\mathscr{D}} Q \mathscr{D}' \mathscr{D} Q = \epsilon \mathscr{D}'(\bar{\mathscr{D}} Q \mathscr{D} Q).
\end{aligned} \tag{19.29}$$

A similar argument yields $\delta(\bar{\mathscr{D}} Q \mathscr{D} Q) = \epsilon \bar{\mathscr{D}}'(\bar{\mathscr{D}} Q \mathscr{D} Q)$ for $\delta Q = \epsilon \bar{\mathscr{D}}' Q$. Moreover, one finds $\delta V = \epsilon \mathscr{D}' V$ and $\delta V = \epsilon \bar{\mathscr{D}}' V$, resp. Since $\mathscr{D}'$ and $\bar{\mathscr{D}}'$ are divergences ($\bar{\theta} \partial/\partial t = \partial/\partial t \bar{\theta}$), the expression under the integral (19.21) differ only by surface terms and thus, yield the same equations of motion.

The supersymmetric Langrangian (19.21) is an example of the BRS supersymmetry, introduced first in the context of gauge theories by Becchi et al. [20, 21] using the Slavnov-Taylor symmetry [241, 250]. It occurs when a constraint is introduced in a path integral, accompanied by a Jacobian, represented by a Gaussian integral over Grassmann variables.

Ward-Takahashi Identities Ward-Takahashi identities are identities which follow from symmetries. The invariance of the action under the transformations induced by $\mathscr{D}'$ and $\bar{\mathscr{D}}'$ will give the fluctuation-dissipation theorem. Let us consider the two-point functions

$$\mathscr{G}_{ij}(t_1, \theta_1^\times, \theta_1; t_2, \theta_2^\times, \theta_2) := \langle Q_i(t_1, \theta_1^\times, \theta_1) Q_j(t_2, \theta_2^\times, \theta_2) \rangle. \tag{19.30}$$

Due to the invariance, they obey

$$\mathscr{D}' \mathscr{G} = 0, \quad \bar{\mathscr{D}}' \mathscr{G} = 0, \tag{19.31}$$

$$\mathscr{D}' = \sum_{i=1}^{2} \frac{\partial}{\partial \theta_i^\times}, \quad \bar{\mathscr{D}}' = \sum_{i=1}^{2} \left(\frac{\partial}{\partial \theta_i} + \frac{1}{k_B T} \theta_i^\times \frac{\partial}{\partial t_i} \right). \tag{19.32}$$

This, also implies, due to $\mathscr{D}'$ translational invariance in $\theta^\times$ and due to the anticommutator $\{\mathscr{D}', \bar{\mathscr{D}}'\}$, translational invariance in time t. Since $\mathscr{G} \in \mathscr{A}_0$, the only possible function, expanded in θ and $\theta^\times$, is

$$\mathscr{G} = C_{ij}(t_1 - t_2) + (\theta_1^\times - \theta_2^\times)[\theta_1 B_{1,ij}(t_1 - t_2) + \theta_2 B_{2,ij}(t_1 - t_2)]. \tag{19.33}$$

We obtain, from $\bar{\mathscr{D}}'\mathscr{G} = 0$,

$$B_{1,ij} + B_{2,ij} = \frac{1}{k_B T} \frac{\partial C_{ij}}{\partial t}. \tag{19.34}$$

We are interested in the correlation of $q = Q(\theta^\times = \theta = 0)$ at different times t. This is given by

$$\langle q_i(t_1)q_j(t_2)\rangle = C_{ij}(t_1 - t_2). \tag{19.35}$$

Another correlation of interest is the response of q to a variation of the potential V at some other time. Suppose we vary the potential by

$$\delta V(Q) = \int d\, t h_j(t) Q_j(t). \tag{19.36}$$

This yields a change of the action

$$-\delta S = - \int d\,\theta d\,\theta^\times d\, t h_j(t) Q_j(t). \tag{19.37}$$

Due to this change of the potential, q_i changes by

$$\langle \delta q_i(t_1)\rangle = - \int d\,\theta_2 d\,\theta_2^\times d\, t_2 h_j(t_2)\mathscr{G}_{ij}(t_1, 0, 0; t_2, \theta_2^\times, \theta_2)$$

$$= \int d\, t_2 h_j(t_2) B_{2,ij}(t_1 - t_2). \tag{19.38}$$

Causality requires that $B_2(t_1 - t_2) = 0$ for $t_2 > t_1$. Similarly $B_1(t_1 - t_2) = 0$ holds for $t_1 > t_2$. Thus,

$$B_{1,ij}(t) = \frac{1}{k_B T}\Theta(-t)\frac{\partial C_{ij}(t)}{\partial t}, \quad B_{2,ij}(t) = \frac{1}{k_B T}\Theta(t)\frac{\partial C_{ij}(t)}{\partial t}. \tag{19.39}$$

and the response reads

$$\langle \delta q_i(t_1)\rangle = \frac{1}{k_B T}\int d\, t_2 \Theta(t_1 - t_2)h_j(t_2)\frac{\partial C_{ij}(t_1 - t_2)}{\partial t_1}. \tag{19.40}$$

One may replace q_i by a general operator $O(q)$ and δq_j by a general potential $\delta V(q)$. Then (19.40) constitutes the fluctuation-dissipation theorem.

The relation between supersymmetry and dissipative Langevin and Fokker-Planck equations was initiated by Parisi and Wu [207] and continued by Zwanziger [305], Feigel'man and Tsvelik [75], Egorian and Kalitzin [70], Nakazato et al. [195], Kirschner [144], Gozzi [98], Chaturvedi et al. [47], and Zinn-Justin [296]. I recommend reading Sects. 16 and 17 of the book by Zinn-Justin [297] and the article by Feigel'man and Tsvelik [75].

19.2 Supersymmetry in High Energy Physics

Supersymmetry adds to the Poincaré Lie-algebra of momentum and angular momentum (including boost) operators four anticommuting operators. The four-dimensional Minkowski space is enlarged by four anticommuting coordinates. If all these fourteen operators are conserved, then bosons and fermions have supersymmetric partners with equal masses.

A supersymmetry relating mesons and baryons was proposed in 1966 by Hironari Miyazawa [190, 191], but went unnoticed. Supersymmetry arose in the context of string theory in the early 1970s: The NSR-model was developed in 1971 by Neveu and Schwarz [198] and by Ramond [221]. The prefix 'super' was first introduced as supergauge by Gervais and Sakita [95]. Similar results were obtained by Golfand and Likhtman [96]. Volkov and Akulov [264, 265], and Wess and Zumino [78, 282].

For more results see the articles by Fayet and Ferrara [74], by Wess [280], by Wess and Bagger [281], and by Martin [176].

Lorentz invariance means that four-momentum P and angular momentum M are constants of motion. They obey the commutator relations

$$[P_\mu, P_\nu] = 0, \quad [M_{\mu\nu}, P_\kappa] = c(g_{\mu\kappa}P_\nu - g_{\nu\kappa}P_\mu), \tag{19.41}$$

$$[M_{\mu\nu}, M_{\kappa\lambda}] = c(g_{\mu\kappa}M_{\nu\lambda} - g_{\mu\lambda}M_{\nu\kappa} - g_{\nu\kappa}M_{\mu\lambda} + g_{\nu\lambda}M_{\mu\kappa}) \tag{19.42}$$

with $[x_\mu, P_\nu] = cg_{\mu\nu}$. We use the conventions by Fayet and Ferrara [74]. This Poincaré algebra can be extended by Grassmannian operators Q and $\bar{Q}$, which are spinors in the representations $(0, \frac{1}{2})$ and $(\frac{1}{2}, 0)$. They obey the commutator relations

$$\{Q_\alpha, Q_\beta\} = \{\bar{Q}_{\dot\alpha}, \bar{Q}_{\dot\beta}\} = 0, \quad \{Q_\alpha, \bar{Q}_{\dot\beta}\} = 2\sigma^\mu_{\alpha\dot\beta}P_\mu, \tag{19.43}$$

$$[Q_\alpha, P_\mu] = [\bar{Q}_{\dot\alpha}, P_\mu] = 0, \tag{19.44}$$

$$[Q_\alpha, M_{\mu\nu}] = K_{\mu\nu,\alpha}{}^\beta Q_\beta, \quad [\bar{Q}_{\dot\alpha}, M_{\mu\nu}] = -\bar{K}_{\mu\nu,\dot\alpha}{}^{\dot\beta}\bar{Q}_{\dot\beta}. \tag{19.45}$$

with

$$K_{\mu\nu,\alpha}{}^{\beta} = \tfrac{1}{4}c(\sigma_{\mu,\alpha\dot\gamma}\,\bar\sigma_\nu{}^{\dot\gamma\beta} - \sigma_{\nu,\alpha\dot\gamma}\,\bar\sigma_\mu{}^{\dot\gamma\beta}), \tag{19.46}$$

$$\bar K_{\mu\nu,\dot\alpha}{}^{\dot\beta} = \tfrac{1}{4}c(\bar\sigma_\mu{}^{\dot\beta\gamma}\,\sigma_{\nu,\gamma\dot\alpha} - \bar\sigma_\nu{}^{\dot\beta\gamma}\,\sigma_{\mu,\gamma\dot\alpha}). \tag{19.47}$$

Equations (19.43)–(19.44) are from (2.19) of [74]. An explicit representation of Q and $\bar Q$ can be given with

$$Q_\alpha = \frac{\partial}{\partial\theta^\alpha} - \sigma^\mu{}_{\alpha\dot\beta}\bar\theta^{\dot\beta}P_\mu, \tag{19.48}$$

$$\bar Q_{\dot\beta} = -\frac{\partial}{\partial\bar\theta^{\dot\beta}} + \sigma^\mu{}_{\alpha\dot\beta}\theta^\alpha P_\mu \tag{19.49}$$

with anticommuting coordinates θ and $\bar\theta$ and

$$M_{\mu\nu} = x_\mu P_\nu - x_\nu P_\mu + \theta^\alpha\,K_{\mu\nu,\alpha}{}^\beta\,\frac{\partial}{\partial\theta^\beta} - \bar\theta^{\dot\alpha}\,\bar K_{\mu\nu,\dot\alpha}{}^{\dot\beta}\,\frac{\partial}{\partial\bar\theta^{\dot\beta}}. \tag{19.50}$$

with the conventions $g_{..} = \mathrm{diag}(-1,1,1,1)$, $\sigma_0 = \mathbf{1}_2$, σ_μ are Pauli-matrices for $\mu = 1,2,3$.

$$\bar\sigma_\mu{}^{\dot\alpha\beta} = \epsilon^{\dot\alpha\dot\gamma}\sigma_{\mu,\delta\dot\gamma}\epsilon^{\beta\delta}, \qquad \epsilon^{..} = \begin{pmatrix} 0 & +1 \\ -1 & 0 \end{pmatrix}, \tag{19.51}$$

which implies

$$\bar\sigma_0^{..} = \mathbf{1}_2, \qquad \bar\sigma_\mu^{..} = -\sigma_{\mu,..}, \qquad \mu = 1,2,3. \tag{19.52}$$

Since Q and $\bar Q$ commute with P, one may choose massive states in the rest frame. Then, apart from normalization, Q and $\bar Q$ act as fermionic annihilation and creation operators, which connect four multiplets, two with angular momentum j, one with $j + \tfrac{1}{2}$, and one with $j - \tfrac{1}{2}$. Thus, two of these multiplets are bosons and two are fermions. If $j = 0$, then there are two scalar bosons and one spin $\tfrac{1}{2}$-fermion. See for example Sect. 2.4 of [74] and Sect. II of [281]. All states of this supermultiplet should have equal mass, provided the symmetry is not spontaneously broken. Until now supersymmetry has not been observed in nature.

Besides the operators Q and $\bar Q$ (19.48), (19.49), one can introduce another set of operators

$$Q'_\alpha = \frac{\partial}{\partial\theta^\alpha} + \sigma^\mu{}_{\alpha\dot\beta}\bar\theta^{\dot\beta}P_\mu, \tag{19.53}$$

$$\bar Q'_{\dot\beta} = +\frac{\partial}{\partial\bar\theta^{\dot\beta}} + \sigma^\mu{}_{\alpha\dot\beta}\theta^\alpha P_\mu. \tag{19.54}$$

They obey

$$\{Q'_\alpha, Q'_\beta\} = \{\bar{Q}'_{\dot\alpha}, \bar{Q}'_{\dot\beta}\} = 0, \quad \{Q'_\alpha, \bar{Q}'_{\dot\beta}\} = -2\sigma^\mu_{\alpha\dot\beta}P_\mu. \tag{19.55}$$

Moreover Q', $\bar{Q}'$ anticommute with Q, Q'. This allows us to characterize scalar chiral superfields ϕ by the condition $Q'\phi = 0$ and $\bar{Q}'\phi = 0$, resp.

References

[20] C. Becchi, A. Rouet, R. Stora, The Abelian Higgs Kibble model, unitarity of the S-operator. Phys. Lett. B **52**, 344 (1974)

[21] C. Becchi, A. Rouet, R. Stora, Renormalization of the abelian Higgs-Kibble model. Commun. Math. Phys. **42**, 127 (1975)

[47] S. Chaturvedi, A.K. Kapoor, V. Srinivasan, Ward Takahashi identities and fluctuation-dissipation theorem in a superspace formulation of the Langevin equation. Z. Phys. B **57**, 249 (1984)

[70] E. Egorian, S. Kalitzin, A superfield formulation of stochastic quantization with fictitious time. Phys. Lett. B **129**, 320 (1983)

[74] P. Fayet, S. Ferrara, Supersymmetry. Phys. Rep. **32**, 249 (1977)

[75] M.V. Feigel'man, A.M. Tsvelik, Hidden supersymmetry of stochastic dissipative dynamics. Sov. Phys. JETP **56**, 823 (1982); Zh. Eksp. Teor. Fiz. **83**, 1430 (1982)

[78] S. Ferrara, J. Wess, B. Zumino, Supergauge multiplets and superfields. Phys. Lett. B **51**, 239 (1974)

[95] J.L. Gervais, B. Sakita, Field theory interpretation of supergauges in dual models. Nucl. Phys. B **34**, 632 (1971)

[96] Y.A. Golfand, E.P. Likhtman, Extension of the Algebra of Poincaré group operators and violation of P-invariance. ZhETF Pis. Red. **12**, 452 (1971); JETP Lett. **13**, 323 (1971)

[98] E. Gozzi, Dimensional reduction in parabolic stochastic equations. Phys. Lett. B **143**, 183 (1984)

[144] R. Kirschner, Quantization by stochastic relaxation processes and supersymmetry. Phys. Lett. B **139**, 180 (1984)

[176] S.P. Martin, A supersymmetry primer. arxiv: hep-ph/9709356 (1997)

[190] H. Miyazawa, Baryon number changing currents. Prog. Theor. Phys. **36**, 1266 (1966)

[191] H. Miyazawa, Spinor currents and symmetries of Baryons and Mesons. Phys. Rev. **170**, 1586 (1968)

[195] H. Nakazato, M. Nakimi, I. Okba, K. Okano, Equivalence of stochastic quantization method to conventional field theories through supertransformation invariance. Prog. Theor. Phys. **70**, 298 (1983)

[198] A. Neveu, J.H. Schwarz, Factorizable dual model of pions. Nucl. Phys. B **31**, 86 (1971)

[207] G. Parisi, Y. Wu, Perturbation theory without gauge fixing. Sci. Sinica **24**, 483 (1981)

[221] P. Ramond, Dual theory for fermions. Phys. Rev. D **3**, 2415 (1971)

[241] A.A. Slavnov, Ward identities in gauge theories. Theor. Math. Phys. **19**, 99 (1972)

[250] J.C. Taylor, Ward identities and charge renormalization of the Yang-Mills field. Nucl. Phys. B **33**, 436 (1971)

[264] D.V. Volkov, V.P. Akulov, Possible universal neutrino interaction. ZhETF Pis. Red. **16**, 621 (1972); JETP Lett. **16**, 438 (1972)

[265] D.V. Volkov, V.P. Akulov, Is the neutrino a Goldstone particle? Phys. Lett. B **46**, 109 (1973)

[280] J. Wess, Fermi-Bose-supersymmetry, in *Trends in Elementary Particle Systems*, edited by H. Rollnik. Lecture Notes in Physics, vol. 37 (Springer, Berlin, 1975), p. 352

[281] J. Wess, J. Bagger, *Supersymmetry and Supergravity*. Princeton Series in Physics (Princeton University Press, Princeton, 1983)

[282] J. Wess, B. Zumino, A Lagrangian model invariant under supergauge transformations. Phys. Lett. B **49**, 52 (1974)

[296] J. Zinn-Justin, Renormalization and stochastic quantization. Nucl. Phys. B **275**, 135 (1986)

[297] J. Zinn-Justin, *Quantum Field Theory and Critical Phenomena* (Clarendon Press, Oxford, 1993)

[305] D. Zwanziger, Covariant quantization of gauge fields without Gribov ambiguity. Nucl. Phys. B **192**, 259 (1981)

Chapter 20
Dimensional Reduction

Abstract An initial section deals with Lie superalgebras, angular momentum and the Laplace operator in superreal spaces. Then several examples of disordered systems are considered, which, due to supersymmetry, can be reduced to pure systems in two fewer dimensions. Among them are the Ising model in a stochastic magnetic field, branched polymers and lattice animals, and electrons in a strong magnetic field. Disorder allows the introduction of a pair of anticommuting coordinates, which cancel against two commuting coordinates, thus reducing the dimension by two. Finally a few remarks are made on critical exponents of isotropic $\phi^{2\sigma}$-theories of fields ϕ with a negative even number of components.

20.1 Rotational Invariance in Superreal Space

We generalize the concept of Lie algebras to Lie superalgebras and use it to introduce the generalization of the operators of angular momentum for the unitary-orthosymplectic group, which acts in superreal space and derive the supersymmetric Laplace operator for this space.

20.1.1 Lie Superalgebra and Jacobi Identity

Lie algebras are characterized by generators G_i, which have the property that the commutator of any pair of generators is a linear combination of the generators again,

$$[G_i, G_j] = f_{ijk} G_k, \tag{20.1}$$

where the coefficients $f_{ijk} = -f_{jik}$ are called structure constants. The Jacobi identity

$$[A, [B, C]] + [B, [C, A]] + [C, [A, B]] = 0 \tag{20.2}$$

can be obtained by expressing the commutators explicitly as a sum of products of the generators.

© Springer-Verlag Berlin Heidelberg 2016

F. Wegner, *Supermathematics and its Applications in Statistical Physics*,
Lecture Notes in Physics 920, DOI 10.1007/978-3-662-49170-6_20

Lie algebras can be derived from continuous groups acting in spaces of real and complex elements in $\mathscr{A}_0$. This concept can be generalized by allowing generators to act in a superspace, that is a space which contains even and odd elements. A generator transforming elements in $\mathscr{A}_{\nu_1}$ to those in $\mathscr{A}_{\nu_2}$ is of Z_2-grade, $\nu_1 + \nu_2$. One defines the supercommutator

$$\{G_i, G_j\} := G_i G_j - (-)^{\nu_i \nu_j} G_j G_i = f_{ijk} G_k, \quad f_{jik} = (-)^{1-\nu_i \nu_j} f_{ijk}, \tag{20.3}$$

where only terms G_k with $\nu_i + \nu_j = \nu_k$ contribute in the sum over k. These supercommutators define a Lie superalgebra. Again, explicit multiplication yields the Jacobi identity

$$(-)^{\nu_A \nu_C}\{A, \{B, C\}\} + (-)^{\nu_B \nu_A}\{B, \{C, A\}\} + (-)^{\nu_C \nu_B}\{C, \{A, B\}\} = 0. \tag{20.4}$$

In Sect. 19.1, on stochastic time-dependent equations, we had examples of Lie superalgebras. The operators $\mathscr{D}$, $\bar{\mathscr{D}}$, Eq. (19.22), and $\frac{\partial}{\partial t}$, constitute a Lie superalgebra, but also the operators $\mathscr{D}'$, $\bar{\mathscr{D}}'$, Eq. (19.23), and $\frac{\partial}{\partial t}$, constitute such an algebra. Also, all five operators constitute a Lie superalgebra.

In Sect. 19.2, on supersymmetry in high energy physics, we had the Lie algebra of the Lorentz group containing the operators M, the Lie algebra of the Poincaré group with operators P and M, and the Lie superalgebras of the operators P, M, Q and $\bar{Q}$. But also the operators P, M, Q', and $\bar{Q}'$ constitute a Lie superalgebra, as well as all the 18 operators.

20.1.2 Unitary-Orthosymplectic Rotations and Supersymmetric Laplace Operator

As a useful example, we consider unitary-orthosymplectic rotations. Thus, we introduce a rotation expressed by the UOSp-matrix $U = \mathbf{1} + W$ with infinitesimal W. Since U is a UOSp-matrix, it obeys $U^\ddagger U = \mathbf{1}$ and thus $W + W^\ddagger = 0$. Since it is superreal, it obeys $\mathsf{C}W + {}^{\mathsf{T}}W\mathsf{C} = 0$ due to (14.5). We introduce $K = \mathsf{C}W = -{}^{\mathsf{T}}W\mathsf{C}$. Then one obtains ${}^{\mathsf{T}}K = -K\sigma$. Thus,

$$f(X + WX) = f(X) + W_{ij} x_j \frac{\partial}{\partial x_i} f = f(X) + K_{lj} x_j \mathsf{C}_{li} \frac{\partial}{\partial x_i} f. \tag{20.5}$$

${}^{\mathsf{T}}K = -K\sigma$ may be rewritten in components $K_{jl} = (-)^{(1+\nu_j)(1+\nu_l)} K_{lj}$. This yields

$$f(X + WX) = f(X) + K_{lj} (-)^{(1+\nu_j)(1+\nu_l)} x_l \mathsf{C}_{ji} \frac{\partial}{\partial x_i} f. \tag{20.6}$$

Thus, K_{lj} appears in two terms and the transformation may be written

$$f(X + WX) = f(X) + \tfrac{1}{2}K_{lj}L_{jl}f(X) \tag{20.7}$$

with

$$L_{jl} = x_j(C\frac{\partial}{\partial x})_l + (-)^{(1+v_j)(1+v_l)}x_l(C\frac{\partial}{\partial x})_j, \quad (C\frac{\partial}{\partial x})_l = C_{li}\frac{\partial}{\partial x_i} \tag{20.8}$$

and the symmetry relation

$$L_{jl} = (-)^{(1+v_j)(1+v_l)}L_{lj}. \tag{20.9}$$

Since $x_j \in \mathscr{A}_{v_j}$ and $x_l \in \mathscr{A}_{v_l}$ in (20.8), we assign the Z_2-degree $v_j + v_l$ to L_{lj}. Equations (20.8) and (20.9) reduce for $v_j = v_l = 0$ to the well-known relations for the components of angular momentum in real space. One obtains the following supercommutators

$$\{x_k, x_l\} = 0, \quad \{(C\frac{\partial}{\partial x})_k, x_l\} = C_{kl}, \quad \{(C\frac{\partial}{\partial x})_k, (C\frac{\partial}{\partial x})_l\} = 0, \tag{20.10}$$

which yield

$$\{L_{jl}, x_k\} = C_{lk}x_j + (-)^{(1+v_j)(1+v_l)}C_{jk}x_l, \tag{20.11}$$

$$\{L_{jl}, (C\frac{\partial}{\partial x})_i\} = (-)^{1+v_j+v_l}C_{ji}(C\frac{\partial}{\partial x})_l + (-)^{v_j+v_l}C_{li}(C\frac{\partial}{\partial x})_j, \tag{20.12}$$

$$\{L_{jl}, L_{ik}\} = C_{li}L_{jk} - (-)^{(v_j+v_l)(v_j+v_i)}C_{kj}L_{il}$$
$$+(-)^{(1+v_j)(1+v_l)}C_{ji}L_{lk} + (-)^{(1+v_i)(1+v_l)}C_{lk}L_{ji}. \tag{20.13}$$

We look now for the bilinear form

$$B = \sum_{ik} x_i B_{ik} x_k, \tag{20.14}$$

invariant under all rotations. Let $B \in \mathscr{A}_{v_B}$, then $B_{ik} \in \mathscr{A}_{v_B+v_i+v_k}$. Moreover

$$B_{ik} = (-)^{(v_B+1)(v_i+v_k)+v_iv_k}B_{ki}. \tag{20.15}$$

Then

$$\{L_{jl}, B\} = 2x_j(CBx)_l + 2(-)^{(1+v_j)(1+v_l)}x_l(CBx)_j. \tag{20.16}$$

The condition $\{L_{jl}, B\} = 0$ is identically fulfilled for $j = l, v_j = 0$. For $j = l, v_j = 1$ it requires $x_j(CBx)_j = 0$. For $j \neq l$ it reduces to $(CBx)_l = a_lx_l, a_l \in \mathscr{A}_{v_B}$. Then one obtains $a_l = c(-)^{v_l(1+v_B)}$, $c \in \mathscr{A}_{v_B}$. Thus, $B_{kl} = c(-)^{v_B v_k}C_{kl}$ and the invariant function is c times the scalar product

$$B = c x_k C_{kl} x_l, \tag{20.17}$$

which, with X, $X^{\ddagger}$, (14.16), (14.17) can be written

$$B = c X^{\ddagger}X, \quad X^{\ddagger}X = \sum_{i=1}^{n} x_i^2 + 2\sum_{i=1}^{r} \xi_i^{\times}\xi_i. \tag{20.18}$$

Similarly we look for a rotational invariant bilinear form of derivatives,

$$D = \sum_{ik}(C\frac{\partial}{\partial x})_i D_{ik}(C\frac{\partial}{\partial x})_k. \tag{20.19}$$

Then

$$\{L_{jl}, D\} = 2(-)^{1+v_j v_l}(C\frac{\partial}{\partial x})_l(CDC\frac{\partial}{\partial x})_j$$
$$+ 2(-)^{v_j+v_k}(C\frac{\partial}{\partial x})_j(CDC\frac{\partial}{\partial x})_l, \tag{20.20}$$

from which we conclude the invariant form

$$D = c\frac{\partial}{\partial x_i}C_{ik}\frac{\partial}{\partial x_k} = c\triangle_{ss}, \quad \triangle_{ss} := \sum_{i=1}^{n}\frac{\partial^2}{\partial x_i^2} + 2\sum_{i=1}^{r}\frac{\partial^2}{\partial \xi_i^{\times}\partial \xi_i}, \tag{20.21}$$

where $\triangle_{ss}$ is the supersymmetric Laplace operator. We observe that

$$\triangle_{ss}(X^{\ddagger}X) = 2(n - 2r). \tag{20.22}$$

20.2 Ising Model in a Stochastic Magnetic Field

Arguments for the dimensional reduction in the problem of the Ising model in a stochastic magnetic field are given and discussed.

We consider an Ising model in a stochastic magnetic field. We first give the argument by Parisi and Sourlas [204]: This disordered system in d dimensions should be equivalent to the Ising model in $d-2$ dimensions without magnetic field. The Landau free energy of the Ising model in a stochastic magnetic field is given by the Lagrangian

$$\mathscr{L} = \int d^d x L(\phi(x)), \quad L(\phi(x)) = \frac{1}{2}(\nabla\phi(x))^2 + V(\phi(x)) - h(x)\phi(x). \tag{20.23}$$

We assume a white-noise distribution for h,

$$\overline{h(x)h(y)} = 2k\delta(x-y), \tag{20.24}$$

which may be written as a probability distribution proportional to

$$[D\,h]\exp\left(-\int d^d x \frac{1}{4k}h^2(x)\right) = [D\,h]\int[D\,\lambda]\exp\left(\int d^d x \frac{1}{k}(\lambda^2(x) + \lambda(x)h(x))\right). \tag{20.25}$$

The Lagrangian assumes its minimum at

$$\frac{\partial L(\phi(x))}{\partial\phi(x)} = -\triangle\,\phi(x) + V'(\phi(x)) - h(x) = 0. \tag{20.26}$$

We proceed now as in Sect. 19.1.2. By means of

$$\frac{\partial h(x)}{\partial\phi(y)} = \delta^d(x-y)\left(-\triangle + V''(\phi(x))\right) \tag{20.27}$$

the most probable averaged expectation value can be expressed by

$$\langle F(\phi_h(x))\rangle = \int[D\,\phi][D\,\lambda][D\,\psi^\times, \psi]F(\phi(x))\exp(-S(\phi(x), \psi^\times(x), \psi(x), \lambda(x))) \tag{20.28}$$

with

$$S = \int d^d x \frac{1}{k}\left(-\lambda^2(x) + \lambda(x)\triangle\,\phi(x)\right.$$
$$\left. - \lambda(x)V'(\phi(x)) + \psi^\times(x)(-\triangle + V''(\phi(x)))\psi(x)\right). \tag{20.29}$$

In terms of a superfield, Φ, depending on coordinates $x \in \mathscr{A}_0$ and $\theta, \theta^\times \in \mathscr{A}_1$

$$\Phi(x, \theta^\times, \theta) = \phi(x) + \psi^\times(x)\theta + \theta^\times\psi(x) + \lambda(x)\theta^\times\theta, \tag{20.30}$$

we can write

$$S = \int d^d x\,d\,\theta\,d\,\theta^\times\bar{L}(\Phi(x, \theta^\times, \theta)),$$
$$\bar{L} = \frac{1}{k}(\tfrac{1}{2}\Phi\,\triangle_{ss}\,\Phi - V(\Phi)) \tag{20.31}$$

with the supersymmetric Laplace operator

$$\triangle_{ss} = \triangle + 2\frac{\partial^2}{\partial\theta^\times\partial\theta}. \tag{20.32}$$

We introduce the Fourier transform

$$\Phi(x, \theta^\times, \theta) = \int d^d q \, d\kappa \, d\kappa^\times e^{i(\sum_{i=1}^{d} q_i x_i + \kappa^\times \theta + \theta^\times \kappa)} \hat{\Phi}_{q, \kappa^\times, \kappa}, \qquad (20.33)$$

which yields

$$\Delta_{ss} \Phi(x, \theta^\times, \theta) = - \int d^d q \, d\kappa \, d\kappa^\times (\sum_{i=1}^{d} q_i^2 + 2\kappa^\times \kappa) e^{i(\sum_{i=1}^{d} q_i x_i + \kappa^\times \theta + \theta^\times \kappa)} \hat{\Phi}_{q, \kappa^\times, \kappa}. \qquad (20.34)$$

The free theory is governed by

$$\bar{L}_0 = \frac{-m^2 \Phi^2 + \Phi \, \Delta_{ss} \, \Phi}{2k} \qquad (20.35)$$

and thus by the action

$$S_0 = -(2\pi)^d \int d^d q \, d\kappa \, d\kappa^\times \hat{\Phi}_{-q, -\kappa^\times, -\kappa} \frac{q^2 + 2\kappa^\times \kappa + m^2}{2k} \hat{\Phi}_{q, \kappa^\times, \kappa}. \qquad (20.36)$$

Let us abbreviate $\tilde{q} = (q, \kappa, \kappa^\times)$ and $[D\tilde{q}] = d^d q \, d\kappa \, d\kappa^\times$. Then we obtain the superpropagator [54, 226]

$$G^{(0)}(x, \theta^\times, \theta; x', \theta'^\times, \theta') = \langle \Phi(x, \theta^\times, \theta) \Phi(x', \theta'^\times, \theta') \rangle_0 \qquad (20.37)$$

$$= \int [D\tilde{q}][D\tilde{q}'] e^{i(qx + q'x')} e^{i(\kappa^\times \theta + \theta^\times \kappa + \kappa'^\times \theta' + \theta'^\times \kappa')} \langle \hat{\Phi}_{\tilde{q}} \hat{\Phi}_{\tilde{q}'} \rangle_0$$

of the free theory (20.36) by considering the integral

$$\int [D\hat{\Phi}] \exp(-\tfrac{1}{2} \int [D\tilde{q}](\hat{\Phi}_{-\tilde{q}} + \hat{a}_{-\tilde{q}}) p_{\tilde{q}}(\hat{\Phi}_{\tilde{q}} + \hat{a}_{\tilde{q}}))$$

$$= \int [D\hat{\Phi}] \exp(-S_0) \exp(-\tfrac{1}{2}(\hat{a}_{\tilde{q}} p_{\tilde{q}} \hat{\Phi}_{-\tilde{q}} - \hat{\Phi}_{\tilde{q}} p_{\tilde{q}} \hat{a}_{-\tilde{q}} - \hat{a}_{\tilde{q}} p_{\tilde{q}} \hat{a}_{-\tilde{q}})), \qquad (20.38)$$

which does not depend on $\hat{a}$. Expansion up to second order in $\hat{a}$ yields the integral, the 'partition function' $\int [D\hat{\Phi}] \exp(-S_0)$, which has to be normalized to unity times

$$1 - \int [D\tilde{q}] \hat{a}_{-\tilde{q}} p_{\tilde{q}} \langle \hat{\Phi}_{\tilde{q}} \rangle - \tfrac{1}{2} \int [D\tilde{q}] \hat{a}_{-\tilde{q}} p \tilde{q} \hat{a}_{\tilde{q}} + \tfrac{1}{2} \langle (\int [D\tilde{q}] \hat{a}_{-\tilde{q}} p_{\tilde{q}} \hat{\Phi}_{\tilde{q}})^2 \rangle + \ldots \qquad (20.39)$$

The terms at second order in $\hat{a}$ can be written as

$$\frac{1}{2} \int \mathrm{d}\tilde{q} \int \mathrm{d}\tilde{q}'\, \hat{a}_{\tilde{q}}\hat{a}_{\tilde{q}'} (p_{\tilde{q}}p_{\tilde{q}'} \langle \hat{\Phi}_{\tilde{q}}\hat{\Phi}_{\tilde{q}'}\rangle_0 - p_{\tilde{q}}\delta_{\tilde{q}+\tilde{q}'}). \tag{20.40}$$

This yields

$$\langle \hat{\Phi}_{\tilde{q}}\hat{\Phi}_{\tilde{q}'}\rangle_0 = \frac{1}{p_{\tilde{q}}}\delta_{\tilde{q}+\tilde{q}'} = \delta_{\tilde{q}+\tilde{q}'}\hat{G}^{(0)}_{q,\kappa^{\times},\kappa} \tag{20.41}$$

with

$$G^{(0)}_{q,\kappa^{\times},\kappa} = -\frac{k}{(2\pi)^d (m^2 + \sum_{i=1}^{d} q_i^2 + 2\kappa^{\times}\kappa)}. \tag{20.42}$$

We show that it has the interesting property of dimensional reduction from d dimensions to $d - 2$ dimensions. Thus, the Grassmann coordinates $\theta^{\times}, \theta$ count as coordinates of negative dimension. More precisely: Let us choose $x = (y, z) \in R^d$, $y \in R^{d-2}$, $z = (0, 0) \in R^2$. We show that the correlations of the d dimensional system with stochastic field is related to the $(d - 2)$-dimensional one without magnetic field by

$$\langle \prod_i \phi(y^{(i)}, 0)\rangle_{\text{stoch } h,d} = \langle \prod_i \phi(y^{(i)})\rangle_{h=0,d-2}. \tag{20.43}$$

The derivation runs like follows

$$\phi(y, 0) = \int \mathrm{d}^{d-2}q e^{\mathrm{i}\sum_{i=1}^{d-2} q_i y_i} \int \mathrm{d}q_{d-1}\mathrm{d}q_d\mathrm{d}\kappa\mathrm{d}\kappa^{\times}\Phi_{q,\kappa^{\times},\kappa}. \tag{20.44}$$

All integrals contributing to diagrams including external legs to $\phi(y, 0)$, contain only functions depending on $q_{d-1}, q_d, \kappa^{\times}, \kappa$ as scalar products $q_{d-1}q'_{d-1} + q_d q'_d + \kappa^{\times}\kappa' + \kappa'^{\times}\kappa$. Thus, the integral theorem of Sect. 16.1 applies: There remain only integrals over the $(d - 2)$-dimensional space, which yields (20.43). Compare with [178], for example. In perturbation theory, one expands in powers of $V - m^2\Phi^2$ and constructs the corresponding diagrams, which involve the propagators G_0.

Discussion The argument given so far has two shortcomings: The configurations $\{\phi\}$ for a given magnetic field $\{h\}$ may not be unique. As a consequence the determinant of the functional derivative (20.27) can be positive or negative, but should always be counted with the same sign. A second shortcoming is that Φ should be superreal. But λ has to be integrated along the imaginary axis, as evident from (20.25). Cardy [45] and Klein and Perez [147] present a non-perturbative argument for the dimensional reduction.

The ϵ-expansion for critical exponents from the upper critical dimension agree for both systems as argued by Aharony, Imry and Ma [4, 120], Grinstein [101], and by Young [293]. The critical behaviour differs at lower dimensions as shown by Imbrie [118, 119] and by Bricmont and Kupiainen [41]. It may either be that at some dimension the determined fixed point becomes unstable and a different one takes over or it may be that both systems differ by a contribution exponentially small at the upper critical dimension.

Brézin and de Dominicis [36, 37] consider the disordered system with a general number of replicas. Then operators appear, which make the system unstable already at dimension 8 and below, which sheds doubt on an expansion around dimension 6. Le Doussal, Wiese, and Chauve [48, 160, 162, 163, 283] argue that with the functional renormalization group a cusp for the disorder distribution function appears and replica symmetry breaking occurs. They argue, whenever several minima appear, this cusp appears. In addition they argue [161] that an n-component disordered ferromagnet has a lower critical dimension 4 for $n > n_c$, but an n-dependent lower critical dimension less than 4 for $n < n_c$, where $n_c = 2.834$ for a random field and $n_c = 9.441$ for random anisotropy. See also the arguments by Fisher [88] and Young and Nauenberg [294]. Tissier and Tarjus [255] argue that dimensional reduction works down to $d \approx 5.1$.

Dimensional reduction leaves open questions for the Ising model in a stochastic magnetic field, but equivalent arguments work correctly for branched polymers and lattice animals, which are considered next.

20.3   Branched Polymers and Lattice Animals

A model of lattice animals, (they serve as models of branched polymers), and of the density of zeroes of the partition function of Ising models as function of a complex magnetic field are compared. Their critical behaviour is connected by dimensional reduction. Similarly models of linear polymers and self-avoiding walks are connected to another model by dimensional reduction.

Animals and Trees   Lattice animals are clusters of connected sites on a regular lattice. Connected means that they are connected by links of the lattice. If any two sites of the cluster have just one connection they are called site trees. Besides these site animals, one also defines bond animals. Bond animals are clusters of bonds between adjacent sites. Bond trees are clusters in which any two bonds are connected by a unique path of bonds. A more detailed explanation can be found in the introduction of Hsu et al. [113]. Branched polymers are defined as sets of site trees and bond trees, resp. The number Z_N and the end-to-end distance R_N of such animals and trees for a given number, N, of sites and bonds, resp., obeys asymptotically, that is, for large N

$$Z_N \sim \mu^N N^\theta, \quad R_N \sim N^\nu. \tag{20.45}$$

Using replica methods Lubensky and Isaacson [170, 171] have described these statistical models in terms of a ϕ^3-theory and obtained ϵ-expansions for the critical exponents θ and ν in $D = 8 - \epsilon$ dimensions.

Lee-Yang Edge Lee and Yang [164] considered the partition function of Ising models with ferromagnetic (or zero) interaction between pairs of spins. They found that as a function of the fugacity, $z = \exp(-2H/k_B T)$, all zeros lie on the circle $|z| = 1$. This is called the Lee-Yang theorem Thus, the zeros are found for purely imaginary magnetic field $H = iH'$. In the thermodynamic limit one can introduce a density of zeros commonly denoted by $\mathscr{G}(H', T)$. This density vanishes above the critical temperature in an interval $-H_0(T) < H' < H_0(T)$ close to the real axis of H. Close to H_0, it shows a power law behavior

$$\mathscr{G}(H', T) \sim (H' - H_0(T))^\sigma . \tag{20.46}$$

The spontaneous magnetization below T_c is proportional to $\mathscr{G}(0, T)$. The critical behaviour around $H_0(T)$ can be described by a ϕ^3-theory with imaginary coupling and upper critical dimension $d_c = 6$ [87, 157]. Earlier papers on this subject are [149, 249]. The identity of the repulsive-core singularity with the Lee-Yang edge criticality has been pointed out by Lai and Fisher [158], and by Park and Fisher [208].

Dimensional Reduction Parisi and Sourlas [206] argue that these two systems are related to each other by dimensional reduction. They find the relations

$$\theta(D) = \sigma(d) + 2, \quad \nu(D) = (\sigma(d) + 1)/d \tag{20.47}$$

for $D = d + 2$. Since $\sigma(0) = -1$ and $\sigma(1) = 1/2$ are exactly known from the zero- and one-dimensional Ising model, one obtains

$$\theta(3) = \tfrac{3}{2}, \quad \nu(3) = \tfrac{1}{2}, \quad \theta(2) = 1. \tag{20.48}$$

The upper critical dimensions are given by $D_c = 8$, $d_c = 6$. Precise analytical arguments for this dimensional reduction have been given by Brydges and Imbrie [42]. Excellent numerical verification has been obtained by Hsu et al. [113] and by Luther and Mertens [172].

Linear Polymers and Self-Avoiding Walks De Gennes [53] has shown that linear polymers and, equivalently, self-avoiding walks can be described by a ϕ^4-theory with $n = 0$ components (replica trick). McKane [180] has given the equivalence to the ϕ^4-theory with the same number of commuting and anticommuting components of ϕ. Parisi and Sourlas [205] have finally mapped this problem to another one in two more dimensions. It seems, however, that this new model has not found much interest. Thus, many calculations are performed on the basis of the $n = 0$, ϕ^4-theory. A review article on self-avoiding walks has been given by Bauerfeind et al. [17]. The static equilibrium properties of polymer solutions and the excluded volume effect are reviewed in the book [230] by L. Schäfer. Sourlas [244] has given a review on the models and their connection, via dimensional reduction, as considered in Sects. 20.2 and 20.3.

20.4   Electron in the Lowest Landau Level

The density of states of electrons moving in a plain in the lowest Landau level of a strong magnetic field in the presence of random point scatterers can be calculated exactly. The corresponding action of the two-dimensional system is supersymmetric and allows a reduction to a zero-dimensional action.

20.4.1   Free Electron in a Magnetic Field

A free electron in two dimensions and in a perpendicular magnetic field B, is governed by the Hamiltonian

$$H_0 = \frac{1}{2m}(\mathbf{p} - \frac{e}{c}\mathbf{A})^2, \quad \mathbf{A} = \tfrac{1}{2}B(-y,x). \tag{20.49}$$

The eigenenergies of H_0 are

$$E_n = \hbar\omega(n + \tfrac{1}{2}), \quad n = 0, 1, 2, \ldots, \tag{20.50}$$

with the cyclotron frequency given by $\omega = eB/mc$. The Hamiltonian reads, with the choice of units for the energy $\hbar\omega = 1$ and for the cyclotron length $l = (\hbar c/eB)^{1/2} = \sqrt{1/2}$,

$$H_0 = -(\partial_z - \tfrac{1}{2}z^*)(\partial_{z^*} + \tfrac{1}{2}z) + \tfrac{1}{2}, \tag{20.51}$$

with the complex coordinate

$$z = x + iy. \tag{20.52}$$

The eigenstates of the lowest Landau level are given by

$$\chi_{0,m}(z) = \frac{1}{\sqrt{\pi m!}}z^m e^{-z^* z/2}, \quad m = 0, 1, 2, \ldots \tag{20.53}$$

In general, the eigenfunctions of the lowest Landau level read

$$\chi(z) = u(z)e^{-z^* z/2}, \quad \partial_{z^*}u = 0, \tag{20.54}$$

with u being holomorphic. Restriction to the lowest Landau level yields the Green's function

$$G_0(z, z', E + i0) = \langle z|P\frac{1}{E - H_0 + i0}P|z'\rangle = \frac{1}{\epsilon}C(z, z') \tag{20.55}$$

where P projects onto the lowest Landau level, and C is the projector in the z-space,

$$\epsilon = E + i0 - \tfrac{1}{2}\hbar\omega,$$ (20.56)

$$C(z, z') = \sum_m \chi_{0,m}(z)\chi^*_{0,m}(z') = \frac{1}{\pi}\exp(-\tfrac{1}{2}z^*z - \tfrac{1}{2}z'^*z' + z'^*z).$$ (20.57)

The probability distribution of the state $\chi_{0,m}$ has its maximum at radius $\sqrt{m}$. Thus, the area covered by $m + 1$ states is πm. The density of the states per area is then in our units,

$$\rho_0 = 1/\pi.$$ (20.58)

The density can also be seen from $G(z, z, E + i0)$, since for large E it decays like ρ_0/E. This density is $1/(2\pi l^2)$ for cyclotron length l.

20.4.2   Random Potential

We add a random potential V to the kinetic energy, Eq. (20.51),

$$H = H_0 + V.$$ (20.59)

This random potential will result in a broadening of the Landau levels. If this broadening is small, in comparison to the spacing $\hbar\omega$ between the Landau levels, then it is a good approximation to restrict the calculation to the Hilbert space of the lowest Landau level. Thus, for the lowest Landau level, we replace (20.59) by

$$H = H_0 + PVP$$ (20.60)

where P is the projector onto the lowest Landau level. If the potential consists of point scatterers, that is, if it is uncorrelated between different points, then the calculation of the averaged one-particle Green's function and thus, the density of states can be reduced to a calculation for a zero-dimensional system. This was shown for a white noise potential

$$\overline{V(r)} = 0, \quad \overline{V(r)V(r')} = W\delta(r - r')$$ (20.61)

by means of a diagrammatic expansion [273]. Brézin et al. [39] have generalized the solution to systems of arbitrary point scatterers characterized by a function g

$$\overline{\exp(-i\int d^2r\alpha(r)V(r))} = \exp(\int d^2r g(\alpha(r))).$$ (20.62)

The function $g(\alpha)$ represents the Fourier transform of the probability distribution P of the potential at a site

$$\exp(g\alpha) = \int \mathrm{d}\,VP(V)\mathrm{e}^{-\mathrm{i}\alpha V}. \tag{20.63}$$

By means of a supersymmetric field theory they could apply dimensional reduction. The derivation will be given in the next subsection. Beforehand we give three examples of stochastic potentials and the corresponding functions g. We also derive the form of the averaged one-particle Green's function.

White Noise Potential The l.h.s. of (20.62) gives

$$1 - \mathrm{i} \int \mathrm{d}^2 r\alpha(r)\overline{V(r)} - \tfrac{1}{2} \int \mathrm{d}^2 r\mathrm{d}^2 r'\alpha(r)\alpha(r')\overline{V(r)V(r')} + \dots$$

$$= 1 - \tfrac{1}{2} \int \mathrm{d}^2 r W\alpha^2(r) + \dots \tag{20.64}$$

The higher-order terms in αV vanish for odd powers. Even powers in αV yield

$$(-)^k \frac{1}{(2k)!} \int \mathrm{d}\,r_1^2 \dots \mathrm{d}\,r_{2k} \overline{\alpha(r_1)V(r_1)\dots\alpha(r_{2k})V(r_{2k})}. \tag{20.65}$$

The factors can be grouped in pairs in $(2k-1)!!$ ways. Hence, (20.65) yields

$$\frac{(-)^k}{k!} \left(\int \mathrm{d}^2 r \tfrac{1}{2} W\alpha^2(r) \right) \tag{20.66}$$

and the corresponding function g reads

$$g(\alpha) = -\tfrac{1}{2} W\alpha^2. \tag{20.67}$$

Poisson Model of Random Impurities They correspond to zero-range scatterers of density $\hat{\rho}$, which are distributed randomly in space

$$V(r) = \lambda \sum_i \delta^2(r - r_i). \tag{20.68}$$

The l.h.s. of (20.62) yields

$$\overline{\exp(-\mathrm{i}\lambda \sum_i \alpha(r_i))}. \tag{20.69}$$

One obtains, in an infinitesimal area of size $\mathrm{d}A$,

$$1 + \hat{\rho}\mathrm{d}A(\mathrm{e}^{-\mathrm{i}\lambda\alpha} - 1), \tag{20.70}$$

which yields the function

$$g(\alpha) = \hat{\rho}(e^{-i\lambda\alpha} - 1) \tag{20.71}$$

Lorentzian Distribution This distribution of the potential is given by (compare to the Lloyd model in Problem 4.1)

$$P(V) = \frac{\Gamma}{\pi(V^2 + \Gamma^2)} \tag{20.72}$$

and yields

$$g(\alpha) = -\Gamma|\alpha|. \tag{20.73}$$

Averaged One-Particle Green's Function We use the translational invariance including gauge-transformation to derive the site dependence of the averaged one-particle Green's function. We generalize the Hamiltonian H_0 to

$$H_a = -(\partial_z - \tfrac{1}{2}z^* + \tfrac{1}{2}a^*)(\partial_{z^*} + \tfrac{1}{2}z - \tfrac{1}{2}a) \tag{20.74}$$

The Hamiltonians H_0 and H_a describe electrons in the same magnetic field B, but with a different gauge, $\mathbf{A}$. They are related by the gauge-transformation

$$H_a e^{(az^* - a^* z)/2} = e^{(az^* - a^* z)/2} H_0. \tag{20.75}$$

Then $G(z, z')$ and $G(z - a, z' - a)$ are related by

$$G(z - a, z' - a) = e^{(az^* - a^* z)/2} G(z, z') e^{(-az'^* + a^* z')/2}. \tag{20.76}$$

Since $G(z, z')$ is a linear combination of $\chi_m(z)\chi_n^*(z')$, (similarly for $G(z - a, z' - a)$) it can be written as

$$G(z, z') = f(z, z'^*)e^{-zz^*/2 - z'z'^*/2},$$
$$G(z - a, z' - a) = f(z - a, z'^* - a^*)e^{-(z-a)(z^*-a^*)/2 - (z'-a)(z'^*-a^*)/2},$$
$$\tag{20.77}$$

where f does not depend on z^* and z'. Insertion of (20.77) into (20.76) yields

$$f(z - a, z'^* - a^*) = e^{-a^* z - a z'^* + a a^*} f(z, z'^*), \tag{20.78}$$

which yields

$$f(z, z'^*) = \text{const } e^{zz'^*}. \tag{20.79}$$

Thus, the one-particle Green's function can be expressed by the on-site Green's function and the projector C from (20.57),

$$G(z, z', E + i0) = \pi C(z, z') G(z, z, E + i0) = \pi C(z, z') G(0, 0, E + i0). \quad (20.80)$$

20.4.3 Supersymmetric Lagrangian

The Green's function can be written as

$$\langle r|\frac{1}{E - H + i0}|r'\rangle = -i \int D\phi D\phi^* D\psi D\psi^* \phi(r)\phi^*(r') \exp(-\mathcal{L}), \quad (20.81)$$

$$\mathcal{L} = -i \int d^2r \left(\phi^*(\epsilon - V)\phi + \psi^*(\epsilon - V)\psi\right), \quad (20.82)$$

where $\epsilon = E - \frac{1}{2}\hbar\omega + i0$, (20.57), and the two fields $\phi \in \mathscr{A}_0$ and $\psi \in \mathscr{A}_1$. They may be thought of as

$$\phi(r) = \sum_m u_m \chi_{0,m}(r) = \frac{e^{-z^*z/2}}{\sqrt{\pi}} u(z), \; u(z) = \sum_m \frac{u_m z^m}{\sqrt{m!}}, \quad \phi, u, u_m \in \mathscr{A}_0, \quad (20.83)$$

$$\psi(r) = \sum_m v_m \chi_{0,m}(r) = \frac{e^{-z^*z/2}}{\sqrt{\pi}} v(z), \; v(z) = \sum_m \frac{v_m z^m}{\sqrt{m!}}, \quad \psi, v, v_m \in \mathscr{A}_1. \quad (20.84)$$

The introduction of the Grassmannian field ψ and v, respectively, guarantees the correct normalization,

$$\int D\phi D\phi^* D\psi D\psi^* \exp(-\mathcal{L}) = 1. \quad (20.85)$$

Now the average over the random potential can be performed by means of (20.62), where $\alpha(r)$ is replaced by $e^{-z^*z}(u^*u + v^*v)$. The averaged Green's function reads

$$\overline{G(r, r', E + i0)} = \overline{\langle r|\frac{1}{E - H + i0}|r'\rangle}$$

$$= -ie^{-z^*z/2 - z'^*z'/2} \int Du Du^* Dv Dv^* u(z)u^*(z')$$

$$\exp\left(-\mathcal{L}_0 - \mathcal{L}'\right), \quad (20.86)$$

where $u(z)u^*(z')$ may be replaced by $v(z)v^*(z)$, and

$$\mathcal{L}_0 = -i\epsilon \int dz dz^* (u^*u + v^*v)e^{-z^*z}, \tag{20.87}$$

$$\mathcal{L}' = -\int dz dz^* g(e^{-z^*z}(u^*u + v^*v)). \tag{20.88}$$

The important observation is that the holomorphic superfield

$$\Phi(z, \theta) = u(z) + \theta v(z) \tag{20.89}$$

allows the representation

$$\mathcal{L}_0 = -i\epsilon \int dz dz^* d\theta d\theta^* e^{-z^*z - \theta\theta^*} \Phi^*\Phi, \tag{20.90}$$

$$\mathcal{L}' = -\int dz dz^* d\theta d\theta^* h(e^{-z^*z - \theta\theta^*} \Phi^*\Phi). \tag{20.91}$$

In order to confirm (20.90), (20.91), one expands $\mathcal{L}_0$ and $\mathcal{L}'$ in v, v^*. The equality of (20.87) and (20.90) can then be seen immediately. The expansion for (20.88) and (20.91) yield, with $\beta = e^{-z^*z}u^*u$,

$$g(e^{-z^*z}(u^*u + v^*v)) = g(\beta) + e^{-z^*z}v^*vg'(\beta), \tag{20.92}$$

$$\int d\theta d\theta^* h(e^{-z^*z - \theta\theta^*} \Phi^*\Phi) = \beta h'(\beta) + e^{-z^*z}v^*v(h'(\beta) + \beta h''(\beta)). \tag{20.93}$$

Thus, g and h have to obey

$$g(\beta) = \beta h'(\beta), \quad g'(\beta) = h'(\beta) + \beta h''(\beta). \tag{20.94}$$

Since the second equation is the derivative of the first one, it is sufficient to solve the first one

$$h(\alpha) = \int_0^\alpha \frac{d\beta}{\beta} g(\beta), \tag{20.95}$$

and $\mathcal{L}_{ss} = \mathcal{L}_0 + \mathcal{L}'$ is invariant under rotation in superspace, which has the two additional Grassmannian coordinates θ and θ^*.

20.4.4   *Dimensional Reduction*

We consider now the dimensional reduction. For this purpose we introduce super-vectors

$$
\tilde{z} := \begin{pmatrix} z \\ \theta \end{pmatrix}, \quad \tilde{u}_m := \begin{pmatrix} u_m \\ v_{m-1} \end{pmatrix}, \quad m > 0. \tag{20.96}
$$

We perform an infinitesimal transformation in the superspace

$$
\delta\tilde{z} = \delta U \tilde{z}, \quad \delta U = \begin{pmatrix} \mathrm{i}p & -\omega^* \\ \omega & \mathrm{i}q \end{pmatrix} \tag{20.97}
$$

and also of $\tilde{u}_m$. Then, one obtains

$$
\delta\Phi = \sum_m (\delta u_m + \mathrm{i}mp u_m - \sqrt{m}\omega v_{m-1}) \frac{z^m}{\sqrt{m!}}
$$
$$
+ \sum_m (\delta v_{m-1} - \omega^* \sqrt{m} u_m + \mathrm{i}(p(m-1) + q) v_{m-1}) \frac{z^{m-1}}{\sqrt{(m-1)!}} \theta, \tag{20.98}
$$

with Φ remaining invariant under this transformation, if one chooses

$$
\delta\tilde{u}_m = \delta U_m \tilde{u}_m, \quad \delta U_m = \begin{pmatrix} -\mathrm{i}mp & -\sqrt{m}\omega \\ \sqrt{m}\omega^* & -\mathrm{i}(m-1)p - \mathrm{i}q \end{pmatrix}. \tag{20.99}
$$

In this derivation p, q, ω, ω^* are infinitesimal. The matrices U and U_m are antihermitian and generate the group $\mathrm{UPL}_1(1,1)$, which can be obtained as

$$
U = \exp \delta U, \quad U_m = \exp \delta U_m, \tag{20.100}
$$

where now the arguments q, etc. are considered finite. Thus, Φ, and consequently Φ^*, are invariant under the transformations

$$
\tilde{u}'_m = U_m \tilde{u}_m, \quad \tilde{z}' = U\tilde{z}, \quad u'_0 = u_0. \tag{20.101}
$$

The partition function can be written as

$$
Z = \int \mathrm{d}u_0^* \mathrm{d}u_0 \mathrm{D}\tilde{u}^\dagger \mathrm{D}\tilde{u} \exp\left(-\int \mathrm{d}\tilde{z}^\dagger \mathrm{d}\tilde{z} \mathscr{L}(\tilde{z}^\dagger \tilde{z}, \Phi^* \Phi) \right) \tag{20.102}
$$

and it has to be shown that Z reduces to the integral, which contains only the field components u_0^* and u_0,

$$Z = \int d u_0^* d u_0 \exp(- \int d \tilde{z}^\dagger d \tilde{z} \mathscr{L}(\tilde{z}^\dagger \tilde{z}, u_0^* u_0)). \qquad (20.103)$$

To show this, we proceed similarly as in Sects. 15.1.3 and 15.1.4. The only difference is that the various supervectors transform with different superunitary transformations U and U_m. But as in Sect. 15.1.4, one shows that

$$Z'(\tilde{u}_m^\dagger, \tilde{u}_m) := \int d u_0^* d u_0 D' \tilde{u}^\dagger D' \tilde{u} \exp\left(- \int d \tilde{z}^\dagger d \tilde{z} \mathscr{L}(\tilde{z}^\dagger \tilde{z}, \Phi^* \Phi)\right)$$

$$(20.104)$$

where over all supervectors, but $\tilde{u}_m$, has been integrated, Z' obeys

$$Z'(\tilde{u}_m^\dagger U_m^\dagger, U_m \tilde{u}_m) = Z'(\tilde{u}_m^\dagger, \tilde{u}_m). \qquad (20.105)$$

Due to the invariance under arbitrary superunitary transformations, the integral reduces to

$$Z = \int D u_m^\dagger D u_m Z'(\tilde{u}_m^\dagger, \tilde{u}_m) = Z'(0, 0). \qquad (20.106)$$

This argument can be used for all $m > 0$, which proves Eq. (20.103). Now $\mathscr{L}$ depends only on $\tilde{z}$, $\tilde{z}^\dagger$, and u_0^*, u_0. The integration over $\tilde{z}^\dagger$, $\tilde{z}$ yields $\pi \mathscr{L}$ at $\tilde{z} = 0$, hence

$$Z = \int d u_0^* d u_0 \exp(-\mathscr{L}(\epsilon, u_0^* u_0)), \quad \mathscr{L}(\epsilon, \alpha) = \pi(i\epsilon\alpha + h(\alpha)). \qquad (20.107)$$

Thus, the integral over a field in two dimensions, plus two anticommuting dimensions, is reduced to a problem for the field $\alpha = u_0^* u_0$ in zero dimensions. One may wonder, why Z, (20.103), is not simply one as in (20.85). We deal with two types of supersymmetry, the one we have used now and where u_m and v_{m-1} are combined to a superfield. Initially, however, we have started with the pairs u_m and v_m. If one takes into account these pairs up to some m, then Z will be one. If, however, we take into account the pairs u_m and v_{m-1} up to some m, then we obtain Z, (20.103).

Since $\Phi^*(0, 0)\Phi(0, 0) = u_0^* u_0$, one obtains, finally, the averaged on-site Green's function

$$G(0, 0, E + i0) = -i \frac{\int d\alpha \, \alpha \exp(-\mathscr{L}(\epsilon, \alpha))}{\int d\alpha \exp(-\mathscr{L}(\epsilon, \alpha))}$$

$$= -\frac{\partial}{\pi \partial \epsilon} \ln \int_0^\infty d\alpha \exp(-\mathscr{L}(\epsilon, \alpha)). \qquad (20.108)$$

Together with (20.80), one obtains the Green's function $G(z, z', E + i0)$. Although the density of states of the lowest Landau level can be calculated exactly for random point-scatterers, there is no generalization known for higher Landau levels. Unfortunately, this type of calculation does not explain the quantized Hall effect, which of course is the most interesting effect of such systems.

The density of states is obtained from (20.108),

$$\rho(E) = -\frac{1}{\pi} \Im G(z, z, E + i0). \tag{20.109}$$

We consider G and ρ for the three examples in Sect. 20.4.2, (20.64)–(20.73).

White Noise Potential For this type of disorder one has

$$g(\alpha) = -\tfrac{1}{2} W \alpha^2, \quad h(\alpha) = -\tfrac{1}{4} W \alpha^2. \tag{20.110}$$

This yields

$$Z = \frac{1}{\sqrt{W}} \exp(-v^2)(1 + iI(v)) \tag{20.111}$$

with

$$v = \sqrt{\frac{\pi}{W}} \epsilon = \sqrt{\frac{\pi}{W}} (E - \tfrac{1}{2} \hbar \omega), \quad I(v) = \frac{2}{\sqrt{\pi}} \int_0^v dx \exp(x^2). \tag{20.112}$$

One concludes that

$$G(0, 0, E + i0) = \frac{2\pi\epsilon}{W} - \frac{2i}{WZ}, \tag{20.113}$$

$$\rho(E) = \frac{2}{\pi^2 \sqrt{W}} \frac{\exp(v^2)}{1 + I^2(v)}. \tag{20.114}$$

Poisson Model of Random Impurities We set the density of the scatterers $\hat{\rho}$ in relation to the density of states ρ_0 in the Landau level, and introduce

$$f := \hat{\rho}/\rho_0 = \pi\hat{\rho}, \quad v := \frac{\pi}{\lambda}(E - \tfrac{1}{2}\hbar\omega). \tag{20.115}$$

The averaged density of states depends strongly on f. In the limit $v \to 0$, and for positive λ, Brézin, Gross, and Itzykson obtain [39], see also [122],

$$\lambda\rho(E) \sim \begin{cases} (1-f)\delta(v) + A(f)v^{-f} + \ldots & 0 < f < 1 \\ \frac{1}{v((\ln(v/v_0))^2 + \pi^2)} + \ldots & f = 1 \\ B(f)v^{f-2} + \ldots & 1 < f. \end{cases} \tag{20.116}$$

with amplitudes $A(f)$ and $B(f)$. There are also weaker non-analyticities of $\rho(E)$ at integer values of ν. For $f < 1$ there are still a finite number of states at $E = \frac{1}{2}\hbar\omega$. Indeed, if within a circle of area πN the Nf scatterers are located at z_i, then the states described by

$$\psi(z) = P(z) \prod_{i=1}^{Nf}(z - z_i)e^{-z*z/2} \tag{20.117}$$

with polynomials P up to order $N(1 - f)$ are states with unchanged energies. These wave-functions vanish at the location of the scatterers and are thus unaffected by them.

Lorentzian Distribution   For this distribution we have

$$g(\alpha) = -\Gamma|\alpha|, \quad h(\alpha) = -\Gamma|\alpha| \tag{20.118}$$

and obtain

$$G(0, 0, E + i0) = \frac{1}{\pi(\epsilon + i\Gamma)}, \quad \rho(E) = \frac{\Gamma}{\pi^2(\Gamma^2 + \epsilon^2)}. \tag{20.119}$$

We remember from (20.58) that the density of states per area $\rho_0 = 1/\pi$. This explains the overall factor $1/\pi$ in G. The density of states is again Lorentzian.

20.5   Isotropic $\phi^{2\sigma}$-Theories with Negative Number of Components

Negative even number of components are equivalent to pairs of anticommuting components. They yield free theories, since these components are nilpotent.
   Formally a ϕ^4-theory with $n = -2$ components can be expressed by $\phi^2 = \bar{\theta}\theta$, which yields $\phi^4 = 0$. Thus, the theory for $n = -2$ is a free theory. This is the reason for factors $(n + 2)$ in expansion coefficients of the ϵ-expansion for the n-component ϕ^4 theory (Balian and Toulouse [15]). Note, however, that this does not apply for exponents of operators, which cannot be expressed in terms of θ and $\bar{\theta}$ (Wegner [276]).
   Similarly, one may ask what happens to a $\phi^{2\sigma}$-theory for $n = -2, -4, \ldots -2\sigma + 2$. They can be expressed by $\phi^2 = \sum_{i=1}^{-n/2} \bar{\theta}_i\theta_i$ and obviously ϕ^{2k} vanishes for $k > -n/2$, in particular, the $\phi^{2\sigma}$-term vanishes. It seems that this has not been fully discussed, but for the tricritical case, $\sigma = 3$, a number of exponents carry the factor $(n + 2)(n + 4)$ at least at order ϵ^2 for $d = 3 - \epsilon$. I refer to [247, 266, 267] and the review by Lawrie and Sarbach [159]. This holds also for the exponent η in order ϵ^3 at the tricritical point [167].

The exponent η carries a factor $(n/2 + \sigma - 1)!(n/2)!$ at order ϵ^2 with dimension $d = 2\sigma/(\sigma - 1) - \epsilon$ [256, 266, 267].

Problems

20.1 Set

$$r^2 = \sum_{i=1}^{n} x_i^2 + c \sum_{i=1}^{r} \theta_i^{\times} \theta_i, \quad \Delta_{\mathrm{ss}} = \sum_{i=1}^{n} \frac{\partial^2}{\partial x_i^2} + c' \sum_{i=1}^{r} \frac{\partial^2}{\partial \theta_i^{\times} \partial \theta_i}$$

Which condition have c, c' to obey such that $\Delta_{\mathrm{ss}} f(r^2)$ is a function of r^2 only?

20.2 Derive Eq. (20.79) from (20.78).

References

[4] A. Aharony, Y. Imry, S. Ma, Lowering of dimensionality in phase transitions with random fields. Phys. Rev. Lett. **37**, 1364 (1976)

[15] R. Balian, G. Toulouse, Critical exponents for transitions with $n = -2$ components of the order parameter. Phys. Rev. Lett. **30**, 544 (1973)

[17] R. Bauerschmidt, H. Duminil-Copin, J. Goodman, G. Slade, Lectures on self-avoiding walks. Clay Math. Proc. **15**, 395 (2012). arXiv:1206.2092

[36] E. Brézin, C. de Dominicis, New phenomena in the random field Ising model. Europhys. Lett. **44**, 13 (1998)

[37] E. Brézin, C. de Dominicis, Interactions of several replicas in the random field Ising model. Eur. Phys. J. B **19**, 467 (2001)

[39] E. Brézin, D.J. Gross, C. Itzykson, Density of states in the presence of a strong magnetic field and random impurities. Nucl. Phys. B **235**, 24 (1984)

[41] J. Bricmont, A. Kupiainen, Lower critical dimension for the random-field Ising model. Phys. Rev. Lett. **59**, 1829 (1987)

[42] D.C. Brydges, J.Z. Imbrie, Branched Polymers and dimensional reduction. Ann. Math. **158**, 1019 (2003)

[45] J. Cardy, Nonperturbative effects in a scalar supersymmetric theory. Phys. Lett. **125B**, 470 (1983)

[48] P. Chauve, P. Le Doussal, K.J. Wiese, Renormalization of pinned elastic systems: How does it work beyond one loop? Phys. Rev. Lett. **86**, 1785 (2001)

[53] P.G. de Gennes, Exponents for the excluded volume problem as derived by the Wilson method. Phys. Lett. **38A**, 339 (1972)

[54] R. Delbourgo, Superfield perturbation theory and renormalization. Nuovo Cimento **25A**, 646 (1975)

[87] M.E. Fisher, Yang-Lee edge singularity and ϕ^3 field theory. Phys. Rev. Lett. **40**, 1610 (1978)

[88] D.S. Fisher, Random fields, random anisotropies, nonlinear σ models, and dimensional reduction. Phys. Rev. B **31**, 7233 (1985)

[101] G. Grinstein, Ferromagnetic phase transitions in random fields: The breakdown of scaling laws. Phys. Rev. Lett. **37**, 944 (1976)

[113] H. Hsu, W. Nadler, P. Grassberger, Statistics of lattice animals. Comput. Phys. Commun. **169**, 114 (2005)

[118] J.Z. Imbrie, Lower critical dimension of the random-field Ising model. Phys. Rev. Lett. **53**, 1747 (1984)

[119] J.Z. Imbrie, The ground state of the three-dimensional random-field Ising model. Commun. Math. Phys. **98**, 145 (1985)

[120] Y. Imry, S.K. Ma, Random-field instability of the ordered state of continuous symmetry. Phys. Rev. Lett. **35**, 1399 (1975)

[122] C. Itzykson, J.-M. Drouffe, *Statistical Field Theory*, two volumes (Cambridge University Press, Cambridge, 1989)

[147] A. Klein, J.F. Perez, Supersymmetry and dimensional reduction: a nonperturbative proof. Phys. Lett. **125B**, 473 (1983)

[149] P.J. Kortmann, R.B. Griffiths, Density of zeroes on the Lee-Yang circle for two Ising ferromagnets. Phys. Rev. Lett. **27**, 1439 (1971)

[157] D.A. Kurtze, M.E. Fisher, Yang-Lee edge singularities at high temperatures. Phys. Rev. **B20**, 2785 (1979)

[158] S. Lai, M.E. Fisher, The universal repulsive-core singularity and Yang-Lee edge criticality. J. Chem. Phys. **103**, 8144 (1995)

[159] I.D. Lawrie, S. Sarbach, Theory of tricritical points, in *Phase Transitions and Critical Phenomena*, vol. 9, ed. by C. Domb, J.L. Lebowitz (Academic Press, London, 1984), p. 1

[160] P. Le Doussal, K.J. Wiese, Functional renormalization group at large N for random manifolds. Phys. Rev. E **67**, 016121 (2003)

[161] P. Le Doussal, K.J. Wiese, Random field spin models beyond one loop: a mechanism for decreasing the lower critical dimension. Phys. Rev. Lett. **96**, 197202 (2006)

[162] P. Le Doussal, K.J. Wiese, Functional renormalization for disordered systems: basic recipes and gourmet dishes. Markov Process. Relat. Fields **13**, 777 (2007)

[164] T.D. Lee, C.N. Yang, Statistical theory of equation of state and phase transitions. II. Lattice gas and Ising model. Phys. Rev. **87**, 410 (1952)

[167] A.L. Lewis, F.W. Adams, Tricritical behavior in two dimensions. II. Universal quantities from the ϵ expansion. Phys. Rev. **B18**, 5099 (1978)

[170] T.C. Lubensky, J. Isaacson, Field theory of statistics of branched polymers, gelation, and vulcanization. Phys. Rev. Lett. **41**, 829 (1978); Erratum Phys. Rev. Lett. **42**, 410 (1979)

[171] T.C. Lubensky, J. Isaacson, Statistics of Lattice animals and branched polymers. Phys. Rev. A **20**, 2130 (1979)

[172] S. Luther, S. Mertens, Counting Lattice animals in high dimensions. J. Stat. Mech. **2011**, P09026 (2011). arXiv: 1106.1078

[178] B. McClain, A. Niemi, C. Taylor, L.C.R. Wijewardhana, Super space, dimensional reduction, and stochastic quantization. Nucl. Phys. B **217**, 430 (1983)

[180] A.J. McKane, Reformulation of $n \to 0$ models using anticommuting scalar fields. Phys. Lett. **76A**, 22 (1980)

[204] G. Parisi, N. Sourlas, Random magnetic fields, supersymmetry, and negative dimensions. Phys. Rev. Lett. **43**, 744 (1979)

[205] G. Parisi, N. Sourlas, Selfavoiding walk and supersymmetry. J. Phys. Lett. **41**, L403 (1980)

[206] G. Parisi, N. Sourlas, Critical behavior of branched polymers and the Lee-Yang edge singularity. Phys. Rev. Lett. **46**, 871 (1981)

[208] Y. Park, M.E. Fisher, Identity of the universal repulsive-core singularity with Yang-Lee edge criticality. Phys. Rev. E **60**, 6323 (1999) [condmat/9907429].

[226] A. Salam, J. Strathdee, Super-gauge transformations. Nucl. Phys. B **76**, 477 (1974)

[230] L. Schäfer, *Excluded Volume Effects in Polymer Solutions as Explained by the Renormalization Group* (Springer, Berlin, 1999)

[244] N. Sourlas, Introduction to Supersymmetry in condensed matter physics. Phys. D **15**, 115 (1985)

[247] M.J. Stephen, J.L. McCauley, Feynman graph expansion for tricritical exponents. Phys. Lett. **44A**, 89 (1973)

[249] M. Suzuki, A theory of the second order phase transition in spin systems. II Complex magnetic field. Prog. Theor. Phys. **38**, 1225 (1967)

[255] M. Tissier, G. Tarjus, Nonperturbative function renormalization group for random field models and related disordered systems. IV. Phys. Rev. B **85**, 104203 (2012)

[256] G.F. Tuthill, J.F. Nicoll, H.E. Stanley, Renormalization-group calculation of the critical-point exponent η for a critical point of arbitrary order. Phys. Rev. B **11**, 4579 (1975)

[266] F.J. Wegner, Exponents for critical points of higher order. Phys. Lett. **54A**, 1 (1975)

[267] F.J. Wegner, The critical state, general aspects in *Phase Transitions and Critical Phenomena*, vol. 6, ed. by C. Domb, M.S. Green (Academic Press, London, 1976), p. 7

[273] F. Wegner, Exact density of states for lowest Landau level in white noise potential. Superfield representation for interacting systems. Z. Phys. B **51**, 279 (1983)

[276] F. Wegner, Anomalous dimensions for the nonlinear sigma-model in $2 + \epsilon$ dimensions (I, II). Nucl. Phys. B **280**, 193, 210 [FS18] (1987)

[283] K.J. Wiese, Disordered systems and the functional renormalization group: a pedagogical introduction. Acta Phys. Slovaca **52**, 341 (2002)

[293] A.P. Young, On the lowering of dimensionality in phase transitions with random fields. J. Phys. C **10**, L257 (1977)

[294] A.P. Young, M. Nauenberg, Quasicritical behavior and first-order transition in the $d = 3$ random field Ising model. Phys. Rev. Lett. **54**, 2429 (1985)

Chapter 21
Random Matrix Theory

Abstract In this chapter the Gaussian random matrix ensembles are investigated. We determine their Green's functions and show that for small energy differences a soft mode appears. As a consequence, the non-linear sigma-model is introduced and the level correlations are determined.

21.1 Green's Functions

Green's functions and their products are introduced.

The Green's function between states $|\alpha\rangle$ and $|\beta\rangle$ is defined as

$$G(\alpha, \beta, z) = \langle\alpha|\frac{1}{z - H}|\beta\rangle \qquad (21.1)$$

It is obtained from the time-integrals ($\Re\eta > 0$)

$$i\int_{-\infty}^{0} d\,t e^{(i\omega+\eta-iH)t} = \frac{1}{\omega - i\eta - H}, \qquad (21.2)$$

$$-i\int_{0}^{+\infty} d\,t e^{(i\omega-\eta-iH)t} = \frac{1}{\omega + i\eta - H}. \qquad (21.3)$$

The upper Green's function is called advanced ($\Im z < 0$) and the lower one retarded ($\Im z > 0$). The density of states per orbital is obtained from the difference of both Green's functions

$$\rho(\alpha, E) = \lim_{\eta\to+0} (G(\alpha, \alpha, E - i\eta) - G(\alpha, \alpha, E + i\eta))/(2\pi i). \qquad (21.4)$$

In the following, we consider averaged products $\prod_{i=1}^{m} G(\alpha_i, \beta_i, z_i)$ of Green's functions of the random matrix model of Sect. 4.4. Since the distribution of matrix-elements (4.13) is invariant under unitary transformations of the matrix-elements $P(f) = P(U^\dagger f U)$, only averaged products of Green's functions differ from zero, if α_i and β_j agree pairwise. Thus, the only averaged one-particle Green's function different from 0 is $\overline{G(\alpha, \alpha, z)}$ and the only two-particle Green's functions different

© Springer-Verlag Berlin Heidelberg 2016
F. Wegner, *Supermathematics and its Applications in Statistical Physics*,
Lecture Notes in Physics 920, DOI 10.1007/978-3-662-49170-6_21

from zero are the averages of $\overline{G(\alpha, \alpha, z)G(\beta, \beta, z')}$ and $\overline{G(\alpha, \beta, z)G(\beta, \alpha, z')}$. Since the distribution of the matrix-elements is invariant under permutation of the αs, the averaged one- and two-particle Green's functions can be expressed in terms of the one-particle Green's function G and two-particle Green's functions K and K'

$$\overline{G(\alpha, \beta, z)} = \delta_{\alpha,\beta}G(z), \tag{21.5}$$

$$\overline{G(\alpha, \beta, z_1)G(\gamma, \delta, z_2)} = \delta_{\alpha\beta}\delta_{\gamma\delta}K(z_1, z_2) + \delta_{\alpha\delta}\delta_{\gamma\beta}K'(z_1, z_2). \tag{21.6}$$

The correlations between the various levels of the eigenstates is described by

$$n^2\tilde{K}(z_1, z_2) := \sum_{\alpha\beta}\overline{G(\alpha, \alpha, z_1)G(\beta, \beta, z_2)} = n^2K(z_1, z_2) + nK'(z_1, z_2) \tag{21.7}$$

with $\alpha, \beta = 1 \ldots n$. The combination

$$n\tilde{K}'(z_1, z_2) := \sum_{\alpha\beta}G(\alpha, \beta, z_1)G(\beta, \alpha, z_2)$$

$$= \sum_{\alpha,\beta}\langle\alpha|\frac{1}{z_1 - H}|\beta\rangle\langle\beta|\frac{1}{z_2 - H}|\alpha\rangle \tag{21.8}$$

$$= \sum_{\alpha}\langle\alpha|\frac{1}{(z_1 - H)(z_2 - H)}|\alpha\rangle$$

$$= \frac{1}{z_1 - z_2}\sum_{\alpha}(\langle\alpha|\frac{1}{z_2 - H}|\alpha\rangle - \langle\alpha|\frac{1}{z_1 - H}|\alpha\rangle)$$

yields

$$\tilde{K}'(z_1, z_2) = K(z_1, z_2) + nK'(z_1, z_2) = \frac{G(z_2) - G(z_1)}{z_1 - z_2} \tag{21.9}$$

and is thus determined by the one-particle Green's function.

21.2 Reduction of the Gaussian Unitary Ensemble to a Matrix Model

The determination of the averaged product of m Green's functions in the Gaussian unitary ensemble is reduced to the determination of correlations in a model of a 2m × 2m super-matrix.

Table 21.1 Gaussian random ensembles

Gaussian	Unitary	Orthogonal	Symplectic	Ensemble
Abbreviation	GUE	GOE	GSE	
Matrix elements	Complex	Real	Quaternion	Numbers
Ensemble invariant under	Unitary	Orthogonal	Symplectic unitary	Transformations

Gaussian Random Matrix Ensembles We consider three types of Gaussian ensembles listed in Table 21.1.

In all cases the matrices are hermitian. The matrix elements are independently distributed apart from the condition of hermiticity. Denoting the matrices by f, the probability distribution is proportional to $\exp(-c\,\mathrm{tr}\,(f^2))$. The average $(UfU^{-1})_{\alpha\beta}(UfU^{-1})_{\gamma\delta}$ does not depend on the transformation matrix U.

We return to the Gaussian unitary ensemble (GUE) whose density of states, or equivalently one-particle Green's functions, we have considered in Sect. 4.4. Here, we will mainly consider the two-particle Green's function, which yields the correlations between the energy levels.

We start from (4.12), but denote the complex components x and the Grassmann components ξ as components of the array S

$$x_\alpha^{(i)} = S_{\mathrm{b}\alpha}^{(i)} \in \mathscr{A}_0, \quad \xi_\alpha^{(i)} = S_{\mathrm{f}\alpha}^{(i)} \in \mathscr{A}_1. \tag{21.10}$$

Instead of the indices $_\mathrm{b}$ and $_\mathrm{f}$ we will sometimes use the Z_2-degree $v = 0, 1$. A product of Green's functions is expressed by

$$\prod_{i=1}^{m} G(\alpha_i, \beta_i, z_i) = \prod_{i=1}^{m} \left(\frac{s_{\mathrm{b}i}}{s_{\mathrm{f}i}}\right)^n \int \prod_{i,\alpha} \left(\frac{\mathrm{d}\,\Re S_{\mathrm{b}\alpha}^{(i)}\,\mathrm{d}\,\Im S_{\mathrm{b}\alpha}^{(i)}}{\pi}\mathrm{d}\,S_{\mathrm{f}\alpha}^{(i)*}\mathrm{d}\,S_{\mathrm{f}\alpha}^{(i)}\right)$$

$$\times \prod_{i}(s_{\mathrm{b}i}S_{\mathrm{b}\alpha_i}^{(i)} S_{\mathrm{b}\beta_i}^{(i)*})\exp(-\mathscr{S}_1(S, 0)), \tag{21.11}$$

$$\mathscr{S}_1(S, A) = \mathrm{str}((sz - \sqrt{s}A\sqrt{s})S^\dagger S - sS^\dagger fS) \tag{21.12}$$

$$= \sum_{i,v,\alpha,\beta} (-)^v s_{v,i} S_{v,\alpha}^{(i)*}(z_i\delta_{\alpha,\beta} - f_{\alpha,\beta})S_{v,\beta}^{(i)}$$

$$- \sum_{i,j} \sqrt{s_{\mathrm{b}i}}A_{i,j}\sqrt{s_{\mathrm{b}j}}S_{\mathrm{b}\alpha}^{(j)*} S_{\mathrm{b}\alpha}^{(i)}.$$

with $s_{\mathrm{b}i} = -\mathrm{isign}(\Im z_i)$, (4.11), where we use (3.35) or (13.33). For the moment we do not decide on the value of $s_{\mathrm{f}i}$. The last term is a source term, which allows one to determine the Green's functions. We introduce the source term only for the even components S_b. This is sufficient to determine the Green's functions.

The field S and the diagonal matrices s and z are given by

$$S = \begin{pmatrix} \cdots & & \cdots & \\ S_{b\alpha}^{(1)} \cdots S_{b\alpha}^{(m)} & S_{f\alpha}^{(1)} \cdots S_{f\alpha}^{(m)} \\ \cdots & & \cdots \end{pmatrix} \in \mathcal{M}(n,0,m,m), \qquad (21.13)$$

$$z = \mathrm{diag}(z_1,\ldots z_m, z_1,\ldots z_m), \quad s = \mathrm{diag}(s_{b1},\ldots s_{bm}, s_{f1},\ldots s_{fm}). \quad (21.14)$$

The indices i and ν of $S_{\nu\alpha}^{(i)}$ number the columns, and the subscript α the rows.

As in Sect. 4.4, we perform the average over the matrix elements f, which are Gaussian distributed with

$$\overline{f_{\alpha\beta}} = 0, \quad \overline{f_{\alpha\beta}f_{\gamma\delta}} = \frac{1}{ng}\delta_{\alpha\delta}\delta_{\beta\gamma}, \qquad (21.15)$$

which yields

$$\overline{\exp(-\mathscr{S}_1(S,A))} = \exp(-\mathrm{str}((sz - \sqrt{s}A\sqrt{s})S^\dagger S) + \frac{1}{2gn}\mathrm{str}((SsS^\dagger)^2)). \qquad (21.16)$$

Next the Hubbard-Stratonovich transformation has to be performed which, formally (i.e. without consideration of convergence requirements) reads (we use $\mathrm{str}((SsS^\dagger)^2) = \mathrm{str}((\sqrt{s}S^\dagger S\sqrt{s})^2)$),

$$\exp(\frac{1}{2gn}\mathrm{str}((\sqrt{s}S^\dagger S\sqrt{s})^2)) = \int [\mathrm{D}R]\exp(-\frac{ng}{2}\mathrm{str}(R^2) + \mathrm{str}(R\sqrt{s}S^\dagger S\sqrt{s})) \qquad (21.17)$$

with a super-matrix $R \in \mathcal{M}(m,m)$. Its precise form will be determined in Sect. 21.4. The Hubbard-Stratonovich transformation allows the reduction of the biquadratic interaction in S to a bilinear expression in S at the expense of a coupling between such a bilinear term in S and the new degree of freedom R. Now the averaged m-particle Green's function reads

$$\overline{\prod_{i=1}^{m} G(\alpha_i,\beta_i,z_i)} = \int [\mathrm{D}S][\mathrm{D}R]\prod_i (s_{bi}S_{b\alpha_i}^{(i)}S_{b\beta_i}^{(i)*})\exp(-\mathscr{S}_2(S,R,0)) \qquad (21.18)$$

with

$$\mathscr{S}_2(S,R,A) = \frac{ng}{2}\mathrm{str}(R^2) - \mathrm{str}(\sqrt{s}(R-z+A)\sqrt{s}S^\dagger S). \qquad (21.19)$$

The Green's functions can be determined from the 'partition function'

$$Z(A) = \prod_{i=1}^{m}\left(\frac{s_{bi}}{s_{fi}}\right)^n \int [\mathrm{D}S][\mathrm{D}R]\exp(-\mathscr{S}_2(S,R,A)) \qquad (21.20)$$

by taking the derivatives with respect to A,

$$\frac{\partial Z}{\partial A_{11}} = nG(z_1), \tag{21.21}$$

$$\frac{\partial^2 Z}{\partial A_{11} \partial A_{22}} = n^2 \tilde{K}(z_1, z_2) = n^2 K(z_1, z_2) + nK'(z_1, z_2), \tag{21.22}$$

$$\frac{\partial^2 Z}{\partial A_{12} \partial A_{21}} = n\tilde{K}'(z_1, z_2) = nK(z_1, z_2) + n^2 K'(z_1, z_2). \tag{21.23}$$

Next, the integration over the fields S is performed. All components, from $\alpha = 1$ to $\alpha = n$, yield the same contribution. Thus,

$$Z(A) = \int [DR] \exp(-n\mathscr{S}_3(R, A)), \quad \mathscr{S}_3(R, A) = \frac{g}{2} \operatorname{str}(R^2) + \ln \operatorname{sdet}(R - z + A). \tag{21.24}$$

We have thus reduced the calculation of the averaged Green's functions to the determination of the expectation values of a model of matrix R. However, we have not yet determined the manifold over which R varies. In order to obtain convergence for the vector field S, it was necessary to choose $s_{bi} = -\mathrm{i} \operatorname{sign} \Im z_i$. s_{fi} is not determined up to now.

21.3 Saddle Point

The saddle point of the matrix R is determined. The averaged one- and two-particle Green's functions are calculated. If the difference of the energies of the retarded and advanced Green's functions are small, then there is a saddle point manifold for the matrix R.

For large n, the integration starts by calculating the saddle point of $\mathscr{S}_3$ at $A = 0$. The saddle point equation reads

$$-gR^{(0)} - \frac{1}{R^{(0)} - z} = 0 \tag{21.25}$$

with a diagonal solution

$$R^{(0)}_{ki,k'i'} = \delta_{kk'} \delta_{ii'} R^{d}_i \tag{21.26}$$

given by

$$\sqrt{g}R^{d}_i = \frac{1}{\sqrt{g}(z_i - R^{d}_i)} = \frac{\sqrt{g}z_i}{2} \pm \mathrm{i}\sqrt{1 - \frac{gz_i^2}{4}} = \mathrm{e}^{\pm \mathrm{i}\phi_i}. \tag{21.27}$$

Including the diagonal source terms in the bosonic sector, one obtains

$$\frac{1}{\text{sdet}(R^d - z + A)} = \prod_{i=1}^{m} \frac{R_i^d - z_i}{R_i^d - z_i + A_{ii}}. \tag{21.28}$$

Thus,

$$\frac{d \frac{1}{\text{sdet}(R^d - z + A)}}{d A_{jj}} \Big|_{A=0} = \frac{1}{z_j - R_j^d} \frac{1}{\text{sdet}(R^d - z)}. \tag{21.29}$$

the saddle point solution for the one-particle Green's function is then

$$G(z_j) = \frac{1}{z_j - R_j^d} = \sqrt{g} e^{s_j \phi_j}. \tag{21.30}$$

The assignment of $s_j = -i\Im z_j$ is due to the discussion in Sect. 4.4. The last equality of (21.27) constitutes a mapping $z \to e^{\pm i\phi}$ with the assignment $\pm i = s$. The real interval $-2/\sqrt{g} \le z \le +2/\sqrt{g}$ is mapped onto the unit circle (ϕ is real): Real $z > 2/\sqrt{g}$ is mapped onto the real interval $(0, +1)$; real $z > -2/\sqrt{g}$ onto $(-1, 0)$. z with $\Im z > 0$ are mapped into the lower half unit circle and z with $\Im z < 0$ into the upper half unit circle.

The saddle point solution (21.30) yields, with (21.4), (21.27),

$$G(E - s_i 0) = g R_i^d = \frac{gE}{2} + s_i \pi \rho^{(0)}(E), \tag{21.31}$$

which gives Wigner's semi-circle law in accordance with (4.24), (4.25).

Now expand $\mathscr{S}_3(R, A)$ around the saddle point with

$$R_{vi,v'j} = \delta_{vv'} \delta_{ij} R_i^d + \frac{1}{\sqrt{g}} X_{vi,v'j}. \tag{21.32}$$

Then

$$\frac{g}{2} \text{str}(R^2) = \frac{1}{2} \text{str}(e^{2s\phi} + 2e^{s\phi} X + X^2), \tag{21.33}$$

$$\text{sdet}(R - z + A) = \text{sdet}(R^d - z)\text{sdet}(1 - e^{s\phi}(X + \sqrt{g}A)), \tag{21.34}$$

with the matrix $e^{s\phi} = \text{diag}(e^{s_i \phi_i})$. The superdeterminant can be rewritten

$$\text{sdet}(1 - \hat{X}) = \text{sdet}(\exp \ln(1 - \hat{X}))$$

$$= \text{sdet}(\exp(-(\hat{X} + \frac{1}{2}\hat{X}^2))) = \exp(-\text{str}(\hat{X} + \frac{1}{2}\hat{X}^2)), \tag{21.35}$$

where we have used (10.35) and expanded up to $\hat{X}^2$. This yields

$$\mathscr{S}_3(R, A) = -\sqrt{g}\,\text{str}(e^{s\phi}A) + \frac{1}{2}\,\text{str}(X^2 - (e^{s\phi}(X - \sqrt{g}A))^2). \qquad (21.36)$$

Completing the square for X, we obtain

$$\exp(-n\mathscr{S}_3(R,0)) = Z(A)\exp\left(-\frac{n}{2}\sum_{vi,v'j}\Pi_{vi,v'j}\tilde{X}_{vi,v'j}\tilde{X}_{v'j,vi}\right) \qquad (21.37)$$

with

$$\tilde{X}_{vi,v'j} = X_{vi,v'j} + \delta_{v0}\delta_{v'0}\frac{e^{s_i\phi_i + s_j\phi_j}}{\Pi_{vi,v'j}}A_{ij}, \qquad (21.38)$$

$$\Pi_{vi,v'j} = (-)^v(1 - e^{s_i\phi_i + s_j\phi_j}), \qquad (21.39)$$

$$Z(A) = \exp\left(n\sqrt{g}\,\text{str}(e^{s\phi}A) + \sum_{ij}\frac{ng}{2(e^{-s_i\phi_i - s_j\phi_j} - 1)}A_{ij}A_{ji}\right). \qquad (21.40)$$

The integral over $\tilde{X}$ yields unity, since the contributions from the bosonic and the fermionic components cancel. Thus, the 'partion function' $Z(A)$ is given by the terms left after completing the square.

The expression for Π vanishes for $s_1 = s_2$ at the band edges. This means, close to the band edge, one has to go beyond second order in X. It turns out that there are tails to the band. We will not pursue this further. If $s_1 = -s_2$, then Π vanishes for $\phi_1 = \phi_2$, i.e. for $z_1 = z_2$. This will lead to special behavior for the correlations of states energetically nearly degenerate.

Let us determine the Green's functions using the saddle point approximation from (21.40),

$$\frac{\partial Z}{\partial A_{ii}} = n\sqrt{g}e^{s_i\phi_i} = nG(z_i), \qquad (21.41)$$

$$\frac{\partial^2 Z}{\partial A_{11}\partial A_{22}} = n^2 g e^{s_1\phi_1 + s_2\phi_2} = n^2\tilde{K}(z_1, z_2) = n^2 G(z_1)G(z_2), \qquad (21.42)$$

$$\frac{\partial^2 Z}{\partial A_{12}\partial A_{21}} = \frac{ng}{e^{-s_1\phi_1 - s_2\phi_2} - 1} = n\tilde{K}'(z_1, z_2). \qquad (21.43)$$

$\tilde{K}$ factorizes in leading order, n^0. In later sections, the corrections to K for $\omega = O(1/n)$ will be calculated. They yield the level correlations. Setting

$$z_1 = E + \tfrac{1}{2}\omega, \quad z_2 = E - \tfrac{1}{2}\omega, \quad E \in \mathbb{R}, \quad \Im\omega > 0, \qquad (21.44)$$

then for small ω, one obtains, from (21.43),

$$\tilde{K}'(z_1, z_2) = \frac{2\pi i \rho(E)}{\omega}. \tag{21.45}$$

The divergence of $\tilde{K}'$ as ω approaches zero indicates a soft mode. Let us rewrite $\mathscr{S}_3(R, A)$ for small ω and A,

$$\mathscr{S}_3(R, A) = \frac{g}{2} \operatorname{str}(R^2) + \ln \operatorname{sdet}(R - E) + \ln \operatorname{sdet}(1 + \frac{1}{R - E}(A - \frac{\omega}{2}A)) \tag{21.46}$$

with $A = \operatorname{diag}(1, -1, 1, -1)$. Here, we have used the multiplication theorem

$$\operatorname{sdet}(R - E + X) = \operatorname{sdet}(R - E)\operatorname{sdet}(1 + (R - E)^{-1}X). \tag{21.47}$$

We learn from (21.46), that for vanishing ω, the saddle point equation reduces to

$$-gR^{(0)} - \frac{1}{R^{(0)} - E} = 0. \tag{21.48}$$

Thus, any matrix $R^{(0)}$ with eigenvalues $E/2 \pm i\sqrt{(1/g) - E^2/4}$ is a solution of the saddle point equation. Only the term containing ω introduces preferred saddle points.

21.4   Convergence and Symmetry

A hermitian matrix R_{bb} in the bosonic sector would yield divergences. Instead following symmetry considerations one integrates over a set of matrices, invariant under pseudounitary transformations.

In the following, we decompose R into submatrices indicated by upper indices ^{R,A} for retarded and advanced contributions, and lower indices $_{\nu=b,f}$ for bosonic and fermionic contributions,

$$R = \begin{pmatrix} R^{RR} & R^{RA} \\ R^{AR} & R^{AA} \end{pmatrix}, R^{\cdot\cdot} = \begin{pmatrix} R^{\cdot\cdot}_{bb} & R^{\cdot\cdot}_{bf} \\ R^{\cdot\cdot}_{fb} & R^{\cdot\cdot}_{ff} \end{pmatrix}, \tag{21.49}$$

$$R = \begin{pmatrix} R_{bb} & R_{bf} \\ R_{fb} & R_{ff} \end{pmatrix}, R_{\cdot\cdot} = \begin{pmatrix} R^{RR}_{\cdot\cdot} & R^{RA}_{\cdot\cdot} \\ R^{AR}_{\cdot\cdot} & R^{AA}_{\cdot\cdot} \end{pmatrix}. \tag{21.50}$$

The dimensions of the matrices $R^{\cdot\cdot}$ depend on the numbers m_R and m_A of retarded and advanced Green's functions considered in (21.11), $m = m_R + m_A$. Then for example

$$R^{RR} \in \mathscr{M}(m_R, m_R), \quad R_{bb} \in \mathscr{M}(m, 0), \quad R^{RR}_{bb} \in \mathscr{M}(m_R, 0). \tag{21.51}$$

Until now, we did not care about convergence requirements. We have to find out how the contours of integration should be chosen. The bosonic sector of the second contribution to $\mathscr{S}_2(S, R, 0)$, (21.19), reads

$$- \mathrm{str}(\sqrt{s}(R - z)\sqrt{s}S^\dagger S)_{\mathrm{bb}} \tag{21.52}$$

$$= - \mathrm{tr} \left(S_{\mathrm{b}}^{\mathrm{R}} \ S_{\mathrm{b}}^{\mathrm{A}} \right) \begin{pmatrix} -iR_{\mathrm{bb}}^{\mathrm{RR}} + i(E + \frac{\omega}{2})\mathbf{1} & R_{\mathrm{bb}}^{\mathrm{RA}} \\ R_{\mathrm{bb}}^{\mathrm{AR}} & iR_{\mathrm{bb}}^{\mathrm{AA}} - i(E - \frac{\omega}{2})\mathbf{1} \end{pmatrix} \begin{pmatrix} S_{\mathrm{b}}^{\mathrm{R}\,\dagger} \\ S_{\mathrm{b}}^{\mathrm{A}\,\dagger} \end{pmatrix},$$

where the trace runs over the indices α of S, (21.13). The S_{b} integrals converge for hermitian $R_{\mathrm{bb}}^{\mathrm{RR}}$ and $R_{\mathrm{bb}}^{\mathrm{AA}}$ or if $R_{\mathrm{bb}}^{\mathrm{RR}}$ is moved into the lower half of the complex plane, similarly if $R_{\mathrm{bb}}^{\mathrm{AA}}$ is moved into the upper half-plane, as suggested by the saddle point solution. There is, however, a problem with $R_{\mathrm{bb}}^{\mathrm{RA}}$ and $R_{\mathrm{bb}}^{\mathrm{AR}}$, if R_{bb} is assumed to be hermitian, since

$$S_{\mathrm{b}}^{\mathrm{R}} R_{\mathrm{bb}}^{\mathrm{RA}} S_{\mathrm{b}}^{\mathrm{A}\,\dagger} + S_{\mathrm{b}}^{\mathrm{A}} R_{\mathrm{bb}}^{\mathrm{AR}} S_{\mathrm{b}}^{\mathrm{R}\,\dagger} = 2\Re(S_{\mathrm{b}}^{\mathrm{R}} R_{\mathrm{bb}}^{\mathrm{RA}} S_{\mathrm{b}}^{\mathrm{A}\,\dagger}) \tag{21.53}$$

can have an arbitrary sign yielding divergent S_{b} integrals. If one multiplies $R_{\mathrm{bb}}^{\mathrm{RR}}$ and $R_{\mathrm{bb}}^{\mathrm{AA}}$ by i, then the integral over R diverges due to the R^2-term in $\mathscr{S}_2(S, R, 0)$.

Instead, one returns to the invariance under transformation of S. One observes

$$\mathrm{str}((SsS^\dagger)^2) = \mathrm{str}((SWsW^\dagger S^\dagger)^2), \quad WsW^\dagger = s, \tag{21.54}$$

from (21.16), under the pseudounitary transformation W. This yields the transformation

$$\mathrm{str}(\sqrt{s}R\sqrt{s}S^\dagger S) \rightarrow \mathrm{str}(W\sqrt{s}R\sqrt{s}W^\dagger S^\dagger S). \tag{21.55}$$

This suggests, to have the saddle point manifold as part of the domain of integration,

$$\sqrt{s}R^{(0)}\sqrt{s} = W\sqrt{s}R^{\mathrm{d}}\sqrt{s}W^\dagger. \tag{21.56}$$

We introduce $T = \sqrt{s}^{-1}W\sqrt{s}$. From $WsW^\dagger = s$ and $T = \sqrt{s}^{-1}W\sqrt{s}$, one derives $TsT^\dagger = s$. Thus, T is also pseudounitary,

$$R^{(0)} = TR^{\mathrm{d}}T^{-1}, \quad TsT^\dagger = s. \tag{21.57}$$

In the RA-representation, we may write

$$R^{\mathrm{d}} = \frac{E}{2} - \frac{i\pi\rho^{(0)}}{g}\Lambda, \quad \Lambda = \mathrm{diag}(\mathrm{sign}(\Im z)) = \begin{pmatrix} \mathbf{1}_{2m_{\mathrm{R}}} & 0 \\ 0 & -\mathbf{1}_{2m_{\mathrm{A}}} \end{pmatrix}. \tag{21.58}$$

Then one obtains

$$R^{(0)} = \frac{E}{2} - \frac{i\pi\rho^{(0)}}{g}Q(T), \quad Q = TAT^{-1}. \tag{21.59}$$

This representation of Q yields

$$Q^2 = \mathbf{1}, \quad Q^\dagger = s^{-1}Qs = sQs^{-1}. \tag{21.60}$$

Since $s^{-1} = -s$, both expressions for $Q^\dagger$ in the above equation apply.

The domain of integration has to be completed. Before Efetov [64, 65] introduced the supersymmetric formulation in 1982, the non-linear σ-model in d dimensions was introduced by means of the replica trick in 1979 and 1980. It was first suggested in [270] and worked out by Schäfer and Wegner [231] (SW-parametrization). Pruisken and Schäfer [219, 220] used a different parametrization (PS-parametrization). Both parametrizations used the pseudounitary (hyperbolic) non-compact symmetry. Efetov et al. [67] used the replica formulation with Grassmann variables, S_f, and compact symmetry, which is appropriate for the fermionic sector, as we will see. Obviously, the S-integrations are without problems in this formulation. Similar approaches are found in [112, 180, 181]. These calculations were applied to extended, that is d-dimensional, systems. We will consider these systems in the next chapter. Here we consider only the zero-dimensional case.

Both SW and PS use a matrix, P, besides the saddle point manifold, which consists of two blocks, $P^{(R)}$ and $P^{(A)}$, for the retarded and the advanced sector, resp.

$$P = \begin{pmatrix} P^{(R)} & 0 \\ 0 & P^{(A)} \end{pmatrix}, \quad P^{(R)} \in \mathcal{M}(m_R, m_R), \quad P^{(A)} \in \mathcal{M}(m_A, m_A) \tag{21.61}$$

with

$$P^{(\cdot)} = \begin{pmatrix} P^{(\cdot)}_{bb} & P^{(\cdot)}_{bf} \\ P^{(\cdot)}_{fb} & P^{(\cdot)}_{ff} \end{pmatrix}, \tag{21.62}$$

but in different ways:

$$R = \frac{E}{2} - \frac{i\pi\rho^{(0)}}{g}Q(T) + \begin{cases} P & \text{SW} \\ TPT^{-1} & \text{PS} \end{cases}. \tag{21.63}$$

In all cases $P^{(\cdot)}_{bb}$ is hermitian and $P^{(\cdot)}_{ff}$ is antihermitian.

Convergence for the S_b The second term of $\mathscr{S}_2$, in (21.19), contains the S_b dependence and may be written

$$- \operatorname{tr}((S\sqrt{s})_b((R-z)s)_{bb}(\sqrt{s^\dagger}S^\dagger)_b). \tag{21.64}$$

Thus, we have to show that the (ordinary part of the) hermitian contribution of the matrix $-((R - z)s)_{bb}$ is positive. s_b has been chosen such that the real part of $(zs)_{bb}$ is positive.

$-Rs$ contains $i\pi\rho Q/g$. With $Q = T\Lambda T^{-1}$, $T^{-1} = sT^\dagger s^{-1}$ and $(\Lambda s)_{bb} = -i\mathbf{1}$, one obtains

$$i(Qs)_{bb} = iT\Lambda sT^\dagger = T_{bb}T_{bb}^\dagger + \text{ nilpotent terms from } T_{bf}. \qquad (21.65)$$

Thus, the ordinary part of $i(Qs)_{bb}$ is positive definite.

The ordinary parts of the contributions $-(Ps)_{bb}$ and $-(TPT^{-1}s)_{bb} = -(TPsT^\dagger)_{bb}$ yield purely imaginary contributions. Thus, convergence of the S_b integrals is guaranteed.

Various Cases If we are interested only in Green's functions or products of Green's functions, which are either all retarded ($m_A = 0$) or all advanced ($m_R = 0$), then R^{AR} and R^{RA} are empty matrices and convergence is guaranteed by the requirements on R^{RR} and R^{AA} given above. The difficult and interesting case is that of averaged products of retarded and advanced Green's functions. A detailed discussion of this will follow. From now on we will restrict to one- and two-particle Green's functions, thus $m = m_R + m_A$ will not exceed 2.

21.5 Nonlinear σ-Model

The steps leading to the nonlinear σ-model are performed. This model is expressed as a model of the soft-mode matrices Q in (21.63), whereas the massive modes described by the matrices P are eliminated. The model is given in 21.5.6, where the parametrization of 21.5.1 and the invariant measure (21.82) are used.

21.5.1 Efetov Parametrization

We first consider the saddle point manifold. Efetov [66] has given a useful parametrization of Q for the interesting case $m_R = m_A = 1$. The choice for s is

$$(s_b^R, s_f^R, s_b^A, s_f^A) = (-i, -i, i, -i). \qquad (21.66)$$

We will explain, after (21.82), why $s_f^R = s_f^A$ has been chosen. Efetov writes

$$T = UV, U^{\pm 1} = \begin{pmatrix} 1 - \frac{1}{2}\alpha^*\alpha & \mp\alpha^* & 0 & 0 \\ \pm\alpha & 1 - \frac{1}{2}\alpha\alpha^* & 0 & 0 \\ 0 & 0 & 1 + \frac{1}{2}\beta^*\beta & \mp i\beta^* \\ 0 & 0 & \pm i\beta & 1 + \frac{1}{2}\beta\beta^* \end{pmatrix},$$

$$V^{\pm 1} = \begin{pmatrix} \lambda_1' & 0 & \mp i\mu_1' & 0 \\ 0 & \lambda_2' & 0 & \mp\mu_2'^* \\ \pm i\mu_1'^* & 0 & \lambda_1' & 0 \\ 0 & \pm\mu_2' & 0 & \lambda_2' \end{pmatrix} \qquad (21.67)$$

with

$$\lambda_1' = \cosh\frac{\theta_1}{2}, \lambda_2' = \cos\frac{\theta_2}{2},$$

$$\mu_1' = \sinh\frac{\theta_1}{2}e^{i\phi_1}, \mu_2' = \sin\frac{\theta_2}{2}e^{i\phi_2}. \qquad (21.68)$$

Then the matrix Q reads

$$Q = T\Lambda T^{-1} = U \begin{pmatrix} \lambda_1 & 0 & i\mu_1 & 0 \\ 0 & \lambda_2 & 0 & \mu_2^* \\ i\mu_1^* & 0 & -\lambda_1 & 0 \\ 0 & \mu_2 & 0 & -\lambda_2 \end{pmatrix} U^{-1} \qquad (21.69)$$

with

$$\lambda_1 = \cosh\theta_1, \lambda_2 = \cos\theta_2,$$

$$\mu_1 = \sinh\theta_1 e^{i\phi_1}, \mu_2 = \sin\theta_2 e^{i\phi_2}. \qquad (21.70)$$

The parameters run in the intervals $\theta_1 = [0..\infty)$, $\theta_2 = [0..\pi]$, $\phi_i = [0..2\pi)$. Thus, the range for the λ_i is $\lambda_1 = [1..\infty)$, $\lambda_2 = [-1..+1]$.

21.5.2   Invariant Measure

The basic aim is to reduce the interaction (21.24) into one in terms of T. The interesting question is: Is $\mathscr{S}_3$, in the limit $\omega \to 0$ and for vanishing source terms A, invariant under rotations by T? Indeed, if we substitute (21.59) into (21.24), then in this limit $\mathscr{S}_3$ does not depend on T. However, we have also to consider what happens to $[DR]$. It has to be replaced by

$$[D\mu(T)] = Id\lambda_1 d\lambda_2 d\phi_1 d\phi_2 d\alpha d\alpha^* d\beta d\beta^*, \qquad (21.71)$$

where I is a function of the variables $\lambda_1 \ldots \beta^*$. This function must have the property that the infinitesimal volume element $[D\,\mu(T)]$, rotated to $T^{(0)} = \mathbf{1}$, does not depend on the variables $\lambda_1 \ldots \beta^*$. Then, $[D\,\mu(T)]$ is the invariant measure. Thus, with

$$Q + dQ = (T + dT)\Lambda(T^{-1} + dT^{-1}) = T(\Lambda + dQ^{(0)})T^{-1} \qquad (21.72)$$

we obtain

$$dQ^{(0)} = T^{-1}dT\Lambda + \Lambda dT^{-1}T = [T^{-1}dT, \Lambda] = 2\begin{pmatrix} 0 & -(T^{-1}dT)^{\mathrm{RA}} \\ (T^{-1}dT)^{\mathrm{AR}} & 0 \end{pmatrix}. \qquad (21.73)$$

since $d(T^{-1}T) = d\mathbf{1} = 0 = T^{-1}dT + dT^{-1}T$. With $T = UV$, one obtains

$$T^{-1}dT = V^{-1}U^{-1}d\,UV + V^{-1}d\,V. \qquad (21.74)$$

With

$$V^{\pm 1} = \begin{pmatrix} V^{\mathrm{D}} & \pm V^{\mathrm{RA}} \\ \pm V^{\mathrm{AR}} & V^{\mathrm{D}} \end{pmatrix}, \qquad (21.75)$$

we obtain

$$V^{-1}d\,V = \begin{pmatrix} \cdots & V^{\mathrm{D}}d\,V^{\mathrm{RA}} - V^{\mathrm{RA}}d\,V^{\mathrm{D}} \\ -V^{\mathrm{AR}}d\,V^{\mathrm{D}} + V^{\mathrm{D}}d\,V^{\mathrm{AR}} & \cdots \end{pmatrix}, \qquad (21.76)$$

with

$$V^{\mathrm{D}}d\,V^{\mathrm{RA}} - V^{\mathrm{RA}}d\,V^{\mathrm{D}} = \begin{pmatrix} -i\lambda_1'd\,\mu_1' + i\mu_1'd\,\lambda_1' & 0 \\ 0 & -\lambda_2'd\,\mu_2'^* + \mu_2'^*d\,\lambda_2' \end{pmatrix}, \qquad (21.77)$$

$$-V^{\mathrm{AR}}d\,V^{\mathrm{D}} + V^{\mathrm{D}}d\,V^{\mathrm{AR}} = \begin{pmatrix} -i\mu_1'^*d\,\lambda_1' + i\lambda_1'd\,\mu_1'^* & 0 \\ 0 & -\mu_2'd\,\lambda_2' + \lambda_2'd\,\mu_2' \end{pmatrix}. \qquad (21.78)$$

Since the off-diagonal matrix elements of these matrices vanish, it is sufficient to determine the off-diagonal elements of $V^{-1}U^{-1}d\,UV$ only. One obtains

$$(V^{-1}U^{-1}d\,UV)^{\mathrm{RA}} = \begin{pmatrix} \cdots & \lambda_1'\mu_2'^*d\,\alpha^* + \mu_1'\lambda_2'd\,\beta^* \\ -i\mu_1'\lambda_2'd\,\alpha - i\lambda_1'\mu_2'^*d\,\beta & \cdots \end{pmatrix}, \qquad (21.79)$$

$$(V^{-1}U^{-1}d\,UV)^{\mathrm{AR}} = \begin{pmatrix} \cdots & i\mu_1^*\lambda_2'd\,\alpha^* - i\lambda_1'\mu_2'd\,\beta^* \\ -\lambda_1'\mu_2'd\,\alpha - \lambda_2'\mu_1'^*d\,\beta & \cdots \end{pmatrix}. \qquad (21.80)$$

The dots ... indicate matrix elements we need not calculate. We denote

$$
d Q^{(0)RA} = \begin{pmatrix} id\,\hat{\mu}_1 & -d\,\kappa_1^* \\ id\,\kappa_2 & d\,\hat{\mu}_2^* \end{pmatrix}, \quad d Q^{(0)AR} = \begin{pmatrix} id\,\hat{\mu}_1^* & id\,\kappa_2^* \\ -d\,\kappa_1 & d\,\hat{\mu}_2 \end{pmatrix}. \tag{21.81}
$$

Then Eqs. (21.77)–(21.80) and Eqs. (21.68), (21.70) allow us to determine the factor I of the invariant measure

$$
\begin{aligned}
I &= \frac{\partial(\hat{\mu}_1, \hat{\mu}_1^*, \hat{\mu}_2, \hat{\mu}_2^*, \kappa_1^*, \kappa_2^*, \kappa_1, \kappa_2)}{\partial(\lambda_1, \phi_1, \lambda_2, \phi_2, \alpha^*, \beta^*, \alpha, \beta)} \\[2mm]
&= \frac{\partial(\hat{\mu}_1, \hat{\mu}_1^*, \hat{\mu}_2, \hat{\mu}_2^*)}{\partial(\lambda_1, \phi_1, \lambda_2, \phi_2)} \frac{\partial(\kappa_1^*, \kappa_2^*, \kappa_1, \kappa_2)}{\partial(\alpha^*, \beta^*, \alpha, \beta)} = \frac{1}{(\lambda_1 - \lambda_2)^2}.
\end{aligned} \tag{21.82}
$$

The invariant measure $[D\,\mu(t)]$, also called Haar-measure, is determined up to a constant factor, which can be chosen arbitrarily. The invariant measure becomes singular at $\lambda_1 = \lambda_2$, that is, for $\lambda_1 = \lambda_2 = +1$. If one had chosen $s_f^R = -s_f^A = \pm i$, then the factor I for invariant measure would also be $(\lambda_1 - \lambda_2)^{-2}$, but both λ_1 and λ_2 would run from 1 to ∞, causing convergence problems for the integrals. This is the reason for the choice $s_f^R = s_f^A$ in (21.66).

21.5.3   Singularity of the Invariant Measure

One observes that the factor I of the invariant measure (21.82) diverges at $\lambda_1 = \lambda_2 = 1$. Moreover, it turns out that, since it does not contain factors of the odd variables, the partition function would vanish. Thus, we have to look closer at this limit. For this purpose, we first express the λs in terms of the matrix

$$
Q^{RR} = \begin{pmatrix} \lambda_1 - \alpha^*\alpha(\lambda_1 - \lambda_2) & \alpha^*(\lambda_1 - \lambda_2) \\ \alpha(\lambda_1 - \lambda_2) & \lambda_2 - \alpha^*\alpha(\lambda_1 - \lambda_2) \end{pmatrix}, \tag{21.83}
$$

which yields

$$
\lambda_1 = Q_{bb}^{RR} + \frac{Q_{bf}^{RR} Q_{fb}^{RR}}{Q_{bb}^{RR} - Q_{ff}^{RR}}, \quad \lambda_2 = Q_{ff}^{RR} + \frac{Q_{bf}^{RR} Q_{fb}^{RR}}{Q_{bb}^{RR} - Q_{ff}^{RR}}. \tag{21.84}
$$

In a next step, we express the Q^{RR} in terms of the Q^{RA} and Q^{AR} by means of $Q^2 = \mathbf{1}$. In a first step, one finds

$$
\begin{aligned}
\gamma_b &:= \mathrm{ord}\,(Q_{bb}^{RR}) = \sqrt{1+A}, \quad A := -Q_{bb}^{RA} Q_{bb}^{AR}, \\
\gamma_f &:= \mathrm{ord}\,(Q_{ff}^{RR}) = \sqrt{1-B}, \quad B := Q_{ff}^{RA} Q_{ff}^{AR}.
\end{aligned} \tag{21.85}
$$

A is positive, since $Q_{bb}^{RA} = -Q_{bb}^{AR}{}^{*}$ One expands in the odd elements Q_{bf} and Q_{fb}. One obtains after some straightforward, but lengthy, calculation

$$\lambda_1 = \gamma_b - \frac{A}{2\gamma_b(A+B)}(K_1 + K_2)$$

$$- \frac{1}{2\gamma_b(A+B)}K_3 - \frac{3A^2 + 2A - AB - 2B}{4\gamma_b^3(A+B)^2}K_1 K_2, \qquad (21.86)$$

$$\lambda_2 = \gamma_f + \frac{B}{2\gamma_f(A+B)}(K_1 + K_2)$$

$$- \frac{1}{2\gamma_f(A+B)}K_3 - \frac{3B^2 - 2B - AB + 2A}{4\gamma_f^3(A+B)^2}K_1 K_2 \qquad (21.87)$$

with

$$K_1 = Q_{fb}^{AR} Q_{bf}^{RA}, \quad K_2 = Q_{fb}^{RA} Q_{bf}^{AR},$$

$$K_3 = -Q_{bb}^{RA} Q_{ff}^{RA} Q_{bf}^{AR} Q_{fb}^{AR} - Q_{bb}^{AR} Q_{ff}^{AR} Q_{bf}^{RA} Q_{fb}^{RA}. \qquad (21.88)$$

The expansions of λ_1 and λ_2 have the form given in Sect. 15.3, where we identify $W = Q^{RA}$, $K_1 = \alpha^* \alpha$, $K_2 = \beta^* \beta$, $K_3 = ab\beta^* \alpha^* + a^* b^* \alpha \beta$, $t = 1$, and $u = -1$. λ_1 and λ_2 obey (15.72). One can show that any function $f(\lambda_1, \lambda_2)$ that can be differentiated twice, obeys this equation. The functions are integrable for positive A and B and thus, yield the value of the function f at $A = B = 0$, that is, at $\lambda_1 = \lambda_2 = 1$. This contribution has to be added to what one obtains from the usual integration over the odd and even variables. As a consequence it yields the correct partition function Z in Sect. 21.5.6.

21.5.4 Schäfer-Wegner Parametrization

The parametrization by Schäfer and Wegner, (21.63), uses

$$R = \tfrac{1}{2}E - \frac{i\pi \rho^{(0)}}{g}Q + P. \qquad (21.89)$$

The real part of the quadratic term $ng \, \mathrm{str}(R^2)/2$ in $\mathscr{S}_2(S, R, A)$ yields, with real $P_{bb}^{(.)}$ and imaginary $P_{ff}^{(.)}$,

$$\Re \mathrm{ord} \, \mathrm{str}(R^2) = (P_{bb}^{(R)} + \tfrac{1}{2}E)^2 + (P_{bb}^{(A)} + \tfrac{1}{2}E)^2 + (iP_{ff}^{(R)} + \frac{\pi \rho^{(0)}}{g}\lambda_2)^2$$

$$+ (iP_{ff}^{(A)} - \frac{\pi \rho^{(0)}}{g}\lambda_2)^2 - \tfrac{1}{2}E^2 - 2(\frac{\pi \rho^{(0)}}{g})^2 \lambda_2^2, \qquad (21.90)$$

which guarantees convergence of the P-integrals. Convergence for the T integration comes about by the imaginary parts of z.

Transformation from R to P and T   Since R^{RA} and R^{AR} do not depend on $P^{(R)}$ and $P^{(A)}$ and $R^{RR} = \frac{1}{2}E - i\pi\rho^{(0)}Q^{RR}/g + R^{(R)}$, similarly for R^{AA}, the Berezinian reduces to

$$\frac{\partial(R)}{\partial(P^{(R)}, P^{(A)}, T)} = \frac{\partial(Q^{RA}, Q^{AR})}{\partial(T)}. \tag{21.91}$$

After some lengthy but straightforward calculation, one finds

$$[\mathrm{D}\,R] = [\mathrm{D}\,P^{(R)}][\mathrm{D}\,P^{(A)}]\mathrm{d}\,\mu_1\mathrm{d}\,\mu_1^*\mathrm{d}\,\mu_2\mathrm{d}\,\mu_2^*\mathrm{d}\,\alpha^*\mathrm{d}\,\alpha\mathrm{d}\,\beta^*\mathrm{d}\,\beta\frac{1}{(\lambda_1^2 - \lambda_2^2)^2}. \tag{21.92}$$

The transformations

$$\frac{\partial(\mu_i^*, \mu_i)}{\partial(\lambda_i, \phi_i)} = 2i\lambda_i \tag{21.93}$$

do not yield the invariant measure.

Fluctuation Contributions   What have to be taken into account are the fluctuation contributions. Up to second order in P, one obtains the contribution

$$\exp\left(-\frac{ng}{2}\left(\mathrm{str}(P^2) - g^{-1}\,\mathrm{str}(P\frac{1}{R^0 - E}P\frac{1}{R^0 - E})\right)\right). \tag{21.94}$$

With

$$gR^0 = \frac{-1}{R^0 - E}, \quad \sqrt{g}R^0 = \tilde{\epsilon} - i\tilde{\rho}Q, \quad \tilde{\epsilon} := \frac{E\sqrt{g}}{2},$$

$$\tilde{\rho} := \frac{\pi\rho^0}{\sqrt{g}}, \quad \tilde{\epsilon}^2 + \tilde{\rho}^2 = 1, \quad \tilde{P} = UPU^{-1}. \tag{21.95}$$

we obtain

$$\mathrm{str}(P^2) - g^{-1}\,\mathrm{str}(P\frac{1}{R^0 - E}P\frac{1}{R^0 - E}) \tag{21.96}$$

$$= \mathrm{str}(\tilde{P}^2) - \mathrm{str}((\tilde{P}(\tilde{\epsilon}\mathbf{1} - i\tilde{\rho}V\Lambda V^{-1}))^2)$$

$$= \left(\tilde{P}_{bb}^{(R)}\ \tilde{P}_{bb}^{(A)}\right)\begin{pmatrix} 1 - (\tilde{\epsilon} - i\tilde{\rho}\lambda_1)^2 & -\tilde{\rho}^2\mu_1\mu_1^* \\ -\tilde{\rho}^2\mu_1\mu_1^* & 1 - (\tilde{\epsilon} + i\tilde{\rho}\lambda_1)^2 \end{pmatrix}\begin{pmatrix} \tilde{P}_{bb}^{(R)} \\ \tilde{P}_{bb}^{(A)} \end{pmatrix} \tag{21.97}$$

$$- \left(\tilde{P}_{ff}^{(R)}\ \tilde{P}_{ff}^{(A)}\right)\begin{pmatrix} 1 - (\tilde{\epsilon} - i\tilde{\rho}\lambda_2)^2 & \tilde{\rho}^2\mu_2\mu_2^* \\ \tilde{\rho}^2\mu_2\mu_2^* & 1 - (\tilde{\epsilon} + i\tilde{\rho}\lambda_2)^2 \end{pmatrix}\begin{pmatrix} \tilde{P}_{ff}^{(R)} \\ \tilde{P}_{ff}^{(A)} \end{pmatrix}$$

$$+2 \left(\tilde{P}_{\mathrm{bf}}^{(R)} \ \tilde{P}_{\mathrm{bf}}^{(A)} \right) \begin{pmatrix} 1 - (\tilde{\epsilon} - i\tilde{\rho}\lambda_1)(\tilde{\epsilon} - i\tilde{\rho}\lambda_2) & i\tilde{\rho}^2 \mu_1^* \mu_2^* \\ i\tilde{\rho}^2 \mu_1 \mu_2 & 1 - (\tilde{\epsilon} + i\tilde{\rho}\lambda_1)(\tilde{\epsilon} + i\tilde{\rho}\lambda_2) \end{pmatrix}$$

$$\times \begin{pmatrix} \tilde{P}_{\mathrm{fb}}^{(R)} \\ \tilde{P}_{\mathrm{fb}}^{(A)} \end{pmatrix}.$$

The determinants of the last three matrices are

$$4\tilde{\rho}^2 \lambda_1^2, \quad 4\tilde{\rho}^2 \lambda_2^2, \quad \tilde{\rho}^2 (\lambda_1 + \lambda_2)^2. \tag{21.98}$$

Thus, the fluctuation contributions yield a factor of

$$\frac{(\lambda_1 + \lambda_2)^2}{4\lambda_1 \lambda_2}. \tag{21.99}$$

Combined with the Berezinians (21.92) and (21.93), we obtain the invariant measure (21.82).

21.5.5 Pruisken-Schäfer Parametrization

Pruisken and Schäfer suggested a different parametrization,

$$R = \frac{E}{2} - \frac{i\pi\rho^{(0)}}{g} Q(T) + TPT^{-1}. \tag{21.100}$$

The invariance under transformations with T is manifest for this parametrization in contrast to the SW-parametrization. However, it has the drawback that the function F becomes singular (see below). Again $P_{\mathrm{bb}}^{(\cdot)}$ is real and $P_{\mathrm{ff}}^{(\cdot)}$ is imaginary.

Variation of T, for small ω, only has a small effect on $\mathscr{S}_3(R, 0)$, since $\mathrm{str}(R^2)$ and $\ln \mathrm{sdet}(R - E)$ of (21.46) are invariant under the similarity transformation by T. Variation of P, however, has a strong effect. Thus, variation of T is related to soft (massless or nearly massless) modes, and variation of P to massive modes.

Berezinian from R to P and T $\quad \mathrm{d}R$ can be written as

$$\mathrm{d}R = T \underbrace{\left([T^{-1}\mathrm{d}T, \tilde{P}] + \mathrm{d}P\right)}_{D} T^{-1}, \quad \tilde{P} = P - \frac{i\pi\rho^{(0)}}{g}\Lambda. \tag{21.101}$$

From (10.2), we learn that the Berezinian from $\mathrm{d}R$ to D equals $\mathrm{sdet}(TT^{-1})^0 = 1$.

Since $[T^{-1}\mathrm{d}\,T, \tilde{P}]$ only has contributions in the sectors ^{RA} and ^{AR}, and P in the sectors ^{RR} and ^{AA}, one has

$$
\begin{pmatrix} D^{\mathrm{RR}} \\ D^{\mathrm{AA}} \\ D^{\mathrm{RA}} \\ D^{\mathrm{AR}} \end{pmatrix} = \begin{pmatrix} 1 & 0 & \cdots & \cdots \\ 0 & 1 & \cdots & \cdots \\ 0 & 0 & M^{\mathrm{RA}} & 0 \\ 0 & 0 & 0 & M^{\mathrm{AR}} \end{pmatrix} \begin{pmatrix} \mathrm{d}\,P^{(\mathrm{R})} \\ \mathrm{d}\,P^{(\mathrm{A})} \\ (T^{-1}\mathrm{d}\,T)^{\mathrm{RA}} \\ (T^{-1}\mathrm{d}\,T)^{\mathrm{AR}} \end{pmatrix} \tag{21.102}
$$

with

$$
M^{\mathrm{RA}} = \begin{pmatrix} \tilde{P}^{(\mathrm{A})}_{\mathrm{bb}} - \tilde{P}^{(\mathrm{R})}_{\mathrm{bb}} & 0 & -P^{(\mathrm{A})}_{\mathrm{fb}} & -P^{(\mathrm{R})}_{\mathrm{bf}} \\ 0 & \tilde{P}^{(\mathrm{A})}_{\mathrm{ff}} - \tilde{P}^{(\mathrm{R})}_{\mathrm{ff}} & -P^{(\mathrm{R})}_{\mathrm{fb}} & -P^{(\mathrm{A})}_{\mathrm{bf}} \\ P^{(\mathrm{A})}_{\mathrm{bf}} & -P^{(\mathrm{R})}_{\mathrm{bf}} & \tilde{P}^{(\mathrm{A})}_{\mathrm{ff}} - \tilde{P}^{(\mathrm{R})}_{\mathrm{bb}} & 0 \\ -P^{(\mathrm{R})}_{\mathrm{fb}} & P^{(\mathrm{A})}_{\mathrm{fb}} & 0 & \tilde{P}^{(\mathrm{A})}_{\mathrm{bb}} - \tilde{P}^{(\mathrm{R})}_{\mathrm{ff}} \end{pmatrix}. \tag{21.103}
$$

The result is

$$
[D\,R] = F^2(P^{(\mathrm{R})}, P^{(\mathrm{A})})[D\,P^{(\mathrm{R})}][D\,P^{(\mathrm{A})}][D\,\mu(T)], \tag{21.104}
$$

$$
[D\,\mu(T)] = -\frac{\mathrm{d}\,\lambda_1 \mathrm{d}\,\lambda_2}{(\lambda_1 - \lambda_2)^2} \frac{\mathrm{d}\,\phi_1 \mathrm{d}\,\phi_2}{(2\pi)^2} \mathrm{d}\,\alpha \mathrm{d}\,\alpha^* \mathrm{d}\,\beta \mathrm{d}\,\beta^*, \tag{21.105}
$$

with

$$
F(P^{(\mathrm{R})}, P^{(\mathrm{A})}) = \mathrm{sdet}(M^{\mathrm{RA}}) = \mathrm{sdet}(M^{\mathrm{AR}}). \tag{21.106}
$$

$F(P)$ can be expressed in terms of the eigenvalues q of the matrices P (for eigenvalues compare with (11.16)),

$$
q^{(\mathrm{R})}_{\mathrm{b}} = \tilde{P}^{(\mathrm{R})}_{\mathrm{bb}} - \frac{P^{(\mathrm{R})}_{\mathrm{bf}} P^{(\mathrm{R})}_{\mathrm{fb}}}{\tilde{P}^{(\mathrm{R})}_{\mathrm{ff}} - \tilde{P}^{(\mathrm{R})}_{\mathrm{bb}}}, \qquad q^{(\mathrm{R})}_{\mathrm{f}} = \tilde{P}^{(\mathrm{R})}_{\mathrm{ff}} - \frac{P^{(\mathrm{R})}_{\mathrm{bf}} P^{(\mathrm{R})}_{\mathrm{fb}}}{\tilde{P}^{(\mathrm{R})}_{\mathrm{ff}} - \tilde{P}^{(\mathrm{R})}_{\mathrm{bb}}}, \tag{21.107}
$$

and similarly for $q^{(\mathrm{A})}_{\mathrm{b,f}}$. Then one obtains

$$
F(P^{(\mathrm{R})}, P^{(\mathrm{A})}) = \frac{(q^{(\mathrm{A})}_{\mathrm{b}} - q^{(\mathrm{R})}_{\mathrm{b}})(q^{(\mathrm{A})}_{\mathrm{f}} - q^{(\mathrm{R})}_{\mathrm{f}})}{(q^{(\mathrm{A})}_{\mathrm{b}} - q^{(\mathrm{R})}_{\mathrm{f}})(q^{(\mathrm{A})}_{\mathrm{f}} - q^{(\mathrm{R})}_{\mathrm{b}})}. \tag{21.108}
$$

(The corresponding F for the orthogonal case was determined in [261]). Thus, the integration over the Ps and T factorizes. This is due to the invariance under the pseudounitary transformations. The measure $[D\,\mu(T)]$ is the invariant measure of the coset space $\mathrm{UPL}(1,1,2,0)/\mathrm{UPL}(1,0,1,0) \times \mathrm{UPL}(0,1,1,0)$. However, the denominator of F vanishes for $q^{(\mathrm{A})}_{\mathrm{b}} = q^{(\mathrm{R})}_{\mathrm{f}}$ and $q^{(\mathrm{A})}_{\mathrm{f}} = q^{(\mathrm{R})}_{\mathrm{b}}$. One cannot escape these points upon integration, since q_{b} is integrated in the real direction, whereas q_{f}

is integrated in the imaginary direction. The question arises, as to whether this yields distributions. Recent investigations were conducted by Fyodorov, Wei, Zirnbauer, and Müller-Hill [91, 92, 194, 277].

21.5.6 The Nonlinear σ-Model Finally

After integration over the massive modes contained in the matrices P, one is left with the action

$$\mathscr{S}_3(R^{(0)}, A) = \frac{g}{2} \operatorname{str}(R^{(0)2}) + \ln \operatorname{sdet}(R^{(0)} - E + \tfrac{1}{2}\omega\Lambda + A). \tag{21.109}$$

The first term vanishes identically. The second term can be rewritten as

$$\ln \operatorname{sdet}(R^{(0)} - E + \tfrac{1}{2}\omega\Lambda + A)$$
$$= \ln \operatorname{sdet}(R^{(0)} - E) + \ln \operatorname{sdet}(1 + \frac{1}{R^{(0)} - E}(\tfrac{1}{2}\omega\Lambda + A)). \tag{21.110}$$

Again the first term vanishes. For further evaluation of the second term, we use (21.48) and (21.59) and keep ω in first order only,

$$\mathscr{S}(Q, A) = -\tfrac{1}{2}i\pi\rho\omega \operatorname{str}(Q\Lambda) + \ln \operatorname{sdet}(1 + (-\tfrac{1}{2}gE1 + i\pi\rho Q)A). \tag{21.111}$$

The first term may be evaluated to $-i\pi\rho\omega(\lambda_1 - \lambda_2)$, where we use (21.69). The partition function is given by

$$Z(A) = \int [\mathrm{D}\,Q] \exp(-n\mathscr{S}(Q, A)), \quad [\mathrm{D}\,Q] = [\mathrm{D}\,\mu(T)]. \tag{21.112}$$

The action $\mathscr{S}(Q, A)$, (21.111), is the action of the non-linear σ-model for the matrix Q, which obeys $Q^2 = 1$. Due to this condition, the model is called non-linear. We observe that the action $\mathscr{S}$ does not depend on odd variables for $A = 0$. Then $Z(0) = 1$ is obtained from the derivation in Sect. 21.5.3.

21.6 Green's Functions

On the basis of the nonlinear-σ model we evaluate the one- and two-particle Green's functions in the large n-limit. We obtain the correlations between different eigenvalues.

From (21.69), (21.70) one obtains the bosonic contributions of Q

$$Q_{bb}^{RR} = \lambda_1 + \alpha^* \alpha (\lambda_2 - \lambda_1), \tag{21.113}$$

$$Q_{bb}^{AA} = -\lambda_1 + \beta^* \beta (\lambda_2 - \lambda_1), \tag{21.114}$$

$$Q_{bb}^{RA} = i\mu_1 (1 - \frac{1}{2}\alpha^* \alpha)(1 + \frac{1}{2}\beta^* \beta) + i\mu_2^* \alpha^* \beta, \tag{21.115}$$

$$Q_{bb}^{AR} = i\mu_1^* (1 - \frac{1}{2}\alpha^* \alpha)(1 + \frac{1}{2}\beta^* \beta) + i\mu_2 \beta^* \alpha. \tag{21.116}$$

The expectation values are evaluated by means of the derivatives of $Z(A)$ as in (21.21)–(21.23). For this purpose, one expands the second term in the action (21.111) in powers of A, where $\ln \mathrm{sdet}(W) = \mathrm{str}\ln(W)$, (10.37), is useful, and uses (21.71), (21.82) and the arguments in Sect. 21.5.3.

Since Q_{bb}^{RR} and Q_{bb}^{AA} do not contain $\alpha^* \alpha \beta^* \beta$-terms, the contributions of the averaged Green's functions come only from the terms not containing any Grassmann variables, at $\lambda_1 = \lambda_2 = 1$,

$$G(z_1) = \sqrt{g} e^{-i\phi}, \quad G(z_2) = \sqrt{g} e^{+i\phi},$$

$$\tilde{K}(z_1, z_1) = G^2(z_1), \quad \tilde{K}(z_2, z_2) = G^2(z_2). \tag{21.117}$$

The one-particle Green's functions agree with (21.30). Only the averaged Green's functions containing both retarded and advanced factors have contributions from the $\alpha^* \alpha \beta^* \beta$-terms,

$$\tilde{K}(z_1, z_2) = G(z_1)G(z_2) + (\pi\rho)^2 \int_1^\infty d\lambda_1 \int_{-1}^{+1} d\lambda_2 e^{i\pi f(\lambda_1 - \lambda_2)}, \tag{21.118}$$

$$\tilde{K}'(z_1, z_2) = (\pi\rho)^2 \int_1^\infty d\lambda_1 \int_{-1}^{+1} d\lambda_2 \frac{\lambda_1 + \lambda_2}{\lambda_1 - \lambda_2} e^{i\pi f(\lambda_1 - \lambda_2)}, \tag{21.119}$$

where $\mu_1 \mu_1^* + \mu_2 \mu_2^* = \lambda_1^2 - \lambda_2^2$ has been used. The mean level spacing in energy between eigenstates is $(n\rho)^{-1}$. Thus,

$$f = n\rho\omega \tag{21.120}$$

increases over this distance by one. Evaluation of the integrals yields

$$\tilde{K}(z_1, z_2) = G(z_1)G(z_2) + (\pi\rho)^2 \frac{2i \sin(\pi f)}{(\pi f)^2} e^{i\pi f}, \tag{21.121}$$

$$\tilde{K}'(z_1, z_2) = (\pi\rho)^2 \frac{2i}{\pi f}. \tag{21.122}$$

For real f, an infinitesimal imaginary part has to be added. Thus,

$$\frac{\sin(\pi f)}{(\pi f)^2} = P\frac{\sin(\pi f)}{(\pi f)^2} - i\delta(f) \tag{21.123}$$

with P indicating the principal value. The correlation of the density of states is given by (compare (21.4), (21.44))

$$(2\pi i)^2 \overline{\rho(E + \frac{\omega}{2})\rho(E - \frac{\omega}{2})} = \tilde{K}(z_1, z_1) + \tilde{K}(z_2, z_2) - (\tilde{K}(z_1, z_2) + (f \to -f))$$

$$= (G(z_1) - G(z_2))^2 - (2\pi\rho)^2\delta(f)$$

$$+ (2\pi\rho)^2 \frac{\sin^2(\pi f)}{(\pi f)^2}. \tag{21.124}$$

Hence, the correlation between the levels reads

$$\overline{\rho(E + \frac{\omega}{2})\rho(E - \frac{\omega}{2})} = \rho^2(E)C_U(f), \tag{21.125}$$

$$C_U(f) = 1 + \delta(f) - \frac{\sin^2(\pi f)}{(\pi f)^2}$$

$$= \delta(f) + \frac{(\pi f)^2}{3} + O(f^4). \tag{21.126}$$

$\delta(f)$ is the self-correlation. Apart from this contribution, the correlation increases for small f proportional to f^2 as a result of level repulsion. At large f, the correlation C approaches the uncorrelated value 1.

This correlation function C describes (after dropping the $\delta(0)$-term) the correlation between a level with any other level. One may also derive the general k-point correlation function

$$\overline{\rho(E + \frac{f_1}{n\rho(E)}) \ldots \rho(E + \frac{f_k}{n\rho(E)})} = \rho^k(E)C_k(f_1, \ldots f_k), \tag{21.127}$$

which can be expressed by the determinant (see for example [183])

$$C_k = \det\left(\frac{\sin \pi(f_i - f_j)}{\pi(f_i - f_j)}\right)^k_{i,j=1}. \tag{21.128}$$

Comment on $[\mathbf{D}\,\mu(t)]$ The factor $1/(2\pi)^2$ has to be explained. The diagonal terms of $d R$ and $d P$ are multiplied by $1/\sqrt{2\pi}$ for correct normalization. They cancel in $R = TPT^{-1}$. Each pair of the two off-diagonal, even elements reads $d\Re R\, d\Im R/\pi = \pm id R\, d R^*/(2\pi)$. Hence the factor $1/(2\pi)^2$.

21.7 Gaussian Orthogonal and Symplectic Ensembles

Until now we have considered the Gaussian unitary ensemble. This ensemble contains complex matrix elements. It is not time-reversal invariant. There are two time-reversal invariant Gaussian ensembles, the orthogonal (GOE) and the symplectic (GSE) Gaussian ensembles.

21.7.1 Gaussian Orthogonal Ensemble

The GOE has real matrix elements, $f_{\alpha,\beta} = f_{\beta,\alpha} \in \mathbb{R}$. It is described by the Gaussian distribution

$$P\{f\} \propto \exp(-\tfrac{1}{4}ng \,\mathrm{tr}\,(f^2)) \tag{21.129}$$

with correlations of the matrix elements

$$\overline{f_{\alpha,\beta}f_{\gamma,\delta}} = \frac{1}{ng}(\delta_{\alpha,\gamma}\delta_{\beta,\delta} + \delta_{\alpha,\delta}\delta_{\beta,\gamma}). \tag{21.130}$$

The number of components of the field S has to be doubled, since now we have to deal with the unitary-orthosymplectic symmetry. Thus, one needs for each α eight independent field components for the two-energy correlations,

$$S = \begin{pmatrix} \cdots & & \cdots & \\ x_\alpha^{(1,1)} & x_\alpha^{(1,2)} & x_\alpha^{(2,1)} & x_\alpha^{(2,2)} & \xi_\alpha^{(1)} & \xi_\alpha^{(1)\times} & \xi_\alpha^{(2)} & \xi_\alpha^{(2)\times} \\ \cdots & & \cdots & \end{pmatrix} \in \mathcal{M}(n,0,4,4), \tag{21.131}$$

where $x \in \mathbb{R}$. We denote the components of S by $S_{\nu\alpha}^{(i,p)}$, with $\nu = 0, 1$ or equivalently $\nu = $ b, f discriminating even and odd elements, $i = 1, 2$ distinguishing two energies z_i (with generalization to more energies being straightforward), and finally $p = 1, 2$,

$$S_{\mathrm{b},}^{(i,p)} = x_\alpha^{(i,p)}, \quad S_{\mathrm{f},}^{(i,1)} = \xi_\alpha^{(i)}, \quad S_{\mathrm{f},}^{(i,2)} = \xi_\alpha^{(i)\times}. \tag{21.132}$$

The indices ν, i, p of $S_{\nu\alpha}^{(i,p)}$ number the columns, the subscript α the rows.

The following steps are similar to those of GUE. A number of factors, however, are different. $\mathscr{S}_1$ reads

$$\mathscr{S}_1(S,A) = \mathrm{str}[\tfrac{1}{2}s(zS^{\ddagger} - S^{\ddagger}f)S - \sqrt{s}A\sqrt{s}S^{\ddagger}S],$$

$$z = \mathrm{diag}(z_1, z_1, z_2, z_2, z_1, z_1, z_2, z_2), \tag{21.133}$$

$$s = \mathrm{diag}(s_{\mathrm{b}1}, s_{\mathrm{b}1}, s_{\mathrm{b}2}, s_{\mathrm{b}2}, s_{\mathrm{f}1}, s_{\mathrm{f}1}, s_{\mathrm{f}2}, s_{\mathrm{f}2}).$$

The average over f yields

$$\overline{\exp(-\mathscr{S}_1(S,A))} = \exp(-\operatorname{str}[(\tfrac{1}{2}sz - \sqrt{s}A\sqrt{s})S^{\ddagger}S] + \frac{1}{4ng}\operatorname{str}[(SsS^{\ddagger})^2]), \quad (21.134)$$

where we use (compare (14.43)) $^{\mathrm{T}}(SsS^{\ddagger}) = CSsS^{\ddagger}C^{-1}$. In the next step, one obtains

$$\mathscr{S}_2(S,R,A) = \tfrac{1}{4}ng\operatorname{str}(R^2) - \operatorname{str}(\tfrac{1}{2}\sqrt{s}(R - z + 2A)\sqrt{s}S^{\ddagger}S), \quad (21.135)$$

$$\frac{\partial Z}{\partial A_{11,11}} = nG(z), \quad (21.136)$$

$$\mathscr{S}_3(R,A) = \tfrac{1}{4}g\operatorname{str}(R^2) + \tfrac{1}{2}\ln\operatorname{sdet}(R - z + 2A^{\mathrm{s}}) \quad (21.137)$$

with $R \in \mathcal{M}(4,4)$. The saddle point is obtained, as for the unitary model in Sects. 21.3 and 21.4. One obtains the same saddle point equation (21.25) and Green's functions (21.30), (21.31). In particular, Eqs. (21.57)–(21.60) hold. Only $Q^{\dagger}$ and $T^{\dagger}$ have to be replaced by $Q^{\ddagger}$ and $T^{\ddagger}$, and $2m_{\mathrm{R}}$ and $2m_{\mathrm{A}}$ by $4m_{\mathrm{R}} = 4$ and $4m_{\mathrm{A}} = 4$.

Since S is real, we obtain (14.43)

$$^{\mathrm{T}}(S^{\ddagger}S) = {}^{\mathrm{T}}S\,{}^{\mathrm{T}}S^{\ddagger} = CS^{\ddagger}SC^{-1}. \quad (21.138)$$

Since (21.135) is obtained from $R - \frac{1}{ng}\sqrt{s}S^{\ddagger}S\sqrt{s}$, we introduce R with the same symmetry as $\sqrt{s}S^{\ddagger}S\sqrt{s}$, that is

$$^{\mathrm{T}}R = CRC^{-1}. \quad (21.139)$$

Indeed, the step from (21.135) to (21.137) is only correct, if R and also A fulfills this condition. Thus, the symmetrized

$$A^{\mathrm{s}}_{ip,i'p'} = \tfrac{1}{2}(A_{ip,i'p'} + A_{i'p',ip}) \quad (21.140)$$

is introduced in (21.137). Compare with the first half of Sect. 14.3. Hence, with (21.63), we first require $^{\mathrm{T}}Q = CQC^{-1}$. This requirement implies

$$^{\mathrm{T}}Q^{\mathrm{RA}} = CQ^{\mathrm{AR}}C^{-1}, \quad ^{\mathrm{T}}Q^{\mathrm{AR}} = CQ^{\mathrm{RA}}C^{-1}, \quad (21.141)$$

where both equalities are equivalent. Then

$$^{\mathrm{T}}(Q^{\mathrm{RA}}Q^{\mathrm{AR}}) = {}^{\mathrm{T}}Q^{\mathrm{AR}}\,{}^{\mathrm{T}}Q^{\mathrm{RA}} = CQ^{\mathrm{RA}}Q^{\mathrm{AR}}C^{-1}, \quad (21.142)$$

$$^{\mathrm{T}}Q^{\mathrm{RR}} = {}^{\mathrm{T}}\sqrt{1 - Q^{\mathrm{RA}}Q^{\mathrm{AR}}} = \sqrt{1 - CQ^{\mathrm{RA}}Q^{\mathrm{AR}}C^{-1}}$$
$$= CQ^{\mathrm{RR}}C^{-1}, \quad (21.143)$$

and similarly for Q^{AA}. Thus, condition (21.141) yields $^{T}Q = CQC^{-1}$ and it is sufficient to require $^{T}P = CPC^{-1}$ in addition to (21.139).

The matrix $P_{bb} \in \mathcal{M}(4,0)$ is symmetric and integrated along the real axis. P_{ff} consists of imaginary quaternions and is antihermitian. It can be written as

$$P_{ff} = \begin{pmatrix} a & 0 & c & d \\ 0 & a & d^{\times} & -c^{\times} \\ -c^{\times} & -d & e & 0 \\ -d^{\times} & c & 0 & e \end{pmatrix} \in \mathcal{M}(0,4), \quad a, e \in \mathbb{I}, \quad c, d \in \mathbb{C}. \tag{21.144}$$

Written in this way, a, e are integrated along the imaginary axis, and the real and imaginary parts of c, d are integrated along the real axis.

In analogy to the derivation in Eqs. (21.109)–(21.111), one obtains from (21.137) the action of the nonlinear sigma-model in the orthogonal case

$$\mathcal{S}(Q,A) = -\tfrac{1}{4}i\pi\rho\omega \, \mathrm{str}(Q\Lambda) + \tfrac{1}{2}\ln \mathrm{sdet}(1 + (-gE + 2i\pi\rho^{(0)}Q)A^{s}). \tag{21.145}$$

The evaluation of the level correlations is more complicated in this orthogonal case since it has to be performed by 8×8 matrices Q. The result [64–66, 182, 183] is

$$C_0(f) = 1 + \delta(f) - \frac{\sin^2(\pi f)}{(\pi f)^2} - \left[\frac{\pi}{2}\mathrm{sign}(f) - \mathrm{Si}(\pi f) \right] \left[\frac{\cos(\pi f)}{\pi f} - \frac{\sin(\pi f)}{(\pi f)^2} \right]$$

$$= \delta(f) + \frac{\pi^2}{6}|f| + O(f^2), \tag{21.146}$$

with the sine integral

$$\mathrm{Si}(x) = \int_0^x dy \frac{\sin y}{y}, \quad \lim_{x \to \infty} \mathrm{Si}(x) = \frac{\pi}{2}. \tag{21.147}$$

The level repulsion yields only a linear term in f in the orthogonal case.

21.7.2   Gaussian Symplectic Ensemble

The Hamiltonian is that of a spin-dependent time-reversal invariant interaction

$$H = \sum_{\alpha m, \beta m'} |\alpha m\rangle f_{\alpha m, \beta m'} \langle \beta m'|. \tag{21.148}$$

Time-reversal invariance of H requires $\Theta H \psi = H \Theta \psi$, with the time-reversal operator $\Theta = -i\sigma^y K$, where K denotes complex conjugation for the representation in real space ($-i\sigma^y = \epsilon$). As a consequence the matrices $f_{\alpha\beta}$ are real quaternions.

$$\begin{pmatrix} f_{\uparrow\uparrow} & f_{\uparrow\downarrow} \\ f_{\downarrow\uparrow} & f_{\downarrow\downarrow} \end{pmatrix} = \epsilon \begin{pmatrix} f^*_{\uparrow\uparrow} & f^*_{\uparrow\downarrow} \\ f^*_{\downarrow\uparrow} & f^*_{\downarrow\downarrow} \end{pmatrix} \epsilon^{-1} = \begin{pmatrix} f^*_{\downarrow\downarrow} & -f^*_{\downarrow\uparrow} \\ -f^*_{\uparrow\downarrow} & f^*_{\uparrow\uparrow} \end{pmatrix} = \begin{pmatrix} f_{\uparrow\uparrow} & f_{\uparrow\downarrow} \\ -f^*_{\uparrow\downarrow} & f^*_{\uparrow\uparrow} \end{pmatrix} \in \mathbb{Q}.$$
(21.149)

If $H\psi = E\psi$, then also $H\Theta\psi = E\Theta\psi$. The wave-functions ψ and $\Theta\psi$ are orthogonal. As a consequence all eigenstates are doubly degenerate (Kramers doublett). We assume Gaussian correlations for f

$$\overline{f_{\alpha m_1, \beta m_2}} = 0, \quad f \in \mathcal{M}(0, 2n),$$
(21.150)

$$\overline{f_{\alpha m_1, \beta m_2} f_{\gamma m_3, \delta m_4}} = \frac{1}{2ng} (\delta_{\alpha\gamma}\delta_{\beta\delta}\epsilon_{m_1 m_3}\epsilon_{m_2 m_4} + \delta_{\alpha\delta}\delta_{\beta\gamma}\delta_{m_1 m_4}\delta_{m_2 m_3}).$$
(21.151)

We describe the system by superreal fields S, which are composed of the two-by-two matrices

$$\Xi_\alpha^{(i)} = \begin{pmatrix} \xi^{(i)}_{\alpha\uparrow} & \xi^{(i)}_{\alpha\downarrow} \\ \xi^{(i)\times}_{\alpha\uparrow} & \xi^{(i)\times}_{\alpha\downarrow} \end{pmatrix}, \quad X_\alpha^{(i)} = \begin{pmatrix} x^{(i)}_{\alpha\uparrow} & -x^{(i)\times}_{\alpha\downarrow} \\ x^{(i)}_{\alpha\downarrow} & x^{(i)\times}_{\alpha\uparrow} \end{pmatrix}.$$
(21.152)

with $x \in \mathbb{C}$. We also denote

$$S^{(i,m)}_{1,\alpha m'} = \Xi^{(i)}_{\alpha m m'} \in \mathcal{A}_1, \quad S^{(i,m)}_{2,\alpha m'} = X^{(i)}_{\alpha m m'} \in \mathcal{A}_0.$$
(21.153)

The indices v, i, m of $S^{(i,m)}_{v\alpha m'}$ denote the columns, and the indices α, m', the rows of S.

$$S = \left(\{\Xi^{(i)}_\alpha\} \ \{X^{(i)}_\alpha\} \right) \in \mathcal{M}(0, 2n, 4, 4).$$
(21.154)

$\mathcal{S}_1$ reads

$$\mathcal{S}_1(S) = -\text{str}(szS^\ddagger S - sS^\ddagger fS + \sqrt{s}S^\ddagger S\sqrt{s}A),$$
(21.155)

$$z = \text{diag}(z_1, z_1, z_2, z_2, z_1, z_1, z_2, z_2),$$
(21.156)

$$s = \text{diag}(s_{f1}, s_{f1}, s_{f2}, s_{f2}, s_{b1}, s_{b1}, s_{b2}, s_{b2}),$$
(21.157)

where we consider only the ff-components of A.

The signs in front of str have changed, since the complex variables f and x are now in the fermionic block, $\text{tr}(f^2) = -\text{str}(f^2)$, etc. The one-particle Green's function can be expressed as

$$G(\alpha m, \beta m', z) = \frac{1}{2} \langle sx_{\alpha,m}x^\times_{\beta,m'} \rangle.$$
(21.158)

The average over f yields

$$\overline{\exp(-\mathscr{S}_1(S))} = \exp(\operatorname{str}(szS^{\ddagger}S) - \frac{1}{2ng}\operatorname{str}((SsS^{\ddagger})^2) + \sqrt{s}S^{\ddagger}S\sqrt{s}A), \quad (21.159)$$

where we use $SsS^{\ddagger} = \epsilon^{-1\,\mathrm{T}}(SsS^{\ddagger})\epsilon$, (14.43). The next steps are

$$\mathscr{S}_2(S,R,A) = -\frac{ng}{2}\operatorname{str}(R^2) + \operatorname{str}(\sqrt{s}(R-z+A)\sqrt{s}S^{\ddagger}S), \quad (21.160)$$

$$\frac{\partial Z}{\partial A_{im,im}} = nG(z_i), \quad (21.161)$$

$$\mathscr{S}_3(R,A) = -\frac{g}{2}\operatorname{str}(R^2) - \ln\operatorname{sdet}(R-z+A). \quad (21.162)$$

The symmetry arguments for R and Q, Eqs. (21.139)–(21.143) apply also for the symplectic case. Now P_{bb} is symmetric, but imaginary, and P_{ff} consists of real quaternions. With $R^{(0)}$, Eq. (21.59), we obtain the nonlinear sigma-model

$$\mathscr{S}(Q,A) = \tfrac{1}{2}i\pi\rho\omega\operatorname{str}(Q\Lambda) - \ln\operatorname{sdet}(1 + (-\tfrac{1}{2}gE + i\pi\rho Q)A^s). \quad (21.163)$$

The level correlation for the symplectic case reads

$$C_S(f) = 1 + \delta(f) - \frac{\sin^2(2\pi f)}{(2\pi f)^2} + \operatorname{Si}(2\pi f)\left[\frac{\cos(2\pi f)}{2\pi f} - \frac{\sin(2\pi f)}{(2\pi f)^2}\right]$$

$$= \delta(f) + \frac{(2\pi f)^4}{135} + O(f^6). \quad (21.164)$$

The double degeneracy of the eigenstates yields a strong repulsion with a probability distribution proportional to f^4, for small energy differences f.

21.8   Circular Ensembles and Level Distributions

In the first subsection the eigenvalues of H along the real axis are replaced by eigenvalues along the unit circle in the complex plane by considering ensembles of unitary matrices. Secondly the distribution of eigenvalues of the Gaussian ensembles are derived.

21.8.1   Circular Ensembles

The half-circle density of states given by the n-orbital model is not typical for most physical applications. However, the correlations between the energies of the eigenstates close by, as given by (21.126), (21.146), and (21.164) are typical. In order to have systems without band edges, Dyson introduced circular ensembles [60], CUE, COE, and CSE, besides the Gaussian ensembles. They are defined as ensembles of unitary matrices. He considers mappings $H \to U$, like

$$U = \frac{1 + iH}{1 - iH}, \quad U = e^{-iHt} \tag{21.165}$$

avoiding a precise definition. The hermiticity $H^\dagger = H$ is equivalent to $U^\dagger = U^{-1}$. Thus, the matrices are unitary. Their eigenvalues can be written $e^{i\phi}$, $\phi \in \mathbb{R}$. Hence, all eigenvalues lie on the unit circle in the complex plane. Since the set of all hermitian H yields all unitary U, CUE is defined as the ensemble of all unitary matrices with invariant measure $d\mu(U) = d\mu(U_l U U_r)$ for any fixed $U_{l,r}$. This ensemble has the property that the density of the angles ϕ of the eigenvalues is constant. Small changes in H correspond to small changes in U. Probabilities of finding angles at ϕ and $\phi + \delta\phi$ correspond to probabilities of finding eigenstates at energies E and $E + \delta E$ for $\delta\phi \ll 1$ and $\delta\phi/2\pi = \rho\delta E$.

The Hamiltonians of the GOE are symmetric, $H = {}^t H$. Thus, the COE is given by unitary matrices obeying $U = {}^t U$. They can be represented by

$$U = {}^t V V, \quad V \in U(n). \tag{21.166}$$

Since U is invariant under the transformation $V \to WV$ with $W \in O(n)$, the matrices U of the COE belong to the coset $U(n)/O(n)$.

Finally H of GSE(2n) obeys $H = \epsilon_n {}^t H \epsilon_n^{-1}$ (for ϵ_n see (12.22)). Consequently $U = \epsilon_n {}^t U \epsilon_n^{-1}$ and U can be represented as

$$U = \epsilon_n {}^t V \epsilon_n^{-1} V \tag{21.167}$$

for the CSE(2n). It is invariant under transformations $V \to WV$, where W is a unitary symplectic matrix. Thus, the matrices U of CSE belong to the coset $U(2n)/USp(2n)$.

The usual Hubbard-Stratonovich transformation is not possible for the circular ensembles. However, an analog yielding a supersymmetric formulation has been given by Zirnbauer [302].

21.8.2 Level Distribution

The probability distribution for the eigenvalues λ_i of the matrices H for the Gaussian ensembles can be given as

$$dp(\{\lambda\}) \propto \exp(-\frac{gn}{2} \sum_k \lambda_k^2) \prod_{l<k} |\lambda_l - \lambda_k|^\beta \prod_k d\lambda_k, \tag{21.168}$$

with $\beta = 1, 2, 4$ for the GOE, GUE, GSE, respectively.

For the Gaussian unitary ensemble it is obtained in the following way. The matrix H can be expressed by the diagonal matrix $\Lambda = \text{diag}(\lambda)$ as

$$H = U\Lambda U^\dagger, \tag{21.169}$$

with a unitary matrix U. Infinitesimal deviations of a unitary matrix from the unit matrix can be written as

$$\mathbf{1} + \sum_m \epsilon_m G^{(m)}, \tag{21.170}$$

where the ϵ are infinitesimal and real. n generators are given by $G^{(\alpha)}_{\alpha,\alpha} = \text{i}$ and all other matrix elements vanish. The other $n(n-1)$ generators $G^{(j)}$, $j = n+1..n^2$, perform transformations between α and β, $\alpha < \beta$. For each pair there are two generators, $G_{\alpha,\beta} = 1$, $G_{\beta,\alpha} = -1$ and $G_{\alpha,\beta} = \text{i}$, $G_{\beta,\alpha} = \text{i}$. All other matrix G elements vanish.

We parametrize the matrices U by $n(n-1)$ variables μ and n variables ϕ. The ϕs are introduced such that

$$U(\{\mu\}, \{\phi\}) = U(\{\mu\}, \{0\})\text{diag}(e^{\text{i}\phi}). \tag{21.171}$$

An infinitesimal variation of the parameters of U yields

$$\text{d}\,U = U\sum_{\alpha=1}^{n} G^{(\alpha)}\text{d}\,\phi_\alpha + U\sum_{i=1}^{n^2-n}\sum_{j=n+1}^{n^2} w_{ij}(\{\mu\})G^{(j)}\text{d}\,\mu_i. \tag{21.172}$$

Then the variation of H reads

$$\text{d}\,H = \text{d}\,U\Lambda U^\dagger + U\text{d}\,\Lambda U^\dagger + U\Lambda\text{d}\,U^\dagger$$

$$= U\left(\sum_{ij} w_{ij}G^{(j)}\Lambda\text{d}\,\mu_i + \text{d}\,\Lambda - \sum_{ij} w_{ij}\Lambda G^{(j)}\text{d}\,\mu_i\right)U^\dagger, \tag{21.173}$$

or in components

$$\text{d}\,H_{\alpha,\beta} = U_{\alpha,k}\left(w_{ij}(\{\mu\})G^{(j)}_{kl}(\lambda_l - \lambda_k)\text{d}\,\mu_i + \delta_{k,l}\text{d}\,\lambda_k\right)U^*_{\beta,l}. \tag{21.174}$$

The variations of the variables ϕ do not contribute. We may integrate over the space of the ϕs and obtain a factor, $(2\pi)^n$, independent of the μs.

The Jacobian reads

$$\frac{\partial H}{\partial(\mu, \lambda)} = \det(U^n)\det(U^{\dagger n})\det(w(\{\mu\}))\prod_{l<k}(\lambda_l - \lambda_k)^2. \tag{21.175}$$

Since $\det(U) \det(U^\dagger) = \det(UU^\dagger) = \det(\mathbf{1}) = 1$, we obtain

$$P(H)\mathrm{d}H \propto \exp(-\frac{gn}{2}\,\mathrm{tr}\,H^2) \prod_{i=1}^{n}\mathrm{d}H_{ii} \prod_{i\leq 1 < j \leq n} \mathrm{d}\,\Re H_{ij}\mathrm{d}\,\Im H_{ij} \qquad (21.176)$$

$$\propto \prod_i \mathrm{d}\mu_i\,\det(w(\{\mu\})) \prod_k \mathrm{d}\lambda_k \exp(-\frac{wN}{2}\sum_k \lambda_k^2) \prod_{l<k}(\lambda_l - \lambda_k)^2.$$

After integration over the μs one is left with the probability distribution (21.168) with $\beta = 2$. Similarly, one obtains the distributions for the other Gaussian ensembles. For the circular ensembles, one can similarly determine the probability distribution of the ϕs,

$$\mathrm{d}p(\{\phi\}) \propto \prod_{l<k}|\sin(\tfrac{1}{2}(\phi_l - \phi_k))|^{\beta} \prod_k \mathrm{d}\phi_k. \qquad (21.177)$$

21.9   Final Remarks

The random matrix ensemble is considered as a model for nuclear levels. Various theoretical approaches are mentioned.

Random matrix ensembles were introduced by Wigner as a stochastic model for nuclear levels [284]. Dyson introduced the three symmetry classes, now called Wigner-Dyson classes [61]. Wigner [286] as well as Dyson and Mehta [60, 62] studied the level distributions. They were intensively studied with various methods. Early reviews were given by Porter [214] and Wigner [287, 288]. A good overview is given in the book by Mehta [182, 183].

Besides the correlation functions for the density of states (21.126), (21.146), (21.164), one is also interested in the correlation between adjacent levels. They were calculated by the method of orthogonal polynomials. See Mehta [182, 183]. The exclusion-inclusion principle [71] allows the determination of the correlations between adjacent levels, if the correlation functions are known for arbitrarily many energies.

Efetov [64, 65] finally introduced the supermatrix non-linear sigma-model considered here. Verbaarschot and Zirnbauer [262] showed that the correct level statistics at energy differences of order $1/n$ cannot be obtained by means of replicas, but by using the supermatrix formulation. Arguments for the combined non-compact—compact symmetry are given in [261, 262]. For the formulation given here I used to some extend the lectures by Mirlin [184]. I also refer to the book by Efetov [66] with many applications.

Wigner Surmise An approximation for the probability distribution of adjacent energy levels was introduced by Wigner for the GOE. It is known as the Wigner surmise [287]. Generalized to arbitrary ensembles one assumes the distribution

$$p(f) = c_\beta f^\beta \exp(-k_\beta f^2). \tag{21.178}$$

The exponent β is given by the probability distribution from level-repulsion at small energy differences, and the constants c and k are determined so that the total probability and the average distance between levels equals unity,

$$\int_0^\infty df\, p(f) = 1, \qquad \int_0^\infty df\, f p(f) = 1. \tag{21.179}$$

The results are shown in (21.180). The correlations vanish in the limit $f \to 0$ as $p(f) = a_\beta f^\beta + O(f^{\beta+1})$ with the exact coefficients a given below. These coefficients are obtained from the correlations (21.126), (21.146), (21.164), since the correlations between the levels further apart decay with higher powers in f as it approaches zero. The ratios a_β/c_β are close to one.

Ensemble	β	c_β	k_β	a_β	a_β/c_β
GOE	1	$\frac{\pi}{2}$	$\frac{\pi}{4}$	$\frac{\pi^2}{6}$	1.047
GUE	2	$\frac{32}{\pi^2}$	$\frac{4}{\pi}$	$\frac{\pi^2}{3}$	1.015
GSE	4	$\left(\frac{64}{9\pi}\right)^3$	$\frac{64}{9\pi}$	$\frac{(2\pi)^4}{135}$	0.995

$$\tag{21.180}$$

More results are given in the papers and books cited above.

Orthogonal Ensemble for Nuclei Nuclear forces obey time-reversal invariance, but do not conserve the spin due to spin-orbit and tensor forces. Thus, one might expect that the level statistics is given by the symplectic ensemble. However, the total angular momentum j is conserved. Thus, j and its z-component m are conserved and $\langle n, j, m | H | n', j', m' \rangle = \delta_{j,j'} \delta_{m,m'} H_{n,n',j}$, where n numbers the basis states for given j, m. Time-reversal maps $|n, j, m\rangle$ into $|n, j, -m\rangle$ if the basis states are appropriately chosen (e.g. real radial wave-functions). Thus, $\langle n, j, m | H | n', j', m' \rangle = \langle n, j, -m | H | n', j', -m' \rangle^*$, which implies $H_{n,n',j} = H_{n,n',j}^*$. Hence, the matrix elements are real and the orthogonal symmetry applies. This has already been pointed out by Dyson [60].

Bohigas-Giannoni-Schmidt Conjecture The matrix elements for a many-particle system, like a nucleus, are far from being invariant under orthogonal or unitary-symplectic transformations between the many-particle states. Assuming a two-particle interaction, many matrix elements vanish. Other matrix elements are not independent from each other. But both experimental observation of levels in nuclei and model calculations show very good agreement with the level statistics of the GOE, if one picks out energy intervals not too close to the ground-state. A recent survey is given in the review by Weidenmüller and Mitchell [279]. Bohigas et al.

[33] conjectured that the spectral fluctuation properties of a quantum system with many-body interactions, that is fully chaotic in the classical limit, coincides with those of the canonical random-matrix ensemble having the same symmetry.

A history of random matrix theory has been given by Bohigas and Weidenmüller [34]. The application of RMT to compound-nucleus reactions has been reviewed in [189, 279]. Guhr et al. [105] review the random matrix ensemble including a number of results given in the next two chapters.

Problems

21.1 Show that (21.43) follows from (21.9) and (21.41). Hint: Express z_i by $e^{\pm i\phi_i}$.

21.2 Derive (21.177) for the circular unitary ensemble.

References

[33] O. Bohigas, M.J. Gianoni, C. Schmitt, Characterization of chaotic quantum spectra and universality of level fluctuation laws. Phys. Rev. Lett. **52**, 1 (1984)

[34] O. Bohigas, H.-A. Weidenmüller, History - an overview, in *Handbook of Random Matrix Theory*, ed. by G. Akeman, J. Baik, P. di Francesco (Oxford University Press, 2011), p. 15

[60] F.J. Dyson, Statistical theory of energy levels of complex systems. I, II, III. J. Math. Phys. **3**, 140, 157, 166 (1962)

[61] F.J. Dyson, The threefold way. Algebraic structure of symmetry groups and ensembles in quantum mechanics. J. Math. Phys. **3**, 1199 (1962)

[62] F.J. Dyson, M.L. Mehta, Statistical theory of energy levels of complex systems. IV, V. J. Math, Phys. **4**, 701, 713 (1963)

[64] K.B. Efetov, Supersymmetry method in localization theory. Zh. Eksp. Teor. Fiz. **82**, 872 (1982); Sov. Phys. JETP **55**, 514 (1982)

[65] K.B. Efetov, Supersymmetry and theory of disordered metals. Adv. Phys. **32**, 53 (1983)

[66] K.B. Efetov, *Supersymmetry in Disorder and Chaos* (Cambridge University Press, Cambridge, 1997)

[67] K.B. Efetov, A.I. Larkin, D.E. Khmel'nitskii, Interaction of diffusion modes in the theory of localization. Zh. Eksp. Teor. Fiz. **79**, 1120 (1980); Sov. Phys. JETP **52**, 568 (1980)

[71] L. Erdös, Universality of Wigner random matrices: a survey of recent results. arXiv 1004.0861 [math-ph]; Russ. Math. Surv. **66**, 507 (2011)

[91] Y.V. Fyodorov, On Hubbard-Stratonovich transformations over hyperbolic domains. J. Phys. Condens. Matter **17**, S1915 (2005)

[92] Y.V. Fyodorov, Y. Wei, M.R. Zirnbauer, Hyperbolic Hubbard-Stratonovich transformations made rigorous. J. Math. Phys. **49**, 053507 (2008)

[105] T. Guhr, A. Müller-Groehling, H.A. Weidenmüller, Random-matrix theories in quantum physics: common concepts. Phys. Rep. **299**, 189 (1998)

[112] A. Houghton, A. Jevicki, R.D. Kenway, A.M.M. Pruisken, Noncompact σ models and the existence of a mobility edge in disordered electronic systems near two dimensions. Phys. Rev. Lett. **45**, 394 (1980)

[180] A.J. McKane, Reformulation of $n \to 0$ models using anticommuting scalar fields. Phys. Lett. **76A**, 22 (1980)

[181] A.J. McKane, M. Stone, Localization as an alternative to Goldstone's theorem. Ann. Phys. **131**, 36 (1981)

[182] M.L. Mehta, *Random Matrices and the Statistical Theory of Energy Levels* (Academic Press, London, 1967)

[183] M.L. Mehta, *Random Matrices* (Academic Press, Boston, 1991)

[184] A.D. Mirlin, Statistics of energy levels and eigenfunctions in disordered and chaotic systems: supersymmetry approach, in *Proceedings of the International School of Physics "Enrico Fermi" on New Directions in Quantum Chaos, Course CXLIII*, ed. by G. Casati, I. Guarneri, U. Smilansky (IOS Press, Amsterdam, 2000), p. 223

[189] G.E. Mitchell, A. Richter, H.A. Weidenmüller, Random matrices and chaos in nuclear physics: nuclear reactions. Rev. Mod. Phys. **82**, 2845 (2010)

[194] J. Müller-Hill, M.R. Zirnbauer, Equivalence of domains for hyperbolic Hubbard-Stratonovich transformations. , J. Math. **22**, 053506 (2011) [arXiv: 1011.1389]

[214] C.E. Porter, *Statistical Theories of Spectra* (Academic Press, New York, 1965)

[219] A.M.M. Pruisken, L. Schäfer, Field theory and the Anderson model for disordered electronic systems. Phys. Rev. Lett. **46**, 490 (1981)

[220] A.M.M. Pruisken, L. Schäfer, The Anderson model for electron localisation non-linear σ model, asymptotic gauge invariance. Nucl. Phys. B **200**, [FS4] 20 (1982)

[231] L. Schäfer, F. Wegner, Disordered system with n orbitals per site: Lagrange formulation, hyperbolic symmetry, and Goldstone modes. Z. Phys. B **38**, 113 (1980)

[261] J.J.M. Verbaarschot, H.A. Weidenmüller, M.R. Zirnbauer, Grassmann integration in stochastic quantum physics: the case of compound-nucleus scattering. Phys. Rep. **129**, 367 (1985)

[262] J.J.M. Verbaarschot, M.R. Zirnbauer, Critique of the replica trick. J. Phys. A **17**, 1093 (1985)

[270] F. Wegner, The mobility edge problem: continuous symmetry and a conjecture. Z. Phys. B **35**, 207 (1979)

[277] Y. Wei, Y.V. Fyodoroy, A conjecture on Hubbard-Stratonovich transformations for the Pruisken-Schäfer parameterizations of real hyperbolic domains. J. Phys. A **40**, 13587 (2007)

[279] H.A. Weidenmüller, G.E. Mitchell, Random matrices and chaos in nuclear physics: nuclear structure. Rev. Mod. Phys. **81**, 539 (2009)

[284] E.P. Wigner, On a class of analytic functions from the quantum theory of collisions. Ann. Math. **53**, 36 (1951)

[286] E.P. Wigner, On the distribution of the roots of certain symmetric matrices. Ann. Math. **67**, 325 (1958)

[287] E.P. Wigner, Results and theory of resonance absorption, in Gatlinburg Conference on Neutron Physics, Oak Ridge Natl. Lab. Rept. No. ORNL-2309, 59 (1957); reprint in [214]

[288] E.P. Wigner, Random matrices in physics. SIAM Rev. **9**, 1 (1967)

[302] M.R. Zirnbauer, Supersymmetry for systems with unitary disorder: circular ensembles. J. Phys. A **29**, 7113 (1996)

Chapter 22
Diffusive Model

Abstract In this chapter, tight-binding models in a d-dimensional space with random on-site and hopping matrix elements are introduced. As in the random matrix models of the preceding chapter, models of three different symmetries are distinguished. Again, the properties of these systems can be described by non-linear σ-models. In Sects. 22.2–22.5, the unitary case, which corresponds to the situation of broken time-reversal invariance, is discussed in some detail, whereas in Sects. 22.6 and 22.7, the properties of the systems obeying time-reversal invariance with and without spin conservation are considered.

22.1 Correlation Functions

The correlation functions for one-particle operators in thermal equilibrium and the linear response to a small perturbation are introduced.

22.1.1 Equilibrium Correlations

We introduce the two-time equilibrium function

$$C_{AB}(t, t') = \langle A(t)B(t') \rangle = \frac{1}{Z} \operatorname{tr} (e^{-\beta(H-\mu N)} A(t)B(t')). \tag{22.1}$$

for one-particle operators A and B with

$$A(t) = e^{iHt/\hbar} A e^{-iHt/\hbar}, \tag{22.2}$$

and similarly for B. Consider A and B in the energy representation,

$$A = |E\rangle A_{E,E'} \langle E'|, \quad H|E\rangle = E|E\rangle. \tag{22.3}$$

© Springer-Verlag Berlin Heidelberg 2016
F. Wegner, *Supermathematics and its Applications in Statistical Physics*,
Lecture Notes in Physics 920, DOI 10.1007/978-3-662-49170-6_22

Then the correlation function is expressed by fermion creation and annihilation operators, $c^\dagger$ and c, resp.

$$C_{AB}(t,t') = A_{E,E'} B_{E'',E'''} \langle c_E^\dagger c_{E'} c_{E''}^\dagger c_{E'''} \rangle e^{i(E-E')t + i(E''-E''')/\hbar}. \tag{22.4}$$

The expectation value can be expressed as

$$\langle c_E^\dagger c_{E'} c_{E''}^\dagger c_{E'''} \rangle = \delta_{E,E'} \delta_{E'',E'''} f_T(E) f_T(E'') + \delta_{E,E'''} \delta_{E',E''} f_T(E)(1 - f_T(E')) \tag{22.5}$$

with the average fermion occupation number

$$f_T(E) = \frac{1}{1 + \exp((E-\mu)/k_B T)}. \tag{22.6}$$

One finally obtains

$$C_{AB}(t,t') = \langle A \rangle \langle B \rangle + \int dE\, dE'\, \text{tr}\,(\delta(H-E)A\delta(H-E')B)$$
$$\times f_T(E)(1 - f_T(E')) e^{i(E-E')(t-t')/\hbar}. \tag{22.7}$$

If for example A and B are given in the position representation, $A = |r\rangle A_{rr'} \langle r'|$, then the trace can be written as

$$\text{tr}\,(\delta(H-E)A\delta(H-E')B) = \langle r'''|\delta(H-E)|r\rangle A_{r,r'} \langle r'|\delta(H-E')|r''\rangle B_{r'',r'''}. \tag{22.8}$$

The δ functions can be expressed as limits of Green's functions [see (22.21)],

$$\delta(H-E) = \frac{1}{2\pi} \lim_{\eta \to +0} \left(\frac{i}{E+i\eta-H} - \frac{i}{E-i\eta+H}\right). \tag{22.9}$$

22.1.2 Linear Response

Next we consider the response to a small perturbation, $V(t)$. We start with the von-Neumann equation for the density matrix ρ,

$$i\hbar \frac{d\rho(t)}{dt} = [H,\rho] \tag{22.10}$$

and assume that the Hamiltonian H can be written

$$H(t) = H_0 + V(t), \tag{22.11}$$

where H_0 is time-independent and $V(t)$ is a small time-dependent perturbation. We transform ρ into the interaction representation

$$\rho_I(t) = e^{iH_0t/\hbar}\rho(t)e^{-iH_0t/\hbar}, \tag{22.12}$$

and similarly for V. Then the von-Neumann equation reads

$$i\hbar\frac{d\rho_I(t)}{dt} = [V_I(t), \rho_I(t)]. \tag{22.13}$$

Starting with thermal equilibrium for H_0, at time $-\infty$,

$$\rho(-\infty) = \rho_0 = e^{-\beta(H_0-\mu N)}/Z, \tag{22.14}$$

one obtains, in first order in the perturbation V,

$$\rho(t) = \rho_0 - \frac{i}{\hbar}\int_{-\infty}^{t} dt'[V_I(t'), \rho_0]. \tag{22.15}$$

Accordingly, the expectation value of A at time t is given by

$$\langle A\rangle(t) = \langle A\rangle_0 - \frac{i}{\hbar}\int_{-\infty}^{t} dt'\, \mathrm{tr}\,(A_I(t)[V_I(t'), \rho_0])$$

$$= \langle A\rangle_0 - \frac{i}{\hbar}\int_{-\infty}^{t} \langle[A_I(t), V_I(t')]\rangle_0. \tag{22.16}$$

For a potential oscillating with frequency ω and increasing slowly in time

$$V(t') = Ve^{(i\omega+\eta)t'}, \tag{22.17}$$

one obtains the response

$$\langle A\rangle(t) = \langle A\rangle_0 - e^{(i\omega+\eta)t}\int dE\,dE'\frac{1}{E-E'+\omega-i\eta}$$

$$\times \mathrm{tr}\,(A\delta(H_0-E')V\delta(H_0-E))(f_T(E) - f_T(E')). \tag{22.18}$$

The Green's functions considered in the next section yield the correlation and response function by use of Eqs. (22.7), (22.9), and (22.18).

22.2   The Unitary Model: Green's Functions and Action

An action $\mathcal{S}$ is introduced, from which averaged products of Green's functions can be expressed as expectation values of matrix variables R.

We consider a tight-binding model on a lattice with N lattice sites. For simplicity, we assume cubic symmetry of the lattice and of the model. Moreover, we assume a unit volume per lattice site. Otherwise a number of factors of the volume per lattice site would have to be carried over. The Hamiltonian of the model reads

$$H = \sum_{r,r'} |r\rangle (t_{r-r'} + f_{r,r'}) \langle r'|. \tag{22.19}$$

We assume a Gaussian distribution of f,

$$\overline{f_{r,r'}} = 0, \quad \overline{f_{r,r'} f_{r'',r'''}} = M_{r-r'} \delta_{r,r'''} \delta_{r',r''}, \tag{22.20}$$

where $M_{r-r'} = M_{r'-r} \geq 0$. Then products of Green's functions

$$G(r, r', z) = \langle r| \frac{1}{z - H} |r'\rangle \tag{22.21}$$

can be written as

$$\prod_{i=1}^{m} G(r_i, r'_i, z_i) = \int [D\,S] \prod_{i=1}^{m} (s_{bi} S_b^{(i)}(r_i) S_b^{(i)*}(r'_i)) \exp(-\mathscr{S}_1(S,0)), \tag{22.22}$$

$$\int [D\,S] = \prod_{i=1}^{m} \left(\frac{s_{bi}}{s_{fi}}\right)^N \int \prod_{j,r} \left(\frac{d\,\Re S_b^{(j)}(r) d\,\Im S_b^{(j)}(r)}{\pi} d\,S_f^{(j)*}(r) d\,S_f^{(j)}(r)\right), \tag{22.23}$$

$$\mathscr{S}_1(S,0) = \text{str}[S^\dagger (sz \otimes \mathbf{1}_N - s \otimes (t+f))S] \tag{22.24}$$

$$= \sum_{i,v,r,r'} s_{vi} S_v^{(i)*}(r)(z_i \delta_{r,r'} - t_{r-r'} - f_{r,r'}) S_v^{(i)}(r'), \tag{22.25}$$

where, as before, v assumes the values $b = 0$ and $f = 1$. We note

$$S(r) \in \mathscr{M}(2,2,1,0), \quad S \in \mathscr{M}(2N,2N,1,0), \quad t,f \in \mathscr{M}(N,0) \tag{22.26}$$

$$s = \text{diag}(s_{b1}, s_{b2}, s_{f1}, s_{f2}) \in \mathscr{M}(2,2), \quad z = \text{diag}(z_1, z_2, z_1, z_2) \in \mathscr{M}(2,2). \tag{22.27}$$

We add a source term

$$\mathscr{S}_1(S,A) = \mathscr{S}_1(S,0) - \sum_{i,j,r,r'} \sqrt{s_{bi} s_{bj}} A_{ij}(r,r') S_b^{(j)*}(r') S_b^{(i)}(r), \tag{22.28}$$

which allows the calculation of the Green's functions from the partition function

$$Z(A) = \int [D\,S] \exp(-\mathscr{S}_1(S,A)). \tag{22.29}$$

$$\prod_{i=1}^{m} G(r_i, r_i', z_i) = \prod_{i=1}^{m} \frac{\partial}{\partial A_{ii}(r, r')} Z(A)|_{A=0}. \tag{22.30}$$

The average over the random terms f yields

$$\overline{\exp(\operatorname{str}(fSS^{\dagger}s))} = \exp(\tfrac{1}{2} \sum_{r,r'} M_{r-r'} \, S^{\dagger}(r) sS(r') \, S^{\dagger}(r') sS(r))$$

$$= \exp(\tfrac{1}{2} \sum_{r,r'} M_{r-r'} \operatorname{str}(T(r)T(r'))),$$

$$T(r) = \sqrt{s} S(r) S^{\dagger}(r) \sqrt{s} \in \mathcal{M}(2,2). \tag{22.31}$$

We use that $S^{\dagger}(r)sS(r')$ and $S^{\dagger}(r')sS(r)$ are in $\mathcal{M}(1,0)$. The Hubbard-Stratonovich transformation yields

$$\overline{\exp(\operatorname{str}(fSS^{\dagger}s))} = \int [D\,R] \exp(-\tfrac{1}{2} \sum_{r,r'} \operatorname{str}(R(r)w_{r-r'}R(r')) + \sum_{r} \operatorname{str}(R(r)T(r))),$$

$$R(r) \in \mathcal{M}(2,2), \tag{22.32}$$

where w is the inverse of M. We assume that the matrices M and thus, also w, is positive. This is surely the case if $M_0 > \sum_{r \neq 0} |M_r|$.

Thus, we obtain

$$\mathscr{S}_2 = \tfrac{1}{2} \sum_{r,r'} \operatorname{str}(R(r)w_{r-r'}R(r')) - \operatorname{str}(S^{\dagger} \sqrt{s}(R - z + t + A)\sqrt{s}S). \tag{22.33}$$

$R \in \mathcal{M}(2N, 2N)$ consists of the block matrices $R(r)$. s and z are multiplied by $\mathbf{1}_N$ and t by $\mathbf{1}_{2,2}$ in this rearrangement. Integration over S yields

$$\mathscr{S}_3(R, A) = \tfrac{1}{2} \sum_{r,r'} \operatorname{str}(R(r)w_{r-r'}R(r')) + \ln \operatorname{sdet}(R - z + t + A). \tag{22.34}$$

This action allows us to give expressions for the one- and two-particle Green's functions. Since

$$\ln \operatorname{sdet}(R - z + t + A) = \ln \operatorname{sdet}(R - z + t) + \ln \operatorname{sdet}(1 - \mathscr{G}A), \quad \mathscr{G} := \frac{1}{z - R - t}, \tag{22.35}$$

we expand in A by use of (10.36),

$$\mathscr{S}_3(R, A) = \mathscr{S}_3(R, 0) + \operatorname{str} \ln(1 - \mathscr{G}A)$$

$$= \mathscr{S}_3(R, 0) - \operatorname{str}(\mathscr{G}A) - \tfrac{1}{2} \operatorname{str}((\mathscr{G}A)^2) + \dots \tag{22.36}$$

Together with (22.30), we obtain the averaged Green's functions

$$G_1(r, r', z_i) := \overline{G(r, r', z_i)} = \overline{\mathscr{G}_{ibr,ibr'}}, \tag{22.37}$$

$$G_2(r_1, r_1', z_1; r_2, r_2', z_2) := \overline{G(r_1, r_1', z_1)G(r_2, r_2', z_2)} = \overline{\mathscr{G}_{1br_1,1br_1'}\mathscr{G}_{2br_2,2br_2'}}$$

$$+ \overline{\mathscr{G}_{1br_1,2br_2'}\mathscr{G}_{2br_2,1br_1'}}, \tag{22.38}$$

One contribution of (22.38) comes from the square of the second term in (22.36), the other from the third term.

22.3 Saddle Point and First Order

The saddle point of the action $\mathscr{S}$ and the corresponding Green's functions are determined.

We determine the saddle point $R^{(0)}$ by setting

$$R = R^{(0)} + \delta R \tag{22.39}$$

and expand up to first order in δR,

$$\mathscr{S}_3(R, 0) = \tfrac{1}{2}\sum_{r,r'} \text{str}(R^{(0)}w_{r-r'}R^{(0)}) + \ln\text{sdet}(R^{(0)} - z + t)$$

$$+ \sum_r \text{str}(\delta R(r) \sum_{r'} w_{r-r'}R^{(0)})$$

$$+ \text{str}\left(\delta R(R^{(0)} - z + t)^{-1}\right) + O((\delta R)^2). \tag{22.40}$$

Fourier Transform We introduce the Fourier transform, where we usually indicate the quantities depending on wave-vectors by a hat. We consider a periodic lattice with N lattice points. The transformation can be easily performed by use of the matrix elements

$$\langle r|q\rangle = e^{iqr}/\sqrt{N}, \quad \langle q|r\rangle = e^{-iqr}/\sqrt{N} \tag{22.41}$$

and the completeness relations

$$\sum_q |q\rangle\langle q| = 1, \quad \sum_r |r\rangle\langle r| = 1, \tag{22.42}$$

where q is confined to the first Brillouin zone. Then t may be expressed as

$$t = \sum_{r,r'} |r\rangle t_{r-r'} \langle r'| = \sum_{r,r',q,q'} |q\rangle \langle q|r\rangle t_{r-r'} \langle r'|q'\rangle \langle q'| = \frac{1}{N} \sum_{r,r',q,q'} |q\rangle e^{-iqr} t_{r-r'} e^{iq'r'} \langle q'|.$$

(22.43)

Setting $r - r' = \tilde{r}$,

$$\sum_{r'} e^{-iq(r'+\tilde{r})+iq'r'} = N\delta_{q,q'} e^{-iq\tilde{r}},$$

(22.44)

yields

$$t = \sum_{q} |q\rangle \hat{t}_q \langle q| \text{ with } \hat{t}_q = \sum_{r} e^{-iqr} t_r,$$

(22.45)

where $\tilde{r}$ has been replaced by r.

The corresponding transformation yields, for M amd w,

$$\hat{M}_q \hat{w}_q = 1.$$

(22.46)

On the other hand, (22.39) can be written

$$R^{(0)} = \sum_{r} |r\rangle R^{(0)} \langle r| = \sum_{q} |q\rangle R^{(0)} \langle q|,$$

(22.47)

$$\delta R = \sum_{r} |r\rangle \delta R(r) \langle r| = \sum_{r,q,q'} |q\rangle \langle q|r\rangle \delta R(r) \langle r|q'\rangle \langle q'| = \frac{1}{N} \sum_{r,q,q'} |q\rangle e^{i(q'-q)r} \delta R(r) \langle q'|.$$

(22.48)

Then

$$\delta R = \sum_{q,q'} |q\rangle \delta \hat{R}_{q-q'} \langle q'| \text{ with } \delta \hat{R}_{q-q'} = \frac{1}{N} \sum_{r} e^{i(q'-q)r} \delta R(r).$$

(22.49)

Thermodynamic Limit and Continuum Limit The sum over the wave-vectors q is in the thermodynamic limit replaced by the integral

$$\lim_{N \to \infty} \frac{1}{N} \sum_{q} = \int_{q} := (2\pi)^{-d} \int d^d q$$

(22.50)

In our explicit calculations, we assume the volume per lattice site v to be unity. If one wishes to perform the continuum limit, then one replaces

$$v \sum_r \leftrightarrow \int d^d r \tag{22.51}$$

and the periodicity volume $V = Nv$ has to be introduced. Then it has to be checked to which extent the various quantities have to be multiplied by powers of v.

Now we rewrite $\mathscr{S}_3$ (22.40), up to first order in δR,

$$\mathscr{S}_3(R,0) = \mathscr{S}_3(R^{(0)},0) + \text{str} \sum_r (\delta R(r)) \left(\hat{w}_0 R^{(0)} + \frac{1}{N} \sum_q \frac{1}{R^{(0)} - z + \hat{t}_q} \right) \tag{22.52}$$

Thus, the saddle-point equation reads

$$\hat{w}_0 R^{(0)} + \frac{1}{N} \sum_q \frac{1}{R^{(0)} - z + \hat{t}_q} = 0. \tag{22.53}$$

Again there are solutions, $R^{(0)}_{iv,i'v'}(r) = R^d_i \delta_{ii'} \delta_{vv'}$, diagonal in iv and $i'v'$. The eqs. for the diagonal solutions have $N-1$ real solutions between the $z - \hat{t}_q$ for real z. The other two solutions, which are relevant for the saddle point, are complex conjugate, if z lies inside the band. Otherwise they are real too. As before (Sects. 4 and 21), the imaginary part of $R^{(0)}_i$ has to be opposite to that of z_i.

One-Particle Green's Function The one-particle Green's function (22.37) is given within the saddle point approximation, $R = R^{(0)}$, by

$$G^{(0)}_1(r,r',z_i) = \langle r | (z_i - R^d_i - t)^{-1} | r' \rangle, \tag{22.54}$$

which reads

$$G^{(0)}_1(r,r',z_i) = \frac{1}{N} \sum_q e^{iq(r-r')} \hat{G}^{(0)}(q,z_i), \quad \hat{G}^{(0)}(q,z_i) = \frac{1}{z_i - R^d_i - \hat{t}_q}, \ . \tag{22.55}$$

and one obtains from (22.53)

$$G^{(0)}_1(r,r,z_i) = \hat{w}^{(0)} R^d_i, \quad \rho(E) = \frac{\hat{w}_0}{\pi} \Im R^d_i \text{ with } z_i = E - i0 \tag{22.56}$$

with ρ the averaged density of states per energy and lattice site. $\rho(E)$ is normalized so that

$$\int_{-\infty}^{+\infty} dE \rho(E) = 1. \tag{22.57}$$

The exact one-particle Green's function, as well as the approximation given here, obeys the symmetry relations

$$G(r, r', z) = G^*(r', r, z^*), \quad \hat{G}(q, z) = \hat{G}^*(q, z^*). \tag{22.58}$$

Two-Particle Green's Function The two-particle Green's function (22.38) factorizes within this approximation,

$$G_2(r_1, r_1', z_1; r_2, r_2', z_2) = G^{(0)}(r_1, r_1', z_1) G^{(0)}(r_2, r_2', z_2). \tag{22.59}$$

Whereas the result for the one-particle Green's function is quite reasonable, we will see in the next section that fluctuations yield an important contribution to the two-particle Green's function.

22.4 Second Order and Fluctuations

Fluctuations around the saddle-point manifold are included in second order. Leading contributions to the diffusion constant and the conductivity are obtained.
   Expand $\mathscr{G}$ (22.35), around $R^{(0)}$,

$$\mathscr{G} = G^{(0)} + G^{(0)} \delta R G^{(0)} + G^{(0)} \delta R G^{(0)} \delta R G^{(0)} + \dots, \tag{22.60}$$

i.e.

$$\langle q | \mathscr{G} | q' \rangle = \hat{G}^{(0)}(q) \delta_{q,q'} + \hat{G}^{(0)}(q) \delta \hat{R}_{q-q'} \hat{G}^{(0)}(q')$$
$$+ \sum_{q''} \hat{G}^{(0)}(q) \delta \hat{R}_{q-q''} \hat{G}^{(0)}(q'') \delta \hat{R}_{q''-q'} \hat{G}^{(0)}(q') + \dots \tag{22.61}$$

Then we obtain, up to second order in δR,

$$\hat{G}_1(q, z_i) = \hat{G}^{(0)}(q, z_i) + (\hat{G}^{(0)}(q, z_i))^2 \sum_{q'jv} \hat{G}^{(0)}(q', z_j) \overline{\delta \hat{R}_{q-q', ib, jv} \delta \hat{R}_{q'-q, jv, ib}}. \tag{22.62}$$

With

$$G_2(r_1, r_1', z_1; r_2, r_2', z_2) = \frac{1}{N^2} \sum_{q_1 q_1', q_2 q_2'} e^{i(q_1 r_1 - q_1' r_1' + q_2 r_2 - q_2' r_2')} \hat{G}_2(q_1, q_1', z_1; q_2, q_2', z_2), \tag{22.63}$$

we obtain, starting from (22.38),

$$\hat{G}_2(q_1,q_1',z_1;q_2,q_2',z_2) = \overline{\langle q_11b|\mathscr{G}|q_1'1b\rangle\langle q_22b|\mathscr{G}|q_2'2b\rangle} \qquad (22.64)$$

$$+\overline{\langle q_11b|\mathscr{G}|q_2'2b\rangle\langle q_22b|\mathscr{G}|q_1'1b\rangle}$$

$$= \delta_{q_1,q_1'}\hat{G}_1(q_1,z_1)\delta_{q_2,q_2'}\hat{G}_1(q_2,z_2) \qquad (22.65)$$

$$+\hat{G}^{(0)}(q_1,z_1)\hat{G}^{(0)}(q_1',z_1)\hat{G}^{(0)}(q_2,z_2)\hat{G}^{(0)}(q_2',z_2)$$

$$\times(\overline{\delta\hat{R}_{q_1-q_1',1b,1b}\delta\hat{R}_{q_2-q_2',2b,2b}} + \overline{\delta\hat{R}_{q_1-q_2',1b,2b}\delta\hat{R}_{q_2-q_1',2b,1b}}).$$

We have omitted the fluctuation terms in first order, since they vanish at the saddle-point. The action yields, up to second order in δR,

$$\mathscr{S}_3(R,0) = \mathscr{S}_3(R^{(0)},0) + \tfrac{1}{2}\sum_{r,r'}\text{str}(\delta R(r)w_{r-r'}\delta R(r')) - \tfrac{1}{2}\text{str}(G^{(0)}\delta RG^{(0)}\delta R)$$

$$= \tfrac{1}{2}N\sum_{q,iv,jv'}(-)^{v}(\hat{w}_q - \hat{\Pi}_q(z_j,z_i))\delta\hat{R}_{-q,iv,jv'}\delta\hat{R}_{q,jv',iv}, \qquad (22.66)$$

with

$$\hat{\Pi}_q(z_i,z_j) = \frac{1}{N}\sum_{q'}\hat{G}^{(0)}(q+q',z_j)\hat{G}^{(0)}(q',z_i). \qquad (22.67)$$

Since $R_{q,iv,jv'} = R^*_{-q,jv',iv}$, we obtain, by means of (13.33),

$$\overline{\delta\hat{R}_{-q_1,i_1v_1,j_1v_1'}\delta\hat{R}_{q_2,j_2v_2',i_2v_2}} = \delta_{q_1,q_2}\delta_{i_1,i_2}\delta_{j_1,j_2}\delta_{v_1,v_2}\delta_{v_1',v_2'}\frac{(-1)^{v_1'}}{N}\hat{\Gamma}_{q_1}(z_{j_1},z_{i_1}) \qquad (22.68)$$

with

$$\hat{\Gamma}_q(z_j,z_i) := \frac{1}{\hat{w}_q - \hat{\Pi}_q(z_j,z_i)}. \qquad (22.69)$$

Thus, the two contributions for $v = 0, 1$ in the one-particle Green's function (22.62) cancel. Only the second fluctuation term $\delta\hat{R}\delta\hat{R}$ in (22.65) contributes to the two-particle Green's function

$$\hat{G}_2(q_1,q_1',z_1;q_2,q_2',z_2) = \delta_{q_1,q_1'}\hat{G}_1(q_1,z_1)\delta_{q_2,q_2'}\hat{G}_1(q_2,z_2) + \frac{\delta_{q_1+q_2-q_1'-q_2',0}}{N}$$

$$\times\hat{\Gamma}_{q_1-q_2'}(z_1,z_2)\hat{G}^{(0)}(q_1,z_1)\hat{G}^{(0)}(q_1',z_1)$$

$$\times\hat{G}^{(0)}(q_2,z_2)\hat{G}^{(0)}(q_2',z_2). \qquad (22.70)$$

This may be rewritten as

$$G_2(r_1, r_1', z_1; r_2, r_2', z_2) = G_1(r_1, r_1', z_1)G_1(r_2, r_2', z_2) + \sum_{\bar{r}, \bar{r}'} G_1(r_1, \bar{r}, z_1)$$

$$\times G_1(\bar{r}, r_2', z_2)G_1(r_2, \bar{r}', z_2)G_1(\bar{r}', r_1', z_1)\Gamma(\bar{r}' - \bar{r}, z_1, z_1)$$

$$(22.71)$$

with

$$\Gamma(\bar{r}' - \bar{r}, z_1, z_2) := \frac{1}{N} \sum_q e^{iq(\bar{r}' - \bar{r})} \hat{\Gamma}_q(z_1, z_2). \tag{22.72}$$

For $q = 0$, one can perform a partial fraction decomposition and obtain by means of (22.53),

$$\hat{\Pi}_0(z_j, z_i) = \frac{1}{N} \sum_{q'} \frac{1}{R_i^d - z_i + \hat{\imath}_{q'}} \frac{1}{R_j^d - z_j + \hat{\imath}_{q'}} = \frac{\hat{w}_0(R_j^d - R_i^d)}{R_j^d - z_j - R_i^d + z_i} \tag{22.73}$$

and thus,

$$\hat{w}_0 - \hat{\Pi}_0(z_j, z_i) = \frac{\hat{w}_0(z_i - z_j)}{R_j^d - z_j - R_i^d + z_i}. \tag{22.74}$$

Expression (22.70) for the two-particle Green's function, with $\hat{\Gamma}$ and $\hat{\Pi}$ given by (22.69) and (22.67), is equivalent to the coherent-potential approximation [258] and the $n = \infty$-limit of the n-orbital model. [269].

22.4.1   Diffusion

In this subsection, we consider the massive and the diffusion modes. From Eq. (22.74), it is obvious that for $q = 0$ and $\omega = z_1 - z_2$, $\Im\omega > 0$, as in (21.44), one obtains, as $\omega \to 0$,

$$\hat{w}_0 - \hat{\Pi}_0(E + \tfrac{1}{2}\omega, E - \tfrac{1}{2}\omega) = \frac{-i\omega\hat{w}_0^2}{2\pi\rho(E)}, \tag{22.75}$$

where we have used (22.53), (22.56). Thus, $\hat{w}_0 - \hat{\Pi}_0$ vanishes in this limit and the modes for $s_2 = -s_1$ are soft. In contrast, if the imaginary parts of z_1 and z_2 have the same sign, $s_1 = s_2$, then $R_1^d - R_2^d$ vanishes proportional to ω, so that $\hat{w}_0 - \hat{\Pi}_0$ remains finite and the modes are massive. Let us consider the soft modes further.

From Eqs. (22.55), (22.67), one deduces

$$\hat{\Pi}_q(z_j, z_i) = \frac{1}{N} \sum_{q'} \hat{G}^{(0)}(q', z_i) \hat{G}^{(0)}(q + q', z_j) = \sum_r e^{iqr} G^{(0)}(r, 0, z_i) G^{(0)}(0, r, z_j).$$

(22.76)

If the hopping matrix elements are real $t_r = t_r^*$, then $\hat{t}_q = \hat{t}_{-q}$. Then the linear term in an expansion of (22.76) in powers of q vanishes. We expand

$$\hat{M}_q = \sum_r e^{-iqr} M_r = \sum_r M_r - \frac{1}{2} \sum_r (qr)^2 M_r + \ldots = \hat{M}_0(1 - \frac{q^2}{2d} r_M^2) + \ldots,$$

(22.77)

where we assume cubic symmetry and define the average range r_M of M

$$r_M^2 := \frac{\sum_r r^2 M_r}{\sum_r M_r}.$$

(22.78)

Then

$$\hat{w}_q = \hat{M}_q^{-1} = \hat{w}_0(1 + \frac{q^2}{2d} r_M^2) + \ldots, \qquad \sum_r r^2 w_r = -\hat{w}_0 r_M^2.$$

(22.79)

One obtains, for small ω and q,

$$\hat{w}_q - \hat{\Pi}_q(E + \tfrac{1}{2}\omega, E - \tfrac{1}{2}\omega) = \frac{-i\omega \hat{w}_0^2}{2\pi\rho(E)} + \frac{\hat{w}_0 q^2 r_M^2}{2d} + \frac{q^2}{2d} \sum_r r^2 G^{(0)}(r, 0, E + i0)$$

$$\times G^{(0)}(0, r, E - i0).$$

(22.80)

Thus, in the hydrodynamic limit, one obtains the diffusion pole

$$\hat{\Gamma}_q(E + \tfrac{1}{2}\omega, E - \tfrac{1}{2}\omega) = \frac{1}{\hat{w}_q - \hat{\Pi}_q(E + \frac{\omega}{2}, E - \frac{\omega}{2})} = \frac{2\pi\rho/\hat{w}_0^2}{-i\omega + D^{(0)} q^2},$$

(22.81)

$$D^{(0)} = \frac{\pi\rho}{d\hat{w}_0} \left(r_M^2 + \frac{1}{\hat{w}_0} \sum_r r^2 G^{(0)}(r, 0, E + i0) G^{(0)}(0, r, E - i0) \right).$$

(22.82)

22.4.2   *Conductivity*

Due to Kubo [156] and Greenwood [100], the frequency dependent conductivity at temperature T can be expressed as

$$\sigma_T(\omega) = \frac{1}{\omega} \int dE (f_T(E - \tfrac{1}{2}\omega) - f_T(E + \tfrac{1}{2}\omega))\sigma(\omega, E) \tag{22.83}$$

with

$$\sigma(\omega, E) = -\frac{e^2 \pi}{2d}\omega^2 \sum_r r^2 S_2(0, r, E + \tfrac{1}{2}\omega, E - \tfrac{1}{2}\omega) \tag{22.84}$$

and the averaged two-particle spectral function

$$S_2(r, r', E_1, E_2) = \overline{\langle r|\delta(H - E_1)|r'\rangle \langle r'|\delta(H - E_2)|r\rangle}, \tag{22.85}$$

where d is the dimension of the lattice and $f_T(E)$ the Fermi function. In the zero-temperature limit, $T = 0$, one obtains

$$\sigma_0(\omega) = \frac{1}{\omega} \int_{\mu - \omega/2}^{\mu + \omega/2} dE \sigma(\omega, E), \tag{22.86}$$

which in the dc-limit, $\omega = 0$, yields $\sigma_0(0) = \sigma(0, \mu)$, with the chemical potential μ. We evaluate the two-particle spectral function S_2 by means of

$$\langle r|\delta(H - E)|r'\rangle = \frac{1}{2\pi i} \left(G(r, r', E - i0) - G(r, r', E + i0) \right). \tag{22.87}$$

From (22.63), we obtain

$$G_2(r, r', z_1; r', r, z_2) = \frac{1}{N} \sum_q e^{iq(r-r')} \hat{K}_q(z_1, z_2), \tag{22.88}$$

$$\hat{K}_q(z_1, z_2) = \frac{1}{N} \sum_{q_1, q_2} \hat{G}_2(q_1, q_2 + q, z_1; q_2, q_1 - q, z_2)$$

$$= \hat{\Pi}_q(z_1, z_2)(1 + \hat{\Pi}_q(z_1, z_2)\hat{\Gamma}_q(z_1, z_2))$$

$$= \hat{w}_q \hat{\Pi}_q(z_1, z_2)\hat{\Gamma}_q(z_1, z_2). \tag{22.89}$$

For small ω, only the averages of products of Green's functions with opposite imaginary parts contribute. Then one obtains

$$\hat{S}_2(q) = \hat{w}_q[(\hat{\Pi}_q\hat{\Gamma}_q)(E+\tfrac{1}{2}\omega+\mathrm{i}0, E-\tfrac{1}{2}\omega-\mathrm{i}0) + (\hat{\Pi}_q\hat{\Gamma}_q)(E-\tfrac{1}{2}\omega+\mathrm{i}0, E+\tfrac{1}{2}\omega-\mathrm{i}0)], \tag{22.90}$$

where we may replace $\hat{w}_q\hat{\Pi}_q = \hat{w}_0^2$ and obtain

$$\hat{S}_2(q) = \hat{w}_0^2(\frac{2\pi\rho/\hat{w}_0^2}{-\mathrm{i}\omega + D^{(0)}q^2} + \frac{2\pi\rho/\hat{w}_0^2}{+\mathrm{i}\omega + D^{(0)}q^2}) = \frac{4\pi\rho D^{(0)}q^2}{\omega^2 + (D^{(0)}q^2)^2}. \tag{22.91}$$

$$\hat{S}_2(q) = 4\pi^2 \sum_r e^{\mathrm{i}qr} S_2(0, r, \ldots) = 4\pi^2 \sum_r S_2(0, r, \ldots)$$

$$-\frac{4\pi^2}{2d}q^2 \sum_r r^2 S_2(0, r, \ldots) + \ldots \tag{22.92}$$

Hence

$$\sum_r r^2 S_2(0, r, \ldots) = -\frac{2d\rho D^{(0)}}{\pi\omega^2}. \tag{22.93}$$

Thus, we obtain the dc-conductivity

$$\sigma(0, E) = e^2\rho D^{(0)}. \tag{22.94}$$

22.5 Nonlinear σ-Model

The model is reduced to the saddle-point manifold. This yields the nonlinear σ-model. Expressions for the averaged Green's functions are given.

As for the random matrix model of Chap. 21, one may introduce the same parametrization for the matrices R and integrate over the massive modes. Then one is left with a model of interacting matrices $Q(r)$. Thus, in generalization of (21.57)–(21.59), we choose

$$R(r) = T(r)P(r)T^{-1}(r), \quad T(r)sT^\dagger(r) = s \tag{22.95}$$

and the saddle-point $P^{(0)} = R^{(0)}$ of $R(r)$,

$$P^{(0)} = \Re R^{(0)} - \frac{\mathrm{i}\pi\rho^{(0)}}{\hat{w}_0}\Lambda. \tag{22.96}$$

Then

$$Q(r) = T(r)\Lambda T^{-1}(r) \tag{22.97}$$

with Λ as in (21.58), thus $\Lambda^2 = 1$. We may insert

$$R(r) = \Re R^{(0)} - \frac{i\pi\rho^{(0)}}{\hat{w}_0}Q(r) \tag{22.98}$$

in $\mathscr{S}_3(R, 0)$, as given in (22.34), and obtain, in the limit $\omega \to 0$, an action invariant under global transformation $T(r) \to T_0 T(r)$.

We argue in the following that the action may be written in the form (22.113). The first term of $\mathscr{S}_3(R, 0)$, (22.34), yields

$$-\frac{1}{2}\left(\frac{\pi\rho^{(0)}}{\hat{w}_0}\right)^2 \sum_{r,r'} \mathrm{str}(Q(r)w_{r-r'}Q(r')). \tag{22.99}$$

For the second term, we consider first the limit $\omega \to 0$. Now suppose that in a certain region in space $Q \approx Q^{\mathrm{loc}} = T_{\mathrm{loc}}\Lambda T_{\mathrm{loc}}^{-1}$. We assume $Q(r)$ close to Q^{loc} and expand $\delta Q(r) = Q(r) - Q^{\mathrm{loc}}$,

$$\delta Q(r) = T_{\mathrm{loc}}\begin{pmatrix} -\frac{1}{2}\tilde{Q}_{12}(r)\tilde{Q}_{21}(r) & \tilde{Q}_{12}(r) \\ \tilde{Q}_{21}(r) & \frac{1}{2}\tilde{Q}_{21}(r)\tilde{Q}_{12}(r) \end{pmatrix}T_{\mathrm{loc}}^{-1}, \tag{22.100}$$

where we have used $Q^2 = 1$. Then the second term of (22.34) yields

$$\ln \mathrm{sdet}(R - E + t) = \ln \mathrm{sdet}(R^{\mathrm{loc}} - E + t) - \frac{i\pi\rho^{(0)}}{\hat{w}_0}\sum_{r} \mathrm{str}(G^{(0)}(r, r, E, Q^{\mathrm{loc}})\delta Q(r))$$

$$+ \frac{\pi^2\rho^{(0)2}}{2\hat{w}_0^2}\sum_{r,r'}\mathrm{str}(G^{(0)}(r, r', E, Q^{\mathrm{loc}})\delta Q(r')$$

$$\times G^{(0)}(r', r, E, Q^{\mathrm{loc}})\delta Q(r)) \tag{22.101}$$

with

$$G^{(0)}(r, r', E, Q^{\mathrm{loc}}) = T_{\mathrm{loc}}\begin{pmatrix} G^{(0)}(r, r', E + i0) & 0 \\ 0 & G^{(0)}(r, r', E - i0) \end{pmatrix}T_{\mathrm{loc}}^{-1}. \tag{22.102}$$

This yields

$$
\ln \mathrm{sdet}(R - E + t)
$$

$$
= \ln \mathrm{sdet}(R^{\mathrm{loc}} - E + t) + \frac{i\pi\rho^{(0)}}{2\hat{w}_0} \sum_r (G^{(0)}(r,r,E+i0) - G^{(0)}(r,r,E-i0))
$$

$$
\times \mathrm{str}(\tilde{Q}_{12}(r)\tilde{Q}_{21}(r)) + \frac{\pi^2\rho^{(0)2}}{2\hat{w}_0^2} G^{(0)}(r,r',E+i0)G^{(0)}(r,r',E-i0)
$$

$$
\times \mathrm{str}(\tilde{Q}_{12}(r)\tilde{Q}_{21}(r') + \tilde{Q}_{21}(r)\tilde{Q}_{12}(r')). \tag{22.103}
$$

Using

$$
\frac{1}{E+i0-R_1^{(0)}-t} - \frac{1}{E-i0-R_2^{(0)}-t} = \frac{R_1^{(0)} - R_2^{(0)}}{(E+i0-R_1^{(0)}-t)(E-i0-R_2^{(0)}-t)} \tag{22.104}
$$

we obtain

$$
G^{(0)}(r,r,E+i0) - G^{(0)}(r,r,E-i0) = \frac{2i\pi\rho^{(0)}}{\hat{w}_0} \sum_{r'} G^{(0)}(r,r',E+i0)G^{(0)}(r',r,E-i0). \tag{22.105}
$$

Thus,

$$
\ln \mathrm{sdet}(R - E + t) = \ln \mathrm{sdet}(R^{\mathrm{loc}} - E + t) + \frac{\pi^2\rho^{(0)2}}{2\hat{w}_0^2} \sum_{r,r'} G^{(0)}(r,r',E+i0)
$$

$$
\times G^{(0)}(r',r,E-i0)\,\mathrm{str}(Q(r)Q(r')), \tag{22.106}
$$

up to bilinear order in $\tilde{Q}$. We combine now (22.99), (22.106) and obtain

$$
\frac{\pi^2\rho^{(0)2}}{2\hat{w}_0^2}(-w_{r-r'} + G^{(0)}(r,r',E+i0)G^{(0)}(r',r,E-i0))\,\mathrm{str}(Q(r)Q(r')). \tag{22.107}
$$

We may replace $Q(r)Q(r') = 1 - \frac{1}{2}(Q(r)-Q(r'))^2$ and $Q(r') = Q(r)+(r'-r)\nabla Q(r)$. This gradient may be understood as the first term in the expansion

$$
Q(r') - Q(r) = \frac{1}{N} \sum_q (e^{iqr'} - e^{iqr})\hat{Q}_q \tag{22.108}
$$

$$
= \frac{i}{N} \sum_q q(r'-r)e^{iqr}\hat{Q}_q - \frac{1}{2N} \sum_q (q(r'-r))^2 e^{iqr}\hat{Q}_q - \cdots
$$

Equation (22.107) reduces with (22.79), (22.82) to

$$-\tfrac{1}{4}\pi\rho^{(0)}D^{(0)}\,\mathrm{str}((\nabla Q)^2). \qquad (22.109)$$

The contribution proportional to ω is obtained from

$$\ln\mathrm{sdet}(R-z+t) = \ln\mathrm{sdet}(R-E-\tfrac{1}{2}\omega\Lambda+t)$$
$$= \ln\mathrm{sdet}(R-E+t) - \mathrm{str}((R^{(0)}-E+t)^{-1}\tfrac{1}{2}\omega\Lambda), \qquad (22.110)$$

where we replace, due to (22.53),

$$(R^{(0)}-E+t)^{-1} = -\hat{w}_0 R^{(0)} = -\hat{w}_0 \Re R^{(0)} + i\pi\rho^{(0)}Q, \qquad (22.111)$$

which yields the contribution,

$$\tfrac{1}{2}i\pi\rho^{(0)}\omega\sum_r \mathrm{str}(\Lambda Q(r)), \qquad (22.112)$$

to the action of the nonlinear σ-model. This is given by

$$\mathcal{S}(Q) = \tfrac{1}{4}\pi\rho^{(0)}\sum_r \mathrm{str}(-D^{(0)}(\nabla Q(r))^2 - 2i\omega\Lambda Q(r)). \qquad (22.113)$$

Correlations   The correlations are given by (22.37), (22.38). One uses (22.53) and puts

$$\langle r|\frac{1}{z-R^{(0)}-t}|r\rangle = \hat{w}_0 R^{(0)}(r), \qquad (22.114)$$

and inserts the saddle point approximation (22.98). Then

$$G_1(r,r,z_i) = \tilde{R}_i - i\pi\rho^{(0)}\overline{Q_{ib,ib}(r)}, \quad \tilde{R}_i := \hat{w}_0\Re R_i^{(0)}. \qquad (22.115)$$

The two-particle Green's functions for $r \neq r'$ are given by

$$G_2(r,r,z_1;r',r',z_2) = \overline{(\tilde{R}_1 - i\pi\rho^{(0)}Q_{1b,1b}(r))(\tilde{R}_2 - i\pi\rho^{(0)}Q_{2b,2b}(r'))} \qquad (22.116)$$
$$= G_1(r,r,z_1)G_1(r',r',z_2)$$
$$-\pi^2\rho^{(0)2}(\overline{Q_{1b,1b}(r)Q_{2b,2b}(r')} - \overline{Q_{1b,1b}(r)}\ \overline{Q_{2b,2b}(r')})$$
$$G_2(r,r',z_1;r',r,z_2) = -\pi^2\rho^{(0)2}\overline{Q_{1b,2b}(r)Q_{2b,1b}(r')}. \qquad (22.117)$$

The two-particle Green's function $G_2(r,r,z_1;r,r,z_2)$ is the sum of (22.116) and (22.117). All other two-particle Green's functions vanish in this approximation of the two-particle Green's functions, since (22.53) holds for an r-independent $R^{(0)}$.

Ward-Takahashi Identity The approximation (22.113) is good in the hydrody-namic limit. Let us consider the expectation value (where we use $\int [D\,Q] \exp(-\mathscr{S}(Q)) = 1$). Let us perform an infinitesimal transformation of Q, $Q' = (1 + \delta T)Q(1 - \delta T) = Q + \delta Q$ and correspondingly

$$\overline{A(Q)} = \int [D\,Q] A(Q) e^{-\mathscr{S}(Q)} = \int [D\,Q] A(Q + \delta Q) e^{-\mathscr{S}(Q + \delta Q)}. \tag{22.118}$$

Then

$$\overline{\delta A - A \delta \mathscr{S}} = 0 \tag{22.119}$$

Let $\hat{Q}_{i,j} = \sum_r Q_{ib,jb}$ and choose $A = \hat{Q}_{1,2}$, then only $\delta T_{1b,2b}$ and $\delta T_{2b,1b}$, which are linearly independent, should differ from zero. The term with $\delta T_{2b,1b}$ yields

$$\delta A = \hat{Q}_{2,2} - \hat{Q}_{1,1}, \quad \delta \mathscr{S} = -i\omega\pi\rho^{(0)}\hat{Q}_{2,1}. \tag{22.120}$$

Thus,

$$\overline{\hat{Q}_{2,2} - \hat{Q}_{1,1}} = -i\omega\pi\rho^{(0)}\overline{\hat{Q}_{1,2}\hat{Q}_{2,1}}, \tag{22.121}$$

which yields

$$\frac{1}{N} \sum_{q,q'} \hat{G}_2(q, q', z_1, q', q, z_2) = \frac{2\pi\rho^{(0)}}{-i\omega}. \tag{22.122}$$

This Ward identity is a consequence of particle number conservation. Generally, if the operator B is conserved, that is, commutes with the Hamiltonian H, then

$$\frac{1}{z_1 - H} B \frac{1}{z_2 - H} = \frac{1}{z_2 - z_1} B \left(\frac{1}{z_1 - H} - \frac{1}{z_2 - H} \right), \tag{22.123}$$

which yields

$$\sum_{q'',q'''} \langle q | \frac{1}{z_1 - H} | q'' \rangle B_{q''q'''} \langle q''' | \frac{1}{z_2 - H} | q' \rangle$$

$$= \frac{1}{z_2 - z_1} \sum_{q''} B_{qq''} \left(\langle q'' | \frac{1}{z_1 - H} | q' \rangle - \langle q'' | \frac{1}{z_2 - H} | q' \rangle \right). \tag{22.124}$$

The exact result (22.122) for the non-linear σ-model (22.113) agrees with (22.70) evaluated with the results in Sect. 22.4. In replacing $\mathscr{S}_3(R, 0)$, (22.34), by the non-linear sigma-model (22.113), it is essential that the invariance under the pseudo-unitary transformations T (for $\omega = 0$) and the effect of the symmetry breaking term proportional to ω remains conserved.

Local Gauge Invariance Since we have used only the contribution of $(z-R-t)^{-1}$ diagonal in r, we obtain only correlations that contain the sites r pairwise, one as first, the other as second argument of $1/(z - H)$. Since the averaged one-particle Green's function $G_1(r, r', z)$ decays rapidly as the distance $|r-r'|$ increases, G_2, (22.71), also decays rapidly as $r_1 - r_2'$ and $r_2 - r_1'$ increase. The extreme case of this behaviour is described by a model in which $t = 0$. Then the only non-vanishing averaged one- and two-particle Green's functions are those given in (22.115)–(22.117). The probability distribution of the eigenfunctions ψ is invariant under arbitrary gauges $\psi(r) \to e^{i\phi(r)}\psi(r)$. Thus, this model is called local gauge invariant model [203, 231, 269].

At first glance, the negative sign in front of $(\nabla Q)^2$ in (22.113) seems to be unexpected. However, using the parametrization (21.69), (21.70), one obtains, for the even elements of Q,

$$\mathrm{str}((\nabla Q)^2) = -2\left((\nabla\theta_1)^2 + \sinh^2\theta_1(\nabla\phi_1^2) + (\nabla\theta_2)^2 + \sin^2\theta_2(\nabla\phi_2)^2\right),$$
(22.125)

so that $-(\nabla Q)^2$ is positive semi-definite.[1]

Diffusons The expansion of $Q(r)$ in Eq. (22.113) up to second order in the transverse components Q^{RA} and Q^{AR} yield the diffusion behaviour. Therefore, these excitations are called diffusons.

The diffusion pole, Eq. (22.81), is obtained since $\hat{w}_q - \hat{\Pi}_q$ approaches zero proportional to ω and to q^2, for small ω and q^2. This pole is obtained by summing all the contributions of

$$\hat{w}_q\hat{\Gamma}_q(z_i, z_j) = 1 + \hat{M}_q\hat{\Pi}_q + (\hat{M}_q\hat{\Pi}_q)^2 + \ldots + (\hat{M}_q\hat{\Pi}_q)^l + \ldots$$
(22.126)

It describes a particle-hole pair multiply scattered by disorder. The term $G^{(0)}G^{(0)}(\hat{M}\hat{\Pi})^l$ describes l scatterings of the pair. In total, the contributions yield the diffusion pole. A word of warning is appropriate: Depending on the dimension of the system under consideration and the strength of disorder, the interaction between the diffusive modes may lead to a loss of the diffusive behaviour and turn the system from a metallic to an insulating behaviour. This is found when one goes beyond bilinear order in the interaction, and will be considered in Chap. 23.

Parity Transposition Instead of defining the fields $S(r) \in \mathcal{M}(2, 2, 1, 0)$ as in (22.26) we could define them as $S(r) \in \mathcal{M}(2, 2, 0, 1)$. The transformation can be simply performed by parity transposition, Sect. 10.5, $S \to {}^{\pi}S$. Then also the matrix Q is replaced by ${}^{\pi}Q$. We realize that ${}^{\pi}\Lambda = \Lambda$ and $s_b^R \leftrightarrow s_f^R, s_b^A \leftrightarrow s_f^A$, since the bosonic and fermionic blocks are exchanged. If we denote ${}^{\pi}Q$ by Q, then the

---

[1] Efetov [66] arranges the fermionic and bosonic variables in different order. Therefore he obtains the opposite sign in front of the supertraces.

action (22.113) reads

$$\mathscr{S}(Q) = \tfrac{1}{4}\pi\rho^{(0)} \sum_r \mathrm{str}(D^{(0)}(\nabla Q(r))^2 + 2i\omega\Lambda Q(r)), \qquad (22.127)$$

since the supertrace changes sign, (10.46). We will return to this representation in Sect. 23.1.

References The nonlinear σ-model was derived in 1979 and 1980 by Wegner [270] and by Schäfer and Wegner [231] and Efetov et al. [67]. Other calculations for the diffusive model and discussions on the lower critical dimension $d = 2$, as well as different behaviour for the three classes unitary, orthogonal and symplectic have been derived by Anderson et al. [1], Gorkov et al. [97], Oppermann and Wegner [203], Hikami et al. [110], and Jüngling and Oppermann [126, 127]. The supersymmetric non-linear sigma-model was introduced by Efetov [64, 65]. Until now the unitary model has been derived; the orthogonal and symplectic one follow in the next two sections. The critical behaviour, that is the behaviour close to the mobility edge, is discussed in Sect. 23.1.

22.6 Orthogonal Case

In this section we derive the expressions for the correlations and the non-linear sigma model for the time-reversal invariant spin-independent random hopping model.

22.6.1 The Lattice Model

Until now we considered the unitary model (22.19). It corresponds to a situation in which time-reversal invariance is broken by magnetic impurities. If the system is time-reversal invariant and no spin dependence is involved, then we choose again the Hamiltonian (22.19), but now with real matrix elements $t_{r,r'}$ and $f_{r,r'}$. Again, f should be Gaussian distributed with zero average,

$$\overline{f_{r,r'}} = 0, \quad \overline{f_{r,r'}f_{r'',r'''}} = (\delta_{r,r'''}\delta_{r',r''} + \delta_{r,r''}\delta_{r',r'''})M_{r-r'}. \qquad (22.128)$$

Since f and t are real and hermitian, they are symmetric. Thus, also the Green's functions have this property

$$G(r, r', z) = G(r', r, z). \qquad (22.129)$$

Similarly, as in (21.131), (21.132), we have to introduce two real and two Grassmannian components, $S_v^{(i,p)}(r)$, for each site r and each energy z_i. Then we may write the product of Green's functions as

$$\prod_{i=1}^{m} G(r_i, r_i', z_i) = \prod_{i=1}^{m} \left(\frac{s_{bi}}{s_{fi}} \right)^N \prod_{i,r} \left(\frac{d\, S_b^{(i,1)}(r) d\, S_b^{(i,2)}(r)}{2\pi} d\, S_f^{(i,1)}(r) d\, S_f^{(i,2)}(r) \right)$$

$$\times \prod_{i} (s_{bi} S_b^{(i,1)}(r_i) S_b^{(i,1)}(r_i')) \exp(-\mathscr{S}_1(S,0)), \qquad (22.130)$$

$$\mathscr{S}_1(S,0) = \tfrac{1}{2} \operatorname{str}[S^{\ddagger}(sz \otimes 1_N - s \otimes (t+f))S]$$

$$= \tfrac{1}{2} \sum_{i,p,v,r,r'} s_{vi} S_v^{\ddagger(i,p)}(r)(z_i \delta_{r,r'} - t_{r-r'} - f_{rr'}) S_v^{(i,p)}(r') \qquad (22.131)$$

with a quasireal field S with (for $m = 2$) dimension

$$S(r) \in \mathscr{M}(4,4,1,0), \quad S \in \mathscr{M}(4N,4N,1,0), \quad t, f \in \mathscr{M}(N,0),$$

$$s = \operatorname{diag}(s_{b1}, s_{b1}, s_{b2}, s_{b2}, s_{f1}, s_{f1}, s_{f2}, s_{f2}) \in \mathscr{M}(4,4),$$

$$z = \operatorname{diag}(z_{b1}, z_{b1}, z_{b2}, z_{b2}, z_{f1}, z_{f1}, z_{f2}, z_{f2}) \in \mathscr{M}(4,4). \qquad (22.132)$$

We can determine the Green's functions, as in (22.29), (22.30), with a source term

$$\mathscr{S}_1(S,A) = \mathscr{S}_1(S,0) - \sum \sqrt{s_{bi} s_{bj}} A_{ij}(r, r') S_b^{(i,1)}(r) S_b^{(j,1)}(r'). \qquad (22.133)$$

The doubling of the number of components is necessary, since the Grassmann variables appear in pairs for real vectors S. The fields S are those of (21.131), (21.132), but now with lattice sites r instead of the states denoted by α.

The average over the random terms f yields

$$\overline{\exp(-\tfrac{1}{2} \operatorname{str}(\sum_{r,r'} f_{r,r'} S(r') S^{\ddagger}(r)s))} = \exp(\tfrac{1}{4} \sum_{r,r'} M_{r-r'} S^{\ddagger}(r)sS(r') S^{\ddagger}(r')sS(r))$$

$$= \exp(\tfrac{1}{4} \sum_{r,r'} M_{r-r'} \operatorname{str}(T(r)T(r'))),$$

$$T(r) = \sqrt{s}S(r)S^{\ddagger}(r)\sqrt{s}, \qquad (22.134)$$

where we use in the first line of (22.134) that $S(r')S^{\ddagger}(r) = S(r)S^{\ddagger}(r')$, and that $S^{\ddagger}(r)sS(r')$ and $S^{\ddagger}(r')sS(r)$ belong to $\mathscr{M}(1,0)$. Thus, there is no need to use the supertrace here.

The Hubbard-Stratonovich transformation yields

$$\overline{\exp(-\tfrac{1}{2}\,\mathrm{str}(\sum_{r,r'} f_{r,r'} S(r')S^{\ddagger}(r)s))} \tag{22.135}$$

$$= \int [\mathrm{D}\,R]\exp(-\tfrac{1}{4}\sum_{r,r'}\mathrm{str}(R(r)w_{r-r'}R(r')) + \tfrac{1}{2}\sum_{r}\mathrm{str}(R(r)T(r)))$$

with $R(r) \in \mathcal{M}(4,4)$. The total matrix $R \in \mathcal{M}(4N,4N)$ consists of the block matrices $R(r)$. Since $T(r)$ is symmetric, and $s^{-1/2}T(r)s^{-1/2}$ is quasireal and quasi-hermitian, $R(r)$ has the same symmetries; compare with Problem 14.1. The partition function now reads

$$Z(A) = \int [\mathrm{D}\,S]\exp(-\mathscr{S}_1(S,A)) = \int [\mathrm{D}\,S][\mathrm{D}\,R]\exp(-\mathscr{S}_2(S,R,A)) \tag{22.136}$$

with

$$\mathscr{S}_2(S,R,A) = \tfrac{1}{4}\sum_{r,r'}\mathrm{str}(R(r)w_{r-r'}R(r')) - \tfrac{1}{2}\,\mathrm{str}(S^{\ddagger}\sqrt{s}(R-z+t+2A)\sqrt{s}S). \tag{22.137}$$

Integration over the fields S yields

$$Z(A) = \int [\mathrm{D}\,R]\exp(-\mathscr{S}_3(R,A)), \tag{22.138}$$

$$\mathscr{S}_3(R,A) = \tfrac{1}{4}\sum_{r,r'}\mathrm{str}(R(r)w_{r-r'}R(r')) + \tfrac{1}{2}\ln\mathrm{sdet}(R-z+t+2A^{\mathrm{s}}). \tag{22.139}$$

Since $S^{\ddagger}S$ obeys

$$^{\mathrm{T}}(S^{\ddagger}(r')S(r)) = \mathrm{C}S^{\ddagger}(r)S(r')\mathrm{C}^{-1}, \tag{22.140}$$

$A, \hat{A}$ have to be replaced by the symmetrized A^{s} and $\hat{A}^{\mathrm{s}}$,

$$A^{\mathrm{s}}_{ip,i'p'}(r,r') = \tfrac{1}{2}(A_{ip,i'p'}(r,r') + A_{i'p',ip}(r',r)),$$

$$\hat{A}^{\mathrm{s}}_{qip,q'i'p'} = \tfrac{1}{2}(\hat{A}_{qip,q'i'p'} + \hat{A}_{-q'i'p',-qip}). \tag{22.141}$$

22.6.2 Saddle Point and Fluctuations, Cooperon

The calculations of the saddle point and the fluctuations is performed similarly as in Sects. 22.3 and 22.4. Additional indices p appear in R, $R^{(0)}$, δR, and A as in (21.132). $R^{(0)}_{ipv,jp'v'}(r)$ is diagonal in p and p' too,

$$R^{(0)}_{ipv,jp'v'}(r) = \delta_{vv'}\delta_{ii'}\delta_{pp'}R^d_i. \tag{22.142}$$

The saddle-point equation agrees with (22.53). The expressions for the one-particle Green's functions hold as in (22.54)–(22.56). The following calculation parallels the calculation in Sect. 22.4, but differs in a number of factors of 2. Instead of (22.66), we now obtain

$$\mathscr{S}_3(R,0) = \mathscr{S}_3(R^{(0)},0) + \tfrac{1}{4}N \sum_{q,ipv,jp'v'} (-)^v(\hat{w}_q - \hat{\Pi}_q(z_j,z_i))\delta\hat{R}_{-q,ipv,jp'v'}\delta\hat{R}_{q,jp'v',ipv}, \tag{22.143}$$

with $\hat{\Pi}$ as in (22.67). Since $S^{\ddagger}(r)S(r)$ obeys (22.140), and since (22.134) is obtained from bilinear terms of $R - M\sqrt{s}S^{\ddagger}S\sqrt{s}$ in the exponent, we require

$$CR(r)C^{-1} = {}^TR(r) \succ C\hat{R}_qC^{-1} = {}^T\hat{R}_q, \tag{22.144}$$

where we use that s commutes with C. Then we obtain

$$\overline{\delta\hat{R}_{-q_1,i_1p_1v_1,i'_1p'_1v'_1}\delta\hat{R}_{q_2,i_2p_2v_2,i'_2p'_2v'_2}}$$
$$= \frac{1}{N}\hat{\Gamma}_{q_1}(z_{i_1},z_{i'_1})\delta_{q_1,q_2}\left((-)^{v'_1}\delta_{i_1i'_2}\delta_{i'_1i_2}\delta_{p_1p'_2}\delta_{p'_1p_2}\delta_{v_1v'_2}\delta_{v'_1v_2}\right.$$
$$\left.+(-)^{v_1v'_1}\delta_{i_1i_2}\delta_{i'_1i'_2}C_{p_1v_1,p_2v_2}C_{p'_1v'_1,p'_2v'_2}\right) \tag{22.145}$$

with $\hat{\Gamma}$ as in (22.69) and [compare with (12.22)]

$$C_{pv,p'v'} = \delta_{vv'}(\delta_{vb}\delta_{pp'} + \delta_{vf}\epsilon_{pp'}). \tag{22.146}$$

Correlations Equation (22.139) yields, up to second order in A^s,

$$Z(A^s) = 1 + \overline{\text{str}(\mathscr{G}A^s)} + \tfrac{1}{2}\overline{(\text{str}(\mathscr{G}A^s))^2} + \overline{\text{str}(\mathscr{G}A^s\mathscr{G}A^s)}. \tag{22.147}$$

To lowest order, that is without fluctuation contributions, we obtain the same results as in (22.55)–(22.59). Including the fluctuation contributions to second order, we now obtain

$$\hat{G}_1(q, z_i) = \hat{G}^{(0)}(q, z_i) + \hat{G}^{(0)2}(q, z_i)\frac{1}{N}\sum_{q'} \hat{G}^{(0)}(q', z_i)\hat{\Gamma}_{q-q'}(z_i, z_i),$$

(22.148)

$$\hat{G}_2(q_1, q_1', z_1; q_2, q_2', z_2) = \delta_{q_1, q_1'}\hat{G}_1(q_1, z_1)\delta_{q_2, q_2'}\hat{G}_1(q_2, z_2) + \frac{\delta_{q_1+q_2-q_1'-q_2'}}{N}\hat{G}^{(0)}(q_1, z_1)$$

$$\times \hat{G}^{(0)}(q_1', z_1)\hat{G}^{(0)}(q_2, z_2)\hat{G}^{(0)}(q_2', z_2)$$

$$\times (\hat{\Gamma}_{q_1-q_2'}(z_1, z_2) + \hat{\Gamma}_{q_1+q_2}(z_1, z_2)),$$

(22.149)

which can be rewritten as

$$G_2(r_1, r_1', z_1; r_2, r_2', z_2) = G_1(r_1, r_1', z_1)G_1(r_2, r_2', z_2) + \sum_{\bar{r}, \bar{r}'} G_1(r_1, \bar{r}, z_1)$$

$$\times G_1(\bar{r}, r_2', z_2)G_1(r_2, \bar{r}', z_2)G_1(\bar{r}', r_1', z_1)\Gamma(\bar{r}'-\bar{r}, z_1, z_2)$$

$$+ \sum_{\bar{r}, \bar{r}'} G_1(r_1, \bar{r}, z_1)G_1(r_2, \bar{r}, z_2)$$

$$\times G_1(\bar{r}', r_2', z_2)G_1(\bar{r}', r_1', z_1)\Gamma(\bar{r}'-\bar{r}, z_1, z_2)$$

(22.150)

with Γ given by (22.69), (22.72).

Cooperon At the end of Sect. 22.5, we realized that two particles running in the opposite direction lead to the long-range behaviour both in space (small q) and time (small ω) called Diffuson. With (22.73), (22.74) and Sect. 22.4.1, we obtain the same diffusion behaviour in the orthogonal case as in the unitary case. The Green's functions are symmetric in the orthogonal case (22.129), which implies

$$\hat{G}_1(q, z) = \hat{G}_1(-q, z)$$

(22.151)

Correspondingly

$$G_2(r_1, r_1', z_1; r_2, r_2', z_2) = G_2(r_1, r_1', z_1; r_2', r_2, z_2),$$

(22.152)

which is reflected in

$$\hat{G}_2(q_1, q_1', z_1; q_2, q_2', z_2) = \hat{G}_2(q_1, q_1', z_1; -q_2', -q_2, z_2),$$

(22.153)

compare to (22.149). Thus, also two particles running in the same direction show long-range behaviour for $q_1 + q_1' = q_2 + q_2' \to 0$. The excitation of such a pair of particles is called the Cooperon, since superconductivity is brought about by the condensate of pairs of particles. However, here we do not deal with such a condensate.

This time-reversal symmetry is also reflected in the extra term $\hat{\Gamma}_{q_1+q_2}(z_1, z_2)$ in (22.150). It yields for the correlation $\hat{K}$ defined in (22.88)

$$\hat{K}_q(z_1, z_2) = \hat{w}_q \hat{\Pi}_q(z_1, z_2) \hat{\Gamma}_q(z_1, z_2) + \frac{1}{N^2} \sum_{q_1, q_2} \hat{G}^{(0)}(q_1, z_1)$$
$$\times \hat{G}^{(0)}(q_2 + q, z_1)\hat{G}^{(0)}(q_2, z_2)\hat{G}^{(0)}(q_1 - q, z_2)\hat{\Gamma}_{q_1+q_2}(z_1, z_2).$$
(22.154)

For fixed q the argument $q_1 + q_2$ of $\hat{\Gamma}$ runs through 0 in the sum over q_1 and q_2 and, equivalently, the integral in the thermodynamic limit. Thus, it diverges for $\omega = 0$ in $d = 1$ and 2 dimensions. This has important consequences, which will be considered in Chap. 23.

Nonlinear σ-Model The nonlinear σ-model is derived in a similar way to the derivation for the unitary case in Sect. 22.5. Obviously, one has to replace † by ‡ and to observe that $\mathscr{S}(R, 0)$ differs by a factor of $\frac{1}{2}$, (22.34), (22.139). Thus, one obtains, instead of (22.113),

$$\mathscr{S} = -\tfrac{1}{8}\pi\rho^{(0)} \sum_r \mathrm{str}(D^{(0)}(\nabla Q(r))^2 + 2i\omega\Lambda Q(r)).$$
(22.155)

From

$$Q(r) = T(r)\Lambda T^{-1}(r), \quad {}^{\mathrm{T}}Q(r) = CQ(r)C^{-1},$$
(22.156)

one obtains

$${}^{\mathrm{T}}T^{-1}\kappa = CT,$$
(22.157)

where κ commutes with Λ. This equation and

$$TsT^\ddagger = s$$
(22.158)

yield

$$\kappa s\kappa^\ddagger = s, \quad T = s^{-1}CT^\times\kappa^{\times -1}s.$$
(22.159)

The choice $\kappa = C$ yields

$$T = s^{-1}CT^\times C^{-1}s.$$
(22.160)

The set of T obeying (22.158), (22.160) constitutes a group and yields a parametrization for the non-linear σ-model (22.155).

The Green's functions G_1 and G_2 in (22.115)–(22.117) are the same if we replace the indices ib by $ib1$. Now $G_2(r, r', z_1; r, r', z_2) = G_2(r, r', z_1; r', r, z_2)$ and $G_2(r, r, z_1; r.r, z_2)$ is the sum of (22.116) and twice (22.117). The discussion of local gauge invariance holds here, with the Z_2 gauge $\psi(r) \to \pm\psi(r)$.

22.7 Symplectic Case

Finally we consider the time-reversal invariant spin-dependent hopping model on a lattice. We consider a rather general model. Special models are mentioned in Sects. 22.7.3 and 22.7.4.

22.7.1 The Lattice Model

Similarly, as in Sect. 21.7.2, we consider a spin-dependent time-reversal invariant interaction, but now on a Bravais lattice. The Hamiltonian reads

$$H = \sum_{rr'mm'} |rm\rangle (t_{r-r',mm'} + f_{rm,r'm'}) \langle r'm'|, \quad f, t \in \mathbb{C}. \tag{22.161}$$

The hopping matrix elements obey

$$t_{r-r',mm'} + f_{rm,r'm'} = \epsilon_{m'm''}(t_{r'-r,m''m'''} + f_{r'm'',rm'''})\epsilon_{mm'''}, \tag{22.162}$$

due to the time-reversal invariance. The Green's function obeys, consequently,

$$G_{mm'}(r, r', z) = \epsilon_{m'm''} G_{m'',m'''}(r', r, z)\epsilon_{mm'''}, \tag{22.163}$$

$$\hat{G}_{mm'}(q, z) = \epsilon_{m'm''} \hat{G}_{m'',m'''}(-q, z)\epsilon_{mm'''}. \tag{22.164}$$

t and f are real quaternions due to time-reversal invariance and hermiticity. We expand $t_{r-r'}$ and $f_{rr'}$ in Pauli matrices and obtain the symmetry relations

$$t_{r-r',mm'} = \sum_k \tilde{i}^k t^k_{r-r'} \sigma^k_{mm'}, \quad t^k_{r'-r} = v^k t^k_{r-r'}, \quad t^k_{r-r'} \in \mathbb{R}, \tag{22.165}$$

$$\tilde{i}^k := \begin{cases} 1 & k = 0, \\ i & k = 1..3. \end{cases}, \quad v^k := \begin{cases} 1 & k = 0, \\ -1 & k = 1..3. \end{cases}, \quad \sigma^0 := 1_2.$$

$$\tag{22.166}$$

We choose the Gaussian distribution

$$f_{rmr'm'} = F_{rr'}\tilde{f}_{r-r'mm'}, \quad \tilde{f}_{r-r'mm'} = \sum_k \tilde{\imath}^k \gamma^k_{r-r'} \sigma^k_{mm'}, \tag{22.167}$$

where $F_{rr'}$ are random variables. F and the γ^k are real,

$$F_{rr'} = F_{r'r}, \quad \gamma^k_{r-r'} = v^k \gamma^k_{r'-r}, \tag{22.168}$$

$$\overline{F_{rr'}} = 0, \quad \overline{F_{r_1 r_1'} F_{r_2 r_2'}} = M_{r_1 - r_1'} (\delta_{r_1 r_2'} \delta_{r_2 r_1'} + \delta_{r_1 r_2} \delta_{r_1' r_2'}). \tag{22.169}$$

The Green's functions can be written

$$\prod_{i=1}^l G_{m_i, m_i'}(r_i, r_i', z_i) = \int [D\,S] \prod_{i=1}^l (s_{fi} S^{(i,m)}_{f,p_i}(r_i) S^{(i,m')\times}_{fp_i}(r_i')) \exp(-\mathscr{S}_1(S,0)), \tag{22.170}$$

$$[D\,S] = \prod_{ir} \frac{d\,\Re a^{(i)}(r)\,d\,\Im a^{(i)}(r)\,d\,\Re b^{(i)}(r)\,d\,\Im b^{(i)}(r)}{\pi^2}$$

$$\times \prod_{ipr} d\,\xi^{(i,p)}(r)\,d\,\xi^{(i,p)\times}(r), \tag{22.171}$$

$$\mathscr{S}_1(S,0) = -\tfrac{1}{2}\,\mathrm{str}[s(zS^{\ddagger} - S^{\ddagger}(t+f))S] \tag{22.172}$$

$$= \tfrac{1}{2}(S^{\ddagger}(r))_{piv,m} s_{iv} (z_i \delta_{rr'} \delta_{mm'} - (t_{r-r',mm'} + f_{rr',mm'})) S_{m',piv}(r')$$

with

$$S^{(i)}_f(r) = \begin{pmatrix} a^{(i)}(r) & -b^{(i)\times}(r) \\ b^{(i)}(r) & a^{(i)\times}(r) \end{pmatrix} \in \mathscr{M}(0,2,0,2), \tag{22.173}$$

$$S_f(r) = \left(S^{(1)}_f(r),\; S^{(2)}_f(r) \right) \in \mathscr{M}(0,2,0,4), \tag{22.174}$$

$$S^{(i,p)}_b(r) = \begin{pmatrix} \xi^{(i,p)}(r) \\ \xi^{(i,p)\times}(r) \end{pmatrix} \in \mathscr{M}(0,2,1,0), \tag{22.175}$$

$$S_b(r) = \left(S^{(1,1)}_b(r)\; S^{(1,2)}_b(r)\; S^{(2,1)}_b(r)\; S^{(2,2)}_b(r) \right) \in \mathscr{M}(0,2,4,0), \tag{22.176}$$

$$S(r) = \left(S_b(r)\; S_f(r) \right) \in \mathscr{M}(0,2,4,4), \tag{22.177}$$

$$S = \begin{pmatrix} \cdots \\ S(r) \\ \cdots \end{pmatrix} \in \mathscr{M}(0,2N,4,4), \tag{22.178}$$

and

$$s = \mathrm{diag}(s_{b,1}, s_{b,1}, s_{b,2}, s_{b,2}, s_{f,1}, s_{f,1}, s_{f,2}, s_{f,2}) \in \mathcal{M}(4,4),$$

$$z = \mathrm{diag}(z_1, z_1, z_2, z_2, z_1, z_1, z_2, z_2) \in \mathcal{M}(4,4). \tag{22.179}$$

We now realize that the bosonic components S_b are in the fermionic sector. The reason is that in the unitary-orthosymplectic group, as defined in Chap. 14, the orthogonal transformations are in the bosonic sector and the unitary-symplectic transformations in the fermionic sector. Indices r, m of S number the rows, and p, i, v the columns. Useful equations for the integrals and averages were derived in the second half of Sect. 14.3. We add the source term in $\mathcal{S}_1$,

$$\mathcal{S}_1(S,A) = \mathcal{S}_1(S,0) - \sum \sqrt{s_{fi} s_{fj}} A_{pim,p'i'm'}(r,r') S_{f,p}^{(i,m)}(r') S_{f,p'}^{(i',m')\times}(r'). \tag{22.180}$$

Then the Green's functions can be obtained by taking derivatives of Z with respect to A. The term of $\mathcal{S}_1$ containing the random potential f reads

$$\tfrac{1}{2}\,\mathrm{str}(sS^{\ddagger}fS) = \tfrac{1}{2}\sum_{rr'} F_{rr'}\,\mathrm{str}(sS^{\ddagger}(r)\tilde{f}_{r-r'}S(r')). \tag{22.181}$$

The average over F yields

$$\overline{\exp(-\tfrac{1}{2}\,\mathrm{str}(sS^{\ddagger}fS))} = \exp(\tfrac{1}{2}\mathcal{D}), \tag{22.182}$$

$$\mathcal{D} = \tfrac{1}{4}\overline{F_{rr'}F_{r''r'''}}\,\mathrm{str}(sS^{\ddagger}(r)\tilde{f}_{r-r'}S(r'))\,\mathrm{str}(sS^{\ddagger}(r'')\tilde{f}_{r''-r'''}S(r'''))$$

$$= \tfrac{1}{4}M_{r-r'}\,\mathrm{str}(sS^{\ddagger}(r)\tilde{f}_{r-r'}S(r'))$$

$$\times(\,\mathrm{str}(sS^{\ddagger}(r')\tilde{f}_{r'-r}S(r)) + \mathrm{str}(sS^{\ddagger}(r)\tilde{f}_{r-r'}S(r'))). \tag{22.183}$$

Since $\mathrm{str}(A) = \mathrm{str}(^{\mathrm{T}}A)$ and since real S and $\tilde{f}$ yield

$$^{\mathrm{T}}S = CS^{\ddagger}C^{-1}, \quad {}^{\mathrm{T}}S^{\ddagger} = CSC^{-1}, \quad {}^{\mathrm{T}}\tilde{f} = C\tilde{f}C^{-1}, \tag{22.184}$$

both terms in the last line of (22.183) are equal. With (22.165), (22.167), we may write

$$\mathcal{D} = \tfrac{1}{2}M_{r-r'}\tilde{\imath}^k\tilde{\imath}^l v^l \gamma_{r-r'}^k \gamma_{r-r'}^l\,\mathrm{str}(S(r')sS^{\ddagger}(r)\sigma^k)\,\mathrm{str}(S(r)sS^{\ddagger}(r')\sigma^l). \tag{22.185}$$

We cannot reduce the supertraces in (22.185) to a scalar, as in the unitary and the orthogonal case, but only to traces over 2×2-matrices. We use the identity

$$\sigma_{mm'}^q \sigma_{m''m'''}^q = 2\delta_{mm'''}\delta_{m'm''}, \tag{22.186}$$

to obtain

$$\mathrm{str}(sS(r')S^{\ddagger}(r)\sigma^k)\,\mathrm{str}(sS(r)S^{\ddagger}(r')\sigma^l)$$

$$= -\tfrac{1}{2}\,\mathrm{str}(S(r')sS^{\ddagger}(r)\sigma^k\sigma^q S(r)sS^{\ddagger}(r')\sigma^l\sigma^q) \tag{22.187}$$

and express $\mathscr{Q}$ as

$$\mathscr{Q} = -\tfrac{1}{2}\tilde{M}_{r-r',kl}\,\mathrm{str}(T_k(r)T_l(r')), \quad T_k(r) = \sqrt{s}S^{\ddagger}(r)\sigma^k S(r)\sqrt{s} \tag{22.188}$$

with

$$\tilde{M}_{r-r'} = M_{r-r'} \cdot \begin{pmatrix} \tilde{\gamma}_+ & 0 & 0 & 0 \\ 0 & \tilde{\gamma}_- + \gamma_1^2 & -\gamma_0\gamma_3 + \gamma_1\gamma_2 & \gamma_0\gamma_2 + \gamma_1\gamma_3 \\ 0 & \gamma_0\gamma_3 + \gamma_1\gamma_2 & \tilde{\gamma}_- + \gamma_2^2 & -\gamma_0\gamma_1 + \gamma_2\gamma_3 \\ 0 & -\gamma_0\gamma_2 + \gamma_1\gamma_3 & \gamma_0\gamma_1 + \gamma_2\gamma_3 & \tilde{\gamma}_- + \gamma_3^2 \end{pmatrix}, \tag{22.189}$$

$$\tilde{\gamma}_+ = \tfrac{1}{2}(\gamma_0^2 + \gamma_1^2 + \gamma_2^2 + \gamma_3^2), \quad \tilde{\gamma}_- = \tfrac{1}{2}(\gamma_0^2 - \gamma_1^2 - \gamma_2^2 - \gamma_3^2) \tag{22.190}$$

where $\gamma_k = \gamma_{r-r'}^k$ and $\tilde{M}$ and w, introduced below, obey

$$\tilde{M}_{r-r',kl} = \tilde{M}_{r'-r,lk}, \quad w_{r-r',kl} = w_{r'-r,lk} \tag{22.191}$$

and $\tilde{M}_{r-r',kl}$ and $w_{r-r',kl}$ are real. The zeros in the upper and the left line of the matrix $\tilde{M}$ indicate that this disorder does not introduce a coupling between density and spin fluctuations. Up to now we have restricted the fluctuations of f by (22.167)–(22.169). One may choose a more general one by superposition of several of these fluctuations. This will not change the symmetry relations (22.191).

Then we perform the Hubbard-Stratonovich transformation

$$\exp(-\tfrac{1}{4}\sum_{rr'kl}\tilde{M}_{r-r',k,l}\,\mathrm{str}(T^k(r)T^l(r'))) \tag{22.192}$$

$$= \int [\mathrm{D}\,R^i]\exp(\tfrac{1}{2}\sum_{rr',kl}w_{r-r',kl}\,\mathrm{str}[R^k(r)R^l(r')] - \tfrac{1}{2}\sum_{rk}\mathrm{str}[R^k(r)T^k(r)]),$$

by multiplication by

$$\int [\mathrm{D}\,R^i]\exp(\tfrac{1}{2}w_{r-r',kl}\,\mathrm{str}[(R^k(r)-\tilde{M}_{r''-r,k'k}T^{k'}(r''))(R^l(r')-\tilde{M}_{r'-r''',ll'}T^{l'}(r'''))])=1, \tag{22.193}$$

where w is one half of the inverse of $\tilde{M}$,

$$\sum_{r'l}w_{r-r',kl}\tilde{M}_{r'-r'',ll'} = \tfrac{1}{2}\delta_{rr''}\delta_{kl'}. \tag{22.194}$$

This yields

$$\mathscr{S}_2(S, R, A) = -\frac{1}{2} \sum_{rr',kl} w_{r-r',kl} \, \mathrm{str}[R^k(r) R^l(r')]$$

$$+ \frac{1}{2} \sum_{rr'k} \mathrm{str}[D^k(r, r') \sqrt{s} S^{\ddagger}(r') \sigma^k S(r) \sqrt{s}] \tag{22.195}$$

with

$$D^k(r, r') = (R^k(r) - z\delta_{k0})\delta_{rr'} + \tilde{\imath}^k t^k_{r'-r} + 2A^{sk}(r, r') \in \mathscr{M}(4, 4), \tag{22.196}$$

Since S is quasireal, $S = C^{(m)} S^{\times} C^{(p)-1}$, we can symmetrize A, as contributions to the source term are pairwise equal,

$$A^s_{pim,p'i'm'}(r, r') = \frac{1}{2}(A_{pim,p'i'm'}(r, r') + (C^{(p)} C^{(m)} A(r', r) C^{(m)-1} C^{(p)-1})_{p'i'm',pim}),$$

$$\hat{A}^s_{pim,p'i'm'}(q, q') = \frac{1}{2}(\hat{A}_{pim,p'i'm'}(q, q') + (C^{(p)} C^{(m)} \hat{A}(-q', -q) C^{(p)-1} C^{(m)-1})_{p'i'm',pim}),$$

$$\tag{22.197}$$

and define $A_{pim,p'i'm'}(r, r') = \sum_k A^k_{pi,p'i'}(r, r') \sigma^k_{mm'}$. $C^{(p)}$ acts on indices ν and p

$$C^{(p)}_{\nu p, \nu' p'} = \delta_{\nu \nu'}(\delta_{\nu b} \delta_{pp'} + \delta_{\nu f} \epsilon_{pp'}) \tag{22.198}$$

and $C^{(m)} = \epsilon$ on indices m.

One obtains the symmetry condition (left equation) by use of ${}^T \sigma^k = v^k C^{(m)} \sigma^k C^{(m)-1}$,

$$C^{(p)} T^k(r) = v^k \, {}^T T^k(r) C^{(p)}, \quad C^{(p)} R^k(r) = v^k \, {}^T R^k(r) C^{(p)}, \quad C^{(p)} D^k = v^k \, {}^T D^k C^{(p)}. \tag{22.199}$$

We require the same symmetry of $R^k(r)$ (middle equation) and since t and A^s obey the same symmetry, it holds for D too (right equation). We also realize that $s^{-1/2} T^k(r) s^{-1/2}$ is hermitian. We require the same of $R^k(r)$.

Next we integrate over S. We introduce a unitary transformation U so that the complex components of S are expressed by real components $\tilde{S}$,

$$S_{m,piv}(r) = \sum_j U^{\nu}_{mpj} \tilde{S}_{jiv}(r) \tag{22.200}$$

and obtain

$$\sum_{rr'k} \mathrm{str}[D^k(r, r') \sqrt{s} S^{\ddagger}(r') \sigma^k S(r) \sqrt{s}] = \mathrm{str}[\tilde{S} \tilde{D} \, {}^T \tilde{S}] \tag{22.201}$$

with

$$\tilde{D}_{jiv,j'i'v'}(r,r') = \sum_{kpmp'm'} U^v_{pmj} \sqrt{s_i} D^k_{piv,p''i'v'}(r,r') C_{p'p''} \sqrt{s_{i'}} \epsilon_{m'm''} \sigma^k_{m''m} U^{v'}_{p'm'j'},$$

(22.202)

which may be written shortly as

$$\tilde{D} = {}^{\mathrm{T}}U \sqrt{s}(D^k C^{(p)-1} \otimes C^{(m)} \sigma^k) \sqrt{s}U,$$

(22.203)

where $\tilde{S} \in \mathcal{M}(0,1,8N,8N)$ and $\tilde{D} \in \mathcal{M}(8N,8N)$. Thus

$${}^{\mathrm{T}}(D^k C^{(p)-1} \otimes C^{(m)} \sigma^k) = \sigma_p(D^k C^{(p)-1} \otimes C^{(m)} \sigma^k)\sigma_m$$

(22.204)

and where $\sigma_p = (-)^v$ and $\sigma_m = -1$,

$${}^{\mathrm{T}}\tilde{D} = -\sigma\tilde{D}.$$

(22.205)

We have to integrate over the term bilinear in S, resp. $\tilde{S}$ in $\exp(-\mathscr{S}_2(S,R,A))$,

$$\int [\mathrm{D}\,\tilde{S}] \exp(-\tfrac{1}{2} \, \mathrm{str}[\tilde{S}\tilde{D}\,{}^{\mathrm{T}}\tilde{S}])$$

(22.206)

We substitute $X = {}^{\mathrm{T}}\tilde{S}$, which implies $\tilde{S} = -{}^{\mathrm{T}}X\sigma$ For further evaluation we use the parity transposed of Sect. 10.5,

$$\mathrm{str}[\tilde{S}\tilde{D}\,{}^{\mathrm{T}}\tilde{S}] = \mathrm{str}[-{}^{\mathrm{T}}X\sigma\tilde{D}X] = \mathrm{str}[{}^{\pi}({}^{\mathrm{T}}X)\sigma\,{}^{\pi}\tilde{D}\,{}^{\pi}X] = \mathrm{str}[{}^{\mathrm{T}}({}^{\pi}X)\,{}^{\pi}\tilde{D}\,{}^{\pi}X].$$

(22.207)

We may now integrate the exponential over ${}^{\pi}X$ instead of $\tilde{S}$. One shows, using (22.205) and the equations in Sect. 10.5, that ${}^{\pi}\tilde{D}$ is symmetric. Thus, the integral (22.206) yields

$$\mathrm{spf}({}^{\pi}\tilde{D}) = \mathrm{sdet}({}^{\pi}\tilde{D})^{-1/2} = \mathrm{sdet}(\tilde{D})^{1/2}.$$

(22.208)

Using the multiplication theorem for superdeterminants we can take away the factors $-{}^{\mathrm{T}}U\sqrt{s}$ and $\sqrt{s}U$ from $\tilde{D}$ in (22.203) and even the factors C_p^{-1} and C_m, since they all yield unit-factors. Then the integral (22.206) reduces to $\mathrm{sdet}(D^k \otimes \sigma^k)^{1/2}$ and $\mathscr{S}_3$ reads

$$\mathscr{S}_3(R,A) = -\tfrac{1}{2}\sum_{rr',kl} w_{r-r',kl} \, \mathrm{str}[R^k(r)R^l(r')] - \tfrac{1}{2} \ln \mathrm{sdet}(\sum_k D^k \otimes \sigma^k).$$

(22.209)

with D^k defined in (22.196).

22.7.2   Saddle Point and Fluctuations

We expand $\mathscr{S}_3$ to first order in δR in order to determine the saddle-point $R^{(0)}$,

$$\mathscr{S}_3(R,0) = \mathscr{S}_3(R^{(0)}0) - \sum_{r,kl} \hat{w}_{0,kl}\, \mathrm{str}[\delta R^k(r)R^{(0)l}]$$

$$- \frac{1}{2N} \sum_{krq} \mathrm{str}[\delta R^k(r) \otimes \sigma^k \frac{1}{\sum_l (R^{(0),l} - z\delta_{l,0} + \hat{t}_q^l) \otimes \sigma^l}] \qquad (22.210)$$

with

$$\hat{w}_{q,kl} = \sum_r w_{r,kl}\mathrm{e}^{-iqr}, \quad \hat{w}_{-q,kl} = \hat{w}_{q,lk} = \hat{w}_{q,kl}^*, \qquad (22.211)$$

$$\hat{t}_q^k = \tilde{t}^k \sum_r \mathrm{e}^{-iqr}t_r^k, \quad \tilde{t}^k t_r^k = \frac{1}{N}\sum_q \mathrm{e}^{iqr}\hat{t}_q^k, \quad \hat{t}_{-q}^k = v^k\hat{t}_q^k, \quad \hat{t}_q^k \in \mathbb{R}. \qquad (22.212)$$

Note the factor $\tilde{t}^k$ in the Fourier transform. The one-particle Green's function is given for R independent of r by

$$\hat{G}^{(0)}(q,z) = \frac{1}{(-R^{(0)l} + z\delta_{l,0} - \hat{t}_q^l) \otimes \sigma^l} \qquad (22.213)$$

and the saddle-point obeys

$$\sum_l \hat{w}_{0,kl}R^{(0)l} = \frac{1}{2N}\, \mathrm{tr}_m(\sigma^k \sum_q \hat{G}^{(0)}(q,z)). \qquad (22.214)$$

If the matrices D^k commute with each other, then

$$\frac{1}{\sum_k D^k \otimes \sigma^k} = \frac{\sum_k v^k D^k \otimes \sigma^k}{\sum_k v^k D^{k2}}, \quad \mathrm{tr}_m[\frac{1}{\sum_k D^k \otimes \sigma^k} \otimes \sigma^l] = 2\frac{v^l D^l}{\sum_k v^k D^{k2}}. \qquad (22.215)$$

Then the saddle-point eqs. read

$$\hat{w}_{0,00}R^{(0)0} = \frac{1}{N}\sum_q \frac{-R^{(0)0} + z - \hat{t}_q^0}{\mathscr{N}_q}, \quad \sum_\lambda \hat{w}_{0,\kappa\lambda}R^{(0)\lambda} = \frac{1}{N}\sum_q \frac{R^{(0)\kappa} + \hat{t}_q^\kappa}{\mathscr{N}_q}, \qquad (22.216)$$

$$\mathscr{N}_q := (R^{(0)0} - z + \hat{t}_q^0)^2 - \sum_\kappa (R^{(0)\kappa} + \hat{t}_q^\kappa)^2. \qquad (22.217)$$

(Greek components run from 1 to 3). Since $t^k_{-r} = v^k t^k_r$, one obtains $\hat{t}^k_{-q} = v^k \hat{t}^k_q$, which allows us to rewrite (22.216)

$$\hat{w}_{0\kappa\lambda} R^{(0)\lambda} = \frac{1}{N} \sum_q \left(\frac{R^{(0)\kappa}}{\mathcal{N}_q} - 2 \frac{R^{(0)\lambda} \hat{t}^\kappa_q \hat{t}^\lambda_q}{\mathcal{N}_q \mathcal{N}_{-q}} \right). \tag{22.218}$$

For small spin-orbit interactions the only small solution for $R^{(0)\alpha}$ is the vanishing one. This means that there is no spontaneous magnetization due to the spin-orbit forces. $R^{(0)0}$ can be chosen diagonal. Thus, the saddle-point is given by

$$R^{(0)0}_{piv,p'i'v'} = R^d_i \delta_{pp'} \delta_{ii'} \delta_{vv'}, \quad R^{(0)\kappa} = 0, \tag{22.219}$$

$$\hat{w}_{0,00} R^d_i = \frac{1}{N} \sum_q \frac{-R^d_i + z_i - \hat{t}^0_q}{(-R^d_i + z_i - \hat{t}^0_q)^2 - \sum_\kappa (\hat{t}^\kappa_q)^2}. \tag{22.220}$$

and $\mathcal{N}$ obeys

$$\mathcal{N}_{-q} = \mathcal{N}_q. \tag{22.221}$$

The Green's functions read

$$\hat{G}^{(0)0}(q,z) = \frac{-R^{(0)0} + z - \hat{t}^0_q}{\mathcal{N}_q}, \quad \hat{G}^{(0)\kappa}(q,z) = \frac{\hat{t}^\kappa_q}{\mathcal{N}_q} \tag{22.222}$$

with the symmetries

$$\hat{G}^{(0)k}(-q,z) = v^k \hat{G}^{(0)k}(q,z), \, G^{(0)k}(r,r',z) = v^k G^{(0)k}(r',r,z),$$
$$\hat{G}^{(0)k}(q,z^*) = \hat{G}^{(0)k}(q,z)^*. \tag{22.223}$$

Fluctuations $\mathscr{S}_3(R,0)$ reads, in second order in δR,

$$\mathscr{S}_3(R,0) - \mathscr{S}_3(R^{(0)},0) = -\frac{N}{2} \sum_{q,kl} \hat{w}_{q,kl} \, \mathrm{str}[\delta \hat{R}^k_q \delta \hat{R}^l_{-q}]$$

$$+ \frac{1}{4} \sum_{qq'} \mathrm{str}[\delta \hat{R}_{q'-q} \hat{G}^{(0)}(q,z) \delta \hat{R}_{q-q'} \hat{G}^{(0)}(q',z)]$$

$$= -\frac{N}{2} \sum_{q,kl,piv,p'jv'} (-)^v (\hat{w}_{q,kl} - \hat{\Pi}_{q,kl}(z_j,z_i))$$

$$\times \delta \hat{R}^k_{-q,piv,p'jv'} \delta \hat{R}^l_{q,p'jv',piv} \tag{22.224}$$

with

$$\hat{\Pi}_{q,kl}(z_j, z_i) = \frac{1}{N} \sum_{k'l'} \frac{1}{2} \operatorname{tr}(\sigma^k \sigma^{k'} \sigma^l \sigma^{l'}) \sum_{q'} \hat{G}^{(0)k'}(q+q', z_j)\hat{G}^{(0)l'}(q', z_i). \quad (22.225)$$

The first supertrace in (22.210), (22.224) does not contain summations over m, whereas the second one does. This explains factors of 2 and $\frac{1}{2}$. The coefficients $\frac{1}{2}\operatorname{tr}(\sigma^k \sigma^{k'} \sigma^l \sigma^{l'})$ equal ± 1, if the indices k, k', l, l' are pairwise equal. They equal $\pm i$, if all indices are different. Otherwise they vanish.

These symmetries and those of $G^{(0)}$, (22.223), yield

$$\hat{\Pi}_{q,kl}(z_j^*, z_i^*) = \hat{\Pi}_{q,lk}^*(z_j, z_i), \quad (22.226)$$

$$\hat{\Pi}_{q,kl}(z_i, z_j) = v^k v^l \hat{\Pi}_{q,kl}(z_j, z_i) = \hat{\Pi}_{-q,lk}(z_j, z_i) = v^k v^l \hat{\Pi}_{-q,lk}(z_i, z_j).$$

Equation (22.226) holds also for $\hat{w} - \hat{\Pi}$ and $\hat{\Gamma}$, defined in (22.235).

In particular, we obtain

$$\hat{\Pi}_{q,00}(z_j, z_i) = \frac{1}{N} \sum_{k,q'} \hat{G}^{(0)k}(q + q', z_j)\hat{G}^{(0)k}(q', z_i). \quad (22.227)$$

Next we consider the sum rules. From

$$\hat{G}^{(0)}(q, z_j)\hat{G}^{(0)}(q, z_i) = (\hat{G}^{(0)k}(q, z_j) \otimes \sigma^k)(\hat{G}^{(0)l}(q, z_i) \otimes \sigma^l) \quad (22.228)$$

$$= \frac{1}{z_j - R_j^{\rm d} - \hat{t}_q} \frac{1}{z_i - R_i^{\rm d} - \hat{t}_q}$$

$$= \frac{1}{z_i - R_i^{\rm d} - z_j + R_j^{\rm d}} \left(\frac{1}{z_j - R_j^{\rm d} - \hat{t}_q} - \frac{1}{z_i - R_i^{\rm d} - \hat{t}_q} \right)$$

$$= \frac{1}{z_i - R_i^{\rm d} - z_j + R_j^{\rm d}} (\hat{G}^{(0)k}(q, z_j) - \hat{G}^{(0)k}(q, z_i)) \otimes \sigma^k$$

we obtain

$$\sum_k \hat{G}^{(0)k}(q, z_j)\hat{G}^{(0)k}(q, z_i)$$

$$= \frac{1}{z_i - R_i^{\rm d} - z_j + R_j^{\rm d}} (\hat{G}^{(0)0}(q, z_j) - \hat{G}^{(0)0}(q, z_i)), \quad (22.229)$$

$$\hat{G}^{(0)0}(q, z_j)\hat{G}^{(0)\kappa}(q, z_i) + \hat{G}^{(0)\kappa}(q, z_j)\hat{G}^{(0)0}(q, z_i)$$

$$= \frac{1}{z_i - R_i^{\rm d} - z_j + R_j^{\rm d}} (\hat{G}^{(0)\kappa}(q, z_j) - \hat{G}^{(0)\kappa}(q, z_i)) \quad (22.230)$$

with

$$\hat{G}^{(0)\alpha}(q, z_j)\hat{G}^{(0)\beta}(q, z_i) = \hat{G}^{(0)\beta}(q, z_j)\hat{G}^{(0)\alpha}(q, z_i). \qquad (22.231)$$

We conclude the sum rules as

$$\hat{\Pi}_{0,00}(z_j, z_i) = \frac{1}{z_i - R_i^{\mathrm{d}} - z_j + R_j^{\mathrm{d}}} \frac{1}{N} \sum_q (\hat{G}^{(0)0}(q, z_j) - \hat{G}^{(0)0}(q, z_i))$$

$$(22.232)$$

$$= \frac{1}{z_i - R_i^{\mathrm{d}} - z_j + R_j^{\mathrm{d}}} (G^{(0)}(0, z_j) - G^{(0)}(0, z_i)),$$

$$\hat{\Pi}_{0,0\kappa}(z_j, z_i) = \frac{1}{z_i - R_i^{\mathrm{d}} - z_j + R_j^{\mathrm{d}}} (G^{(0)\kappa}(0, z_j) - G^{(0)\kappa}(0, z_i)) = 0.$$

$$(22.233)$$

Correlations We introduce the fluctuation contributions for the correlations. Since $\hat{w} - \hat{\Pi}$ appears in (22.224), we introduce its inverse,

$$\hat{\Gamma}_q(z_j, z_i) := (\hat{w}_q - \hat{\Pi}_q(z_j, z_i))^{-1}, \qquad (22.234)$$

that is

$$\hat{\Gamma}_{q,kl}(z_j, z_i)(\hat{w}_{q,lm} - \hat{\Pi}_{q,lm}(z_j, z_i)) = \delta_{km}. \qquad (22.235)$$

Then similar to (22.143)–(22.145), and due to (22.224), we obtain

$$\overline{\delta \hat{R}^k_{-q_1, i_1 p_1 v_1, i'_1 p'_1 v'_1} \delta \hat{R}^l_{q_2, i_2 p_2 v_2, i'_2 p'_2 v'_2}}$$

$$= -\frac{1}{N} \hat{\Gamma}_{q_1, lk}(z_{i'_1}, z_{i_1}) \delta_{q_1, q_2} \qquad (22.236)$$

$$\times \left((-)^{v'_1} \delta_{i_1 i'_2} \delta_{i'_1 i_2} \delta_{p_1 p'_2} \delta_{p'_1 p_2} \delta_{v_1 v'_2} \delta_{v'_1 v_2} + (-)^{v_1 v'_1} v^l \delta_{i_1 i_2} \delta_{i'_1 i'_2} C_{p_2 v_2, p_1 v_1} C_{p'_2 v'_2, p'_1 v'_1} \right).$$

The one-particle Green's function reads

$$\hat{G}_1(q, z_i)_{mm'} = \hat{G}^{(0)}(q, z_i)_{mm'} \qquad (22.237)$$

$$+ \frac{1}{N} \sum_{q'kl} (\hat{G}^{(0)}(q, z_i)\sigma^k \hat{G}^{(0)}(q + q', z_i)\sigma^l \hat{G}^{(0)}(q, z_i))_{mm'} \hat{\Gamma}_{q', lk}(z_i, z_i)$$

(no summation over q, i, p). The two-particle correlation function is given by

$$\hat{G}_{2,m_1 m_1', m_2 m_2'}(q_1, q_1', z_i; q_2 q_2', z_j)$$

$$= \overline{\hat{\mathscr{G}}_{q_1 ipm_1 \mathrm{f}, q_1' ipm_1' \mathrm{f}} \hat{\mathscr{G}}_{q_2 jp' m_2 \mathrm{f}, q_2' jp' m_2' \mathrm{f}}}$$

$$+ \overline{\hat{\mathscr{G}}_{q_1 ipm_1 \mathrm{f}, q_2' jp' m_2' \mathrm{f}} \hat{\mathscr{G}}_{q_2 jp' m_2 \mathrm{f}, q_1' ipm_1' \mathrm{f}}}$$

$$+ C_{p'\bar{p}'} C_{m_2 \bar{m}_2} C_{p\bar{p}} C_{m_1' \bar{m}_1'} \overline{\hat{\mathscr{G}}_{q_1 ipm_1 \mathrm{f}, -q_2 j\bar{p}' \bar{m}_2 \mathrm{f}} \hat{\mathscr{G}}_{-q_1' i\bar{p}\bar{m}_1' \mathrm{f}, q_2' jp' m_2' \mathrm{f}}} \qquad (22.238)$$

$$= \hat{G}_{1,m_1,m_1'}(q_1, z_i) \delta_{q_1, q_1'} \hat{G}_{1, m_2, m_2'}(q_2, z_j) \delta_{q_2, q_2'} + \frac{1}{N} \delta_{q_1 + q_2, q_1' + q_2'}$$

$$\times \left(\left((\hat{G}^{(0)}(q_1, z_i) \sigma^k \hat{G}^{(0)}(q_2', z_j))_{m_1, m_2'} (\hat{G}^{(0)}(q_2, z_j) \right. \right.$$

$$\times \sigma^l \hat{G}^{(0)}(q_1', z_i))_{m_2, m_1'} \hat{\Gamma}_{q_2 - q_1', lk}(z_j, z_i)$$

$$+ (\hat{G}^{(0)}(q_1, z_i) \sigma^k \hat{G}^{(0)}(-q_2, z_j) \epsilon^{-1})_{m_1 m_2} (\epsilon \hat{G}^{(0)}(-q_1', z_i) \sigma^l \hat{G}^{(0)}(q_2', z_j))_{m_1' m_2'}$$

$$\left. \times v^l \hat{\Gamma}_{-q_1 - q_2, lk}(z_j, z_i) \right) \qquad (22.239)$$

(no summation over i, p, j, p').

Then the two-particle Green's function between spins at sites r and r', where the occupation number is included for σ^0, can be written as

$$\langle \operatorname{tr}(\sigma^k(r) \frac{1}{z_i - H} \sigma^l(r') \frac{1}{z_j - H}) \rangle = \overline{\operatorname{tr}(\sigma^k G(r, r', z_i) \sigma^l G(r', r, z_j))}$$

$$= \frac{1}{N} \sum_q e^{iq(r - r')} \hat{K}_q^{k,l}(z_i, z_j) \qquad (22.240)$$

with

$$\hat{K}_q^{k,l}(z_i, z_j) = \frac{1}{N} \sum_{q_1, q_2} \sigma^k_{m_2' m_1} \sigma^l_{m_1' m_2} \hat{G}_{2, m_1 m_1', m_2, m_2'}(q_1, q_2 + q, z_i; q_2, q_1 - q, z_j)$$

$$= (\hat{w} \hat{\Gamma} \hat{\Pi})_{q, kl}(z_i, z_j) + \frac{1}{N^2} \sum_{q_1 q_2} \operatorname{tr}(\sigma^{p_1} \sigma^{\bar{k}} \sigma^{p_2} \sigma^l \sigma^{p_1'} \sigma^{\bar{l}} \sigma^{p_2'} \sigma^k)$$

$$\times \hat{G}^{(0) p_1}(q_1, z_i) \hat{G}^{(0) p_2}(-q_2, z_j) \hat{G}^{(0) p_1'}(-q_2 - q, z_i) \hat{G}^{(0) p_2'}(q_1 - q, z_j) v^l v^{\bar{l}}$$

$$\times \hat{\Gamma}_{-q_1 - q_2, \overline{lk}}(z_j, z_i). \qquad (22.241)$$

Cooperon   Time-reversal invariance

$$G_1(r', r, z) = \epsilon \, {}^{\mathrm{T}} G_1(r, r', z) \epsilon^{-1} \qquad (22.242)$$

yields

$$\overline{G_{mm'}(r, r', z_1) G_{m'm}(r', r, z_2)} = \epsilon_{m_1 m_2} \overline{G_{m_1 m_1'}(r, r', z_1) G_{m_2 m_2'}(r, r', z_2)} \epsilon_{m_1' m_2'}^{-1}$$
(22.243)

that is

$$\epsilon_{m_1 m_2} \overline{G_{2,m_1 m_1', m_2 m_2'}(r, r', z_1; r, r', z_2)} \epsilon_{m_1' m_2'}^{-1} = \overline{G_{2,mm',m'm}(r, r', z_1; r', r, z_2)}$$
(22.244)

Thus, two particles running in the same direction show similar long-range behavior to the time-reversal invariant spin-independent (orthogonal) case. The factors ϵ take care of the antisymmetrization of the particles.

Hydrodynamic Behaviour In the hydrodynamic limit we may expand

$$\hat{w}_{q,00} - \hat{\Pi}_{q,00}(E + \tfrac{1}{2}\omega, E - \tfrac{1}{2}\omega) = \frac{\hat{w}_{0,00}^2}{2\pi\rho^{(0)}}(-i\omega + D^{(0)}q^2 + \ldots),$$
(22.245)

$$\hat{w}_{q,0\alpha} - \hat{\Pi}_{q,0\alpha}(E + i0, E - i0) = i(b_\alpha \cdot q) + \ldots$$
(22.246)

$$\hat{w}_{q,\alpha 0} - \hat{\Pi}_{q,\alpha 0}(E + i0, E - i0) = i(b_\alpha \cdot q) + \ldots$$
(22.247)

$$\hat{w}_{0,\alpha\beta} - \hat{\Pi}_{0,\alpha\beta}(E + \tfrac{1}{2}\omega, E - \tfrac{1}{2}\omega) = a_{\alpha\beta} - ic_{\alpha\beta}\omega + \ldots,$$
(22.248)

where the 3×3 matrices a and c are real and symmetric. $(b_\alpha \cdot q)$ is the scalar product of the real vector b_α with q.

The lowest eigenvalue of the 4×4 matrix $\hat{w}_q - \hat{\Pi}_q(E + i0, E - i0)$, up to order q^2 and ω, is given by

$$\frac{\hat{w}_{0,00}^2}{2\pi\rho^{(0)}}(-i\omega + D^{(0)}q^2) + (b_\alpha \cdot q)(a^{-1})_{\alpha\beta}(b_\beta \cdot q),$$
(22.249)

Thus, the terms $b \cdot q$ yield additional contributions to the diffusion constant. The other eigenvalues of $\hat{w}_0 - \hat{\Pi}_q(E + i0, E - i0)$ do not vanish. The matrix $a - ic\omega$ describes the decay of the total spin due to the spin-orbit interaction, which violates spin conservation. The spin-orbit relaxation times, τ_{so}, of the decay proportional to $\exp(-t/\tau_{so})$ are obtained from $\det(a\tau_{so} - c) = 0$. Thus, the hydrodynamics of the density and spin fluctuations are obtained.

We show that indeed the matrix a in (22.248) is positive. Using the symmetry $M_{-r} = M_r$ and of γ_r^k, Eqs. (22.168), one obtains the average of $\tilde{M}_r$ and $\tilde{M}_{-r}$ by setting the terms $\gamma_0\gamma_k$ in (22.189) equal to zero. Then the matrix has two eigenvalues $M_r\tilde{\gamma}_+$ and two eigenvalues $M_r\tilde{\gamma}_-$. The eigenvalues $M_r\tilde{\gamma}_+$ belong to the eigenvectors $(1, 0, 0, 0)$ and $(0, \gamma_1, \gamma_2, \gamma_3)$. If the ratios $\gamma_1 : \gamma_2 : \gamma_3$ are equal for all distance vectors r, then the spin direction given by $\gamma_i\sigma^i$ is conserved as well as the particle

298 22 Diffusive Model

number. Otherwise, the eigenvalues of the 3×3-matrix are less than $\sum_r \tilde{M}_{r,00}$. Since $\sum_r \tilde{M}_{r,kl}\hat{w}_{0,ll'} = \frac{1}{2}\delta_{kl'}$, compare to (22.194), the eigenvalues of the 3×3 matrix $\hat{w}_0$ are larger. We have assumed that the fluctuations of the spin-dependent components are small in comparison to the spin-independent fluctuations, so that the 3×3-matrix $\sum_r \tilde{M}_{r,\kappa\lambda}$ is positive.

Further, we have to consider $\hat{\Pi}_0$. This matrix is hermitian due to (22.226). Suppose it is diagonalized. Then (22.225) yields

$$\hat{\Pi}_{0,kk}(E+\mathrm{i}0,E-\mathrm{i}0) = \sum_l c_{kl}\frac{1}{N}\sum_q \hat{G}^{(0)l}(q,E+\mathrm{i}0)\hat{G}^{(0)l}(q,E-\mathrm{i}0),$$

$$c_{kl} = \begin{cases} +1 & \text{if } k = 0 \text{ or } l = 0 \text{ or } k = l \\ -1 & \text{otherwise} \end{cases} \tag{22.250}$$

Since $\hat{G}^{(0)l}(q,E+\mathrm{i}0) = \hat{G}^{(0)l*}(q,E-\mathrm{i}0)$, the sums over q are positive and thus, $\hat{\Pi}_{0,00} > \hat{\Pi}_{0,\kappa\kappa}$. Consequently the matrix a is positive.

Nonlinear σ-Model Now only the component R^k with $k=0$ is taken into account. As for the orthogonal case, R has to obey (22.144). Again we write

$$R(r) = R^{(0)}(T) + \begin{cases} P & SW \\ TPT^{-1} & PS \end{cases}, \quad R^{(0)}(T) = \Re R^{\mathrm{d}} - \frac{\mathrm{i}\pi\rho^{(0)}}{\hat{w}_0}Q(T),$$

$$\tag{22.251}$$

$$Q(T) = TAT^{-1}, \quad TsT^{\ddagger} = s \tag{22.252}$$

with T,P,Q functions of r. P decays in blocks as given by (21.61), (21.62). This time, however, P^{bb} is antihermitian and P^{ff} is hermitian to guarantee convergence. Consider now (22.144) $\mathsf{C}R\mathsf{C}^{-1} = {}^{\mathsf{T}}R$. Obviously it has to hold for P^{R} and P^{A}, but also for Q. Similarly, as in (22.156)–(22.159), we obtain, as in (22.160),

$$T = \mathsf{C}s^{-1}T^{\times}s\mathsf{C}^{-1}. \tag{22.253}$$

This choice has the nice property that the matrices T constitute a group under multiplication. Comparing with (14.5), we see that T is pseudoreal with $g = \mathrm{i}s$. Similar to the discussion in the paragraph 'Convergence for S_{b}' in Sect. 21.4, we have to consider here the convergence of S_{f}. Basically, one has to replace the indices ${}_{\mathrm{b}}$ by ${}_{\mathrm{f}}$ in (21.64), (21.65) and then the same discussion carries through. This requires that now the real part of $(zs)_{\mathrm{ff}}$ has to be positive and $(\Lambda s)_{\mathrm{ff}} = -\mathrm{i}\mathbf{1}$. Thus,

$$s_{\mathrm{fR}} = -\mathrm{i}, \quad s_{\mathrm{fA}} = +\mathrm{i}, \quad s_{\mathrm{bR}} = s_{\mathrm{bA}}. \tag{22.254}$$

This choice, different from that of the orthogonal case, yields different T and Q.

The derivation of the non-linear σ-model is performed as for the unitary case. $\hat{w}_q$ is replaced by $\hat{w}_{q,00}$ and $\hat{\Pi}_q$ by $\hat{\Pi}_{q,00}$. Due to the different signs of $\mathcal{S}_3$ in (22.34) and (22.209), one obtains the action

$$\mathcal{S} = \tfrac{1}{4}\pi\rho^{(0)} \sum_r \mathrm{str}(D^{(0)}(\nabla Q(r))^2 + 2i\omega\Lambda Q(r)). \tag{22.255}$$

22.7.3   Some Simplifications

If we choose $t^\alpha = 0$, as many authors do, and consider only the spin-orbit effects in the fluctuating part f, then many expressions simplify, in particular,

$$G^{(0)}_{mm'}(r,r',z) = \delta_{mm'}G^{(0)0}(r,r',z), \quad G^{(0)k}(r,r',z) = \delta_{k0}G^{(0)0}(r,r',z), \tag{22.256}$$

and similarly for $\hat{G}^{(0)}(q,z)$. $\hat{\Pi}$ simplifies to

$$\hat{\Pi}_{q,kl}(z_i,z_j) = \delta_{kl}\frac{1}{N}\sum_{q'} \hat{G}^{(0)0}(q+q',z_i)\hat{G}^{(0)}(q',z_j). \tag{22.257}$$

Further,

$$\hat{K}^{kl}_q(z_i,z_j) = (\hat{w}\hat{\Gamma}\hat{\Pi})_{q,kl}(z_i,z_j) + \frac{1}{N^2}\sum_{q_1,q_2} \mathrm{tr}\,(\sigma^{\bar{k}}\sigma^l\sigma^{\bar{l}}\sigma^k)v^l v^{\bar{l}}$$
$$\times \hat{G}^{(0)0}(q_1,z_i)\hat{G}^{(0)0}(q_2,z_j)$$
$$\times \hat{G}^{(0)0}(q_2+q,z_i)\hat{G}^{(0)0}(q_1-q,z_j)\hat{\Gamma}_{q_1+q_2,\bar{k},\bar{l}}(z_i,z_j). \tag{22.258}$$

22.7.4   The Extreme and Pure Case

Coming from the Gaussian symplectic ensemble suggests the choice

$$\overline{f_{rmr'm'}} = 0, \tag{22.259}$$
$$\overline{f_{rmr'm'}f_{r''m''r'''m'''}} = M_{r-r'}(\delta_{rr'''}\delta_{r'r''}\delta_{mm'''}\delta_{m'm''} + \delta_{rr''}\delta_{r'r'''}\epsilon_{mm''}\epsilon_{m'm'''}).$$

The expressions depending on the spin-components may be rewritten as

$$\delta_{mm'''}\delta_{m'm''} = \tfrac{1}{2}\sum_k \sigma^k_{mm'}\sigma^k_{m''m'''},$$

$$\epsilon_{mm''}\epsilon_{m'm'''} = \tfrac{1}{2}\sum_k v^k \sigma^k_{mm'}\sigma^k_{m''m'''}. \tag{22.260}$$

This leads to a superposition of four contributions as mentioned after (22.191),

$$\tilde{M}_{r-r',kl} = \tfrac{1}{2}M_{r-r'}\delta_{k0}\delta_{l0}. \tag{22.261}$$

Then also,

$$\hat{\Gamma}_{q,kl}(z_i,z_j) = \delta_{k0}\delta_{l0}\,\hat{\Gamma}_{q,00}(z_i,z_j) \tag{22.262}$$

and we can leave the indices k and l in $\hat{G}^{(0)}$ and $\hat{\Gamma}$ aside and obtain

$$\hat{K}_{q,kl}(z_i,z_j) = \delta_{k0}\delta_{l0}\hat{w}_q\hat{\Gamma}_q(z_i,z_j)\hat{\Pi}_q(z_i,z_j) + \delta_{kl}v'\frac{2}{N^2}\sum_{q_1,q_2}\hat{G}^{(0)}(q_1,z_i)\hat{G}^{(0)}(q_2,z_j)$$

$$\times\hat{G}^{(0)}(q_2+q,z_i)\hat{G}^{(0)}(q_1-q,z_j)\hat{\Gamma}_{q_1+q_2}(z_i,z_j). \tag{22.263}$$

Summary   We can describe the diffusive models in three symmetry classes as in the preceding chapter on random matrix theory. The occurrence of these classes depends on the symmetry of the system, as has been summarized in Table 21.1. We will see in the following chapter that there are more symmetry classes, if certain energies are singled out.

We have considered only pure cases. If, for example, a small term breaking time-reversal invariance is added to a time-reversal invariant Hamiltonian then it will be governed by the unitary case at a frequency of the order of the inverse mean scattering time caused by the perturbation and below.

Problems

22.1   Denote $D_{q,kl}(E\pm,E\mp) := \hat{w}_{q,kl} - \hat{\Pi}_{q,kl}(E\pm i\eta, E\mp i\eta)$ with real E and positive η and $D_{q,kl}(E+,E-) = e_{kl} + o_{kl}$, where e and o are even and odd functions in q. Determine by means of the symmetry relations (22.211), (22.226) $D_{q,kl}(E-,E+)$ and the relations between e_{kl} and e_{lk}, and o_{kl} and o_{lk}. Are e_{kl} and o_{kl} real or imaginary?

22.2   Replace $i\eta$ in the previous problem by $\tfrac{1}{2}\omega + i\eta$. Derive Eqs. (22.245)–(22.248).

References

[1] E. Abrahams, P.W. Anderson, D.C. Licciardello, T.V. Ramakrishnan, Scaling theory of localization: absence of quantum diffusion in two dimensions. Phys. Rev. Lett. **42**, 673 (1979)

[64] K.B. Efetov, Supersymmetry method in localization theory. Zh. Eksp. Teor. Fiz. **82**, 872 (1982); Sov. Phys. - JETP **55**, 514 (1982)

[65] K.B. Efetov, Supersymmetry and theory of disordered metals. Adv. Phys. **32**, 53 (1983)

[66] K.B. Efetov, *Supersymmetry in Disorder and Chaos* (Cambridge University Press, Cambridge, 1997)

[67] K.B. Efetov, A.I. Larkin, D.E. Khmel'nitskii, Interaction of diffusion modes in the theory of localization. Zh. Eksp. Teor. Fiz. **79**, 1120 (1980); Sov. Phys. JETP **52**, 568 (1980)

[97] L.P. Gorkov, A.I. Larkin, D.E. Khmelnitskii, Particle conductivity in a two-dimensional random potential. Pisma Zh. Eksp. Teor. Fiz. **30**, 248 (1979); JETP Lett. **30**, 228 (1979)

[100] D.A. Greenwood, The Boltzmann equation in the theory of electrical conduction in metals. Proc. Phys. Soc. Lond. **71**, 585 (1958)

[110] S. Hikami, A.I. Larkin, Y. Nagaoka, Spin-orbit interaction and magnetoresistance in the two dimensional random system. Prog. Theor. Phys. **63**, 707 (1980)

[126] K. Jüngling, R. Oppermann, Random electronic models with spin-dependent hopping. Phys. Lett. **76A**, 449 (1980)

[127] K. Jüngling, R. Oppermann, Effects of spin-interactions in disordered electronic systems: loop expansions and exact relations among local gauge invariant models. Z. Phys. B **38**, 93 (1980)

[156] R. Kubo, A general expression for the conductivity tensor. Canad. J. Phys. **34**, 1274 (1956)

[203] R. Oppermann, F.J. Wegner, Disordered systems with n orbitals per site: $1/n$ expansion. Z. Phys. B **34**, 327 (1979)

[231] L. Schäfer, F. Wegner, Disordered system with n orbitals per site: Lagrange formulation, hyperbolic symmetry, and Goldstone modes. Z. Physik B **38**, 113 (1980)

[258] B. Velicky, Theory of electronic transport in disordered binary alloys: coherent-potential approximation. Phys. Rev. **184**, 614 (1969)

[269] F. Wegner, Disordered systems with n orbitals per site: $n = \infty$ limit. Phys. Rev. **B19**, 783 (1979)

[270] F. Wegner, The mobility edge problem: continuous symmetry and a conjecture. Z. Phys. **B35**, 207 (1979)

Chapter 23
More on the Non-linear σ-Model

Abstract In the first section of this chapter we go beyond the harmonic approximation of the nonlinear sigma-model. This allows the description of the behaviour close to the mobility edge, in particular the scaling theory of the conductivity, and the multifractality of the wave-functions. Besides the three classes, called Wigner-Dyson classes, there are seven more classes, known as chiral classes and Bogolubov-de Gennes classes. They are listed in the second section together with their relation to topological insulators and superconductors. The physics of two-dimensional disordered systems is particularly rich. Some aspects are mentioned in the third section. Finally we mention superbosonization, which in certain cases can replace the Hubbard-Stratonovich transformation. It has been particularly useful for the n-orbital model.

23.1 Beyond the Saddle-Point Solution

Until now we have only considered the saddle point. Within this approximation we obtain diffusion within the whole band. The situation changes drastically, when we take fluctuations into account. Diffusion is still possible in $d = 3$ dimensions, whereas in $d = 1$ dimension there is no longer diffusion. The eigenstates are localized. $d = 2$ is a borderline dimension, where weak localisation appears.

In the preceding chapter, we have derived the nonlinear σ-models for the three Dyson classes. The action of the non-linear σ-models can be written as

$$\mathscr{S}(Q) = \frac{\alpha}{t} \int d^d r \, \text{str}[\tfrac{1}{2}(\nabla Q(r))^2 + i\tilde{\omega} \Lambda Q(r)], \quad \tilde{\omega} = \omega/D^{(0)} \tag{23.1}$$

with

ensemble	α	$1/t$	Eq.
unitary	-1	$\tfrac{1}{2}\pi\rho^{(0)}D^{(0)}$	(22.113)
unitary	$+1$	$\tfrac{1}{2}\pi\rho^{(0)}D^{(0)}$	(22.127),
orthogonal	-1	$\tfrac{1}{4}\pi\rho^{(0)}D^{(0)}$	(22.155)
symplectic	$+1$	$\tfrac{1}{2}\pi\rho^{(0)}D^{(0)}$	(22.255)

(23.2)

© Springer-Verlag Berlin Heidelberg 2016
F. Wegner, *Supermathematics and its Applications in Statistical Physics*,
Lecture Notes in Physics 920, DOI 10.1007/978-3-662-49170-6_23

where $\rho^{(0)}$ is the density of states per energy and volume. Depending on α we have chosen for energies according to (21.44),

$$
\begin{array}{cccccl}
\alpha & s_{\mathrm{b}}^{\mathrm{R}} & s_{\mathrm{f}}^{\mathrm{R}} & s_{\mathrm{b}}^{\mathrm{A}} & s_{\mathrm{f}}^{\mathrm{A}} & \text{Eq.} \\
+1 & -i & \mp i & +i & \mp i & (21.82) \\
-1 & \mp i & -i & \mp i & +i & (22.254)
\end{array}
\tag{23.3}
$$

It is important to consider the fluctuations. The soft modes due to variations of T still implemented in the matrix Q have a decisive effect. They already appeared in the corrections to the pure diffusive behaviour in the last term of (22.154) for the orthogonal case and in the last terms of (22.258), (22.263) for the symplectic case. We will consider their consequences here.

The mobility edge behaviour, that is, the transition between extended and localized states, attracted many physicists. The first theoretic explanation was probably by Anderson [10]. He and S. Edwards introduced the replica trick for disordered systems [63]. In 1979, Anderson et al. [1] developed renormalization for the conductance. Basically, at the same time the description of this transition by means of the nonlinear sigma-model was developed [64, 65, 67, 110, 203, 231, 270] with some precursors [3, 97, 203, 269]. Very useful was Polyakov's [212] renormalization of n-vector models in $2 + \epsilon$ dimensions, which could be transferred to the nonlinear sigma-matrix-models.

23.1.1 Symmetry and Correlations

We start by considering symmetries between non-linear σ-models. Formal expansions in t yield symmetry relations for partition functions, Eqs. (23.22) and (23.30), and for correlations, Eq. (23.37). We write the action in local space

$$
\mathcal{S} = -\frac{\alpha}{t} \left(\sum_{r,r'} \tfrac{1}{2} J_{rr'} \, \mathrm{str}(Q(r)Q(r')) + \sum_{r} h(r) \, \mathrm{str}(\Lambda Q(r)) \right),
$$

$$
Q^{\mathrm{RR}} \in \mathcal{M}(n^{\mathrm{R}}, m^{\mathrm{R}}), \quad Q^{\mathrm{AA}} \in \mathcal{M}(n^{\mathrm{A}}, m^{\mathrm{A}}), \quad Q \in \mathcal{M}(n^{\mathrm{R}} + n^{\mathrm{A}}, m^{\mathrm{R}} + m^{\mathrm{A}}),
$$

$$
\tag{23.4}
$$

and assume J positive semi-definite and $\Re h(r) > 0$. Since $Q^2(r) = 1$, we express

$$
Q^{\mathrm{RR}} = (\mathbf{1} - Q^{\mathrm{RA}} Q^{\mathrm{AR}})^{1/2}, \quad Q^{\mathrm{AA}} = -(\mathbf{1} - Q^{\mathrm{AR}} Q^{\mathrm{RA}})^{1/2}.
\tag{23.5}
$$

With $Q^{\dagger}(r) = sQ(r)s^{-1}$, we insert

$$
Q^{\mathrm{RA}} = s^{\mathrm{R}\,-1} Q^{\mathrm{AR}\,\dagger} s^{\mathrm{A}}
\tag{23.6}
$$

and obtain, in harmonic approximation,

$$\mathscr{S}_0 = \frac{\alpha}{t} \sum_{rr'} \hat{J}_{rr'} \, \text{str}(s^{R\,-1}\, Q^{AR\,\dagger}(r)s^A Q^{AR}(r')) = \frac{\alpha}{t} \sum_{rr'} \hat{J}_{rr'} \, \text{str}(Q^{RA}(r)Q^{AR}(r')),$$

$$\hat{J}_{rr'} = -J_{rr'} + \delta_{rr'}(h(r) + \sum_{r''} J_{rr''}). \tag{23.7}$$

Convergence requires

$$\alpha \, s_\nu^{R\,-1} \, s_\nu^A = (-)^{1-\nu}. \tag{23.8}$$

This condition is fulfilled for the cases we consider (23.3). This yields the expectation values with respect to $\exp(-\mathscr{S}_0)$,

$$\langle Q_{ab}^{RA}(r)Q_{cd}^{AR}(r')\rangle_0 = \delta_{ad}\delta_{bc}(-)^{v_b}\alpha G_{rr'}^{(0)}, \quad G^{(0)} = t\hat{J}^{-1}. \tag{23.9}$$

The complete action reads

$$\mathscr{S} = \mathscr{S}_0 + \mathscr{S}', \quad \mathscr{S}' = \frac{\alpha}{t} \sum_r (h(r) + \sum_{r'} J_{rr'}) \sum_{k>1} \gamma_k \, \text{str}((Q^{RA}(r)Q^{AR}(r))^k)$$

$$- \frac{\alpha}{2t} \sum_{rr',kl} \gamma_k \gamma_l J_{rr'} \big(\, \text{str}((Q^{RA}(r)Q^{AR}(r))^k (Q^{RA}(r')Q^{AR}(r'))^l)$$

$$+ (^{RA} \leftrightarrow {}^{AR})\big) \tag{23.10}$$

with

$$\sqrt{1-x} = 1 - \sum_k \gamma_k x^k, \quad \gamma_k = \frac{(2k-3)!!}{2^k k!}. \tag{23.11}$$

Invariant Measure   An infinitesimal rotation yields

$$dQ^{AR} = dT^{AR}Q^{RR} - Q^{AA}dT^{AR}. \tag{23.12}$$

By means of similarity transformations, we introduce

$$\tilde{Q}^{RR} = U^R Q^{RR} \, U^{R\,-1}, \quad \tilde{Q}^{AA} = U^A Q^{AA} \, U^{A\,-1},$$

$$d\tilde{Q}^{AR} = U^A dQ^{AR} \, U^{R\,-1}, \quad d\tilde{T}^{AR} = U^A dT^{AR} \, U^{R\,-1}. \tag{23.13}$$

Then we obtain

$$\frac{\partial \tilde{Q}^{RR}}{\partial Q^{RR}} = \frac{\partial \tilde{Q}^{AA}}{\partial Q^{AA}} = 1, \quad \frac{\partial \tilde{Q}^{AR}}{\partial Q^{AR}} = \frac{\partial \tilde{T}^{AR}}{\partial T^{AR}} = \mathrm{sdet}(U^A)^{n^R - m^R} \mathrm{sdet}(U^R)^{m^A - n^A}.$$
$$(23.14)$$

We choose the similarity transformation so that $\tilde{Q}^{RR}$ and $\tilde{Q}^{AA}$ are diagonal with eigenvalues λ^R and λ^A, resp.,

$$\mathrm{d}\,\tilde{Q}^{AR}_{ij} = (\lambda^R_i - \lambda^A_j)\mathrm{d}\,\tilde{T}^{AR}_{ij}, \quad I := \frac{\partial Q^{AR}}{\partial T^{AR}} = \frac{\partial \tilde{Q}^{AR}}{\partial \tilde{T}^{AR}} = \prod_{i,j}(\lambda^R_i - \lambda^A_j)^{\sigma^R_i \sigma^A_j}, \quad (23.15)$$

where

$$\prod_r [\mathrm{D}\, T^{AR}(r)][\mathrm{D}\, T^{RA}(r)] = \prod_r [\mathrm{D}\, Q^{AR}(r)][\mathrm{D}\, Q^{RA}(r)]I^2(r) \qquad (23.16)$$

is the invariant measure. The square appears since the transformation of $\mathrm{d}\, Q^{RA}$ with $\mathrm{d}\, T^{RA}$ yields the same factor and T^{AR} and T^{RA} are linearly independent. We rewrite

$$\ln I = \sum_{i,j} \sigma^R_i \sigma^A_j \ln(\lambda^R_i - \lambda^A_j) = \sum_{i,j}(-)^{\nu^R_i + \nu^A_j}(\ln 2 + \ln(1 - \tfrac{1}{2}(\Delta^R_i + \Delta^A_j)))$$

$$= d^R d^A \ln 2 - \sum_{ij}(-)^{\nu^R_i + \nu^A_j} \sum_{n \geq 1} \frac{1}{n}(\tfrac{1}{2}(\Delta^R_i + \Delta^A_j))^n$$

$$= d^R d^A \ln 2 - (d^R + d^A) \sum_{n \geq 1} \frac{1}{n} L_n - \sum_{k \geq 1, l \geq 1} \frac{(k + l - 1)!}{k! l!} L_k L_l, \qquad (23.17)$$

$$\lambda^R_i := 1 - \Delta^R_i, \quad \lambda^A_j := -1 + \Delta^A_j,$$

$$d^R := n^R - m^R = \mathrm{str}(\mathbf{1}^{RR}), \quad d^A := n^A - m^A = \mathrm{str}(\mathbf{1}^{AA}) \qquad (23.18)$$

with

$$L_k := \sum_i (-)^{\nu^R_i} (\tfrac{1}{2}\Delta^R_i)^k = \sum_i (-)^{\nu^R_i} (\tfrac{1}{2}(1 - \lambda^R_i))^k = (\tfrac{1}{2})^k \mathrm{str}((\mathbf{1} - \sqrt{1 - Q^{RA}Q^{AR}})^k)$$

$$= \sum_j (-)^{\nu^A_j} (\tfrac{1}{2}\Delta^A_j)^k = \sum_j (-)^{\nu^A_j} (\tfrac{1}{2}(\lambda^A_j - 1))^k. \qquad (23.19)$$

Elimination In order to integrate over $[\mathrm{D}\, Q^{RA}]$ and $[\mathrm{D}\, Q^{AR}]$, one expands $\exp(-\mathscr{S}')$ multiplied by the Haar measure in a polynomial of supertraces and evaluates them with respect to $\exp(-\mathscr{S}_0)$,

$$\langle A \rangle = \int [D\,Q]e^{-\mathscr{S}}A \Big/ \int [D\,Q]e^{-\mathscr{S}} = \langle e^{-\mathscr{S}'} \prod_r I^2(r)A \rangle_0 \Big/ \langle e^{-\mathscr{S}'} \prod_r I^2(r) \rangle_0.$$

(23.20)

The last denominator equals Z/Z_0. In order to eliminate $Q^{RA}(r)$ we have to contract it with any $Q^{AR}(r')$. Both factors may be in the same supertrace or in two different ones. One obtains

$$\langle \,\mathrm{str}(Q^{RA}(r)AQ^{AR}(r')B) \rangle_0 = \alpha G^{(0)}(r, r')\,\mathrm{str}A\,\mathrm{str}B,$$

$$\langle \,\mathrm{str}(Q^{RA}(r)A)\,\mathrm{str}(Q^{AR}(r')B) \rangle_0 = \alpha G^{(0)}(r, r')\,\mathrm{str}(AB).$$

(23.21)

Now it is essential that the average over such a pair $Q^{RA}(r)$, $Q^{AR}(r')$, adds one factor of α and changes the number of supertraces by one. All supertraces of $\mathscr{S}'$ carry one factor of α, the supertraces of $I(r)$ appear always pairwise including the supertraces $\mathrm{str}(\mathbf{1})$ expressed by d^R and d^A. Thus, after integration over all degrees of freedom we find that the partition function obeys, in a formal expansion in t,

$$\frac{Z_U}{Z_{U0}}(t, d^R, d^A) = \frac{Z_U}{Z_{U0}}(-t, -d^R, -d^A).$$

(23.22)

Expectation values can be obtained from

$$\langle \prod_i \mathrm{str}(A_i^{\rho_i,\rho_i'} Q^{\rho_i',\rho_i}(r_i))) \rangle,$$

(23.23)

where ρ, ρ' stand for R and A. They evaluate to linear combinations of products $\prod_k \mathrm{str}(\prod_j A_{i(k,j)})$, where each A_i appears once in the product and $\rho_{i(k,j)} = \rho'_{i(k,j-1)}$. This expression reflects the invariance of the action and the correlations under similarity transformations of the Q.

Orthogonal and Symplectic Ensembles In these ensembles T^{AR} and T^{RA} are linearly dependent (22.157), (22.159)

$$\mathrm{d}\,T^{AR} = -\mathbf{C}^{-1}\mathrm{d}\,{}^T T^{RA}\mathbf{C}$$

(23.24)

and thus, the Haar measure contains only one factor $I(r)$ and not two. The expectation values read

$$\langle Q_{ab}^{RA}(r)Q_{cd}^{AR}(r') \rangle_0 = \delta_{ad}\delta_{bc}(-)^{v_b}\alpha G_{rr'}^{(0)}, \qquad G^{(0)} = \tfrac{1}{2}t\hat{J}^{-1}$$

(23.25)

and the contractions between Q^{RA} and Q^{AR} are the same as in (23.21). We have moreover due to

$$CQ^{\mathrm{AR}}C^{-1} = {}^{\mathrm{T}}Q^{\mathrm{RA}}, \tag{23.26}$$

the expectation values

$$\langle Q^{\mathrm{RA}}_{ab}(r)Q^{\mathrm{RA}}_{cd}(r')\rangle_0 = C_{ac}C_{bd}(-)^{\nu_a+\nu_b+\nu_a\nu_b}\alpha G^{(0)}(r,r') \tag{23.27}$$

This yields the contractions

$$\begin{aligned}
\langle\, \mathrm{str}(Q^{\mathrm{RA}}(r)AQ^{\mathrm{RA}}(r')B)\rangle_0 &= \langle\, \mathrm{str}(Q^{\mathrm{RA}}(r)A)\,\mathrm{str}(Q^{\mathrm{RA}}(r')B)\rangle_0 \\
&= \alpha G^{(0)}(r,r')\,\mathrm{str}(AC^{-1}\,{}^{\mathrm{T}}BC) \\
&= \alpha G^{(0)}(r,r')\,\mathrm{str}(BC^{-1}\,{}^{\mathrm{T}}AC). \tag{23.28}
\end{aligned}$$

Again expectation values can be obtained from (23.23), but now also factors $C^{-1}\,{}^{\mathrm{T}}AC$, instead of A, appear in the products mentioned after (23.23).

Symmetry Relations The low-temperature expansion of Z/Z_0 and of expectation values are Taylor expansions in powers of t. The partition function and correlations appear as functions of αt, but not of α or t separately. Thus, we obtain, formally, symmetry relations between the partition functions of non-compact and compact systems

$$\frac{Z}{Z_0}_{\,\mathrm{n-c}}(t) = \frac{Z}{Z_0}_{\,\mathrm{comp}}(-t), \tag{23.29}$$

and similarly for expectation values for functions of $Q^{\rho,\rho'}$. Since the orthogonal and symplectic ensemble are related by parity transposition (Sect. 10.5) and exchange of non-compact with compact, one obtains

$$\frac{Z_{\mathrm{Sp}}}{Z_{\mathrm{Sp}0}}(t,d_{\mathrm{R}},d_{\mathrm{A}}) = \frac{Z_{\mathrm{O}}}{Z_{\mathrm{O}0}}(-t,-d_{\mathrm{R}},-d_{\mathrm{A}}) \tag{23.30}$$

and similarly for expectation values.

Before the introduction of supersymmetry, only the bosonic or the fermionic sector were considered in the replica limit $d^{\mathrm{R}} = d^{\mathrm{A}} = 0$.

The matrices $\tilde{Q}$ of these models did not have Grassmann variables and one used

$$\begin{aligned}
\tilde{Q}^{\mathrm{RA}} &= \tilde{Q}^{\mathrm{AR}\,\dagger}, && \text{unitary} \\
\tilde{Q}^{\mathrm{RA}} &= \tilde{Q}^{\mathrm{AR}\,\ddagger}, && \tilde{Q} = {}^{\mathrm{t}}\tilde{Q} && \text{orthogonal} \\
\tilde{Q}^{\mathrm{RA}} &= \tilde{Q}^{\mathrm{AR}\,\ddagger}, && C\tilde{Q}C^{-1} = {}^{\mathrm{t}}\tilde{Q} && \text{symplectic}
\end{aligned} \tag{23.31}$$

$$\tilde{Q}^{\mathrm{RR}} = (\mathbf{1}^{\mathrm{R}} - \tilde{\alpha}\tilde{Q}^{\mathrm{RA}}\tilde{Q}^{\mathrm{AR}})^{1/2}, \quad \tilde{Q}^{\mathrm{AA}} = -(\mathbf{1}^{\mathrm{A}} - \tilde{\alpha}\tilde{Q}^{\mathrm{AR}}\tilde{Q}^{\mathrm{RA}})^{1/2} \tag{23.32}$$

with $\tilde{\alpha} = 1$ for the compact and $\tilde{\alpha} = -1$ for the non-compact manifold. The action read

$$\mathscr{S} = -\frac{\tilde{\alpha}}{t} \left(\sum_{r,r'} \tfrac{1}{2} J_{rr'} \, \mathrm{tr} \, (\tilde{Q}(r)\tilde{Q}(r')) + \sum_r h(r) \, \mathrm{tr} \, (\Lambda \tilde{Q}(r)) \right). \tag{23.33}$$

Cycles We compare the expectation values of products of traces and supertraces of cycles

$$B = Q^{p_1,p_2} \cdots Q^{p_{k-1},p_k} Q^{p_k,p_1}, \tag{23.34}$$

and similarly for $\tilde{B}$ expressed by factors $\tilde{Q}$. Moreover,

$$Q^{RA} Q^{AR} = \tilde{\alpha} \tilde{Q}^{RA} \tilde{Q}^{AR}. \tag{23.35}$$

Thus,

$$\mathrm{tr} \, (\tilde{B}) = (-)^v \tilde{\alpha}^{n_Q} \, \mathrm{str}(B), \tag{23.36}$$

where n_Q is the number of pairs $\tilde{Q}^{RA} \tilde{Q}^{AR}$ in $\tilde{B}$. Then we obtain the relations

$$\langle \prod \mathrm{tr} \, (\tilde{B}) \rangle_{\mathrm{O},n-c}(t, n) = (-)^{\sum n_Q} \langle \prod \mathrm{str}(B) \rangle_{\mathrm{OSp}}(-t, n),$$

$$\langle \prod \mathrm{tr} \, (\tilde{B}) \rangle_{\mathrm{Sp,comp}}(t, n) = (-)^{n_B} \langle \prod \mathrm{str}(B) \rangle_{\mathrm{OSp}}(t, -n),$$

$$\langle \prod \mathrm{tr} \, (\tilde{B}) \rangle_{\mathrm{O,comp}}(t, n) = \langle \prod \mathrm{str}(B) \rangle_{\mathrm{OSp}}(-t, n),$$

$$\langle \prod \mathrm{tr} \, (\tilde{B}) \rangle_{\mathrm{Sp},n-c}(t, n) = (-)^{n_B + \sum n_Q} \langle \prod \mathrm{str}(B) \rangle_{\mathrm{OSp}}(t, -n), \tag{23.37}$$

where n_B is the number of cycles B. n is shorthand for n^R, n^A. The arguments indicate on the l.h.s. the number of components, on the r.h.s. the differences d^R, d^A. Thus, the supersymmetric correlations can be obtained from the results of the replica symmetry. The groups O and Sp may be replaced by the unitary group U.

One has to be aware that these results are only good for small t, where the averages (23.9), (23.25) are so small that the restriction $|Q^{RA}| < 1$, in the compact sector, is negligible.

23.1.2   Scaling Theory of Conductivity

Due to the propagator $\tilde{G}^{(0)}(q) \propto 1/q^2$, the perturbation expansion of Z in powers of t will diverge for $\Im \omega = 0$ in $d \le 2$ dimensions. It will diverge in $d \ge 2$ dimensions, if there is no upper cut-off in wave-vector space which, however, is provided for finite lattice spacing a. Thus, one introduces either an ultraviolet cut-off Λ or an infrared cut-off λ by replacing $1/q^2$ by $1/(q^2 + \lambda^2)$. Efetov et al. [67], see also [65, 66], followed the renormalization group procedure by Polyakov [212] for the nonlinear vector models and eliminated the fast varying components of Q down to the wave-vector λ. Then the action keeps its form, but t becomes a function of λ. This function obeys the renormalization group equation

$$\frac{d\tilde{t}}{d \ln \lambda} = \tilde{\beta}(\tilde{t}), \quad \tilde{t} = \frac{t\lambda^\epsilon}{2^{d+1}\pi d\Gamma(d/2)}, \quad \epsilon = d - 2 \qquad (23.38)$$

with

$$\tilde{\beta}(\tilde{t}) = \begin{cases} \epsilon\tilde{t} - \tilde{t}^2 - \frac{3}{4}\zeta(3)\tilde{t}^5 + \frac{27}{64}\zeta(4)\tilde{t}^6 + O(\tilde{t}^7) & \text{orthogonal ensemble} \\ \epsilon\tilde{t} - \frac{1}{2}\tilde{t}^3 - \frac{3}{8}\tilde{t}^5 + O(\tilde{t}^7) & \text{unitary ensemble} \\ \epsilon\tilde{t} + \tilde{t}^2 - \frac{3}{4}\zeta(3)\tilde{t}^5 - \frac{27}{64}\zeta(4)\tilde{t}^6 + O(\tilde{t}^7) & \text{symplectic ensemble} \end{cases}$$
$$(23.39)$$

and the Riemann ζ-function, $\zeta(3) = 1.202$, $\zeta(4) = \pi^4/90 = 1.0823$ [31, 109]. The β-function of the non-linear σ-model describes the change of $\tilde{t}$ under variation of the length-scale. Due to (23.22) and (23.30), one has the symmetry relations

$$\tilde{\beta}_U(-\tilde{t}) = -\tilde{\beta}_U(\tilde{t}), \quad \tilde{\beta}_{Sp}(-\tilde{t}) = -\tilde{\beta}_O(\tilde{t}). \qquad (23.40)$$

The conductivity σ is related to t and $\tilde{t}$ by (22.94)

$$\sigma = e^2\rho D = (2e^2)(\pi t)^{-1} \propto \tilde{t}^{-1}\lambda^\epsilon. \qquad (23.41)$$

The conductance g (inverse resistance) of a cylindric body of length L and cross-section L^{d-1} is

$$g = \sigma L^\epsilon \propto \sigma\lambda^{-\epsilon} \propto \tilde{t}^{-1}. \qquad (23.42)$$

Thus, the $\tilde{\beta}$-function for $\tilde{t}$ may be rewritten as the β-function for the conductance g,

$$\frac{d \ln g}{d \ln L} = \frac{d \ln \tilde{t}}{d \ln \lambda} = \beta(g(L)) \qquad (23.43)$$

with

$$\beta(g) = \tilde{t}^{-1}\tilde{\beta}(\tilde{t}) = \frac{g}{g_0}\tilde{\beta}(\frac{g_0}{g}) \qquad (23.44)$$

$$= \begin{cases} \epsilon - \frac{g_0}{g} - \frac{3}{4}\zeta(3)(\frac{g_0}{g})^4 + \frac{27}{64}\zeta(4)(\frac{g_0}{g})^5 + O(g^{-6}) & \text{orthogonal ensemble} \\ \epsilon - \frac{1}{2}(\frac{g_0}{g})^2 - \frac{3}{8}(\frac{g_0}{g})^4 + O(g^{-6}) & \text{unitary ensemble} \\ \epsilon + \frac{g_0}{g} - \frac{3}{4}\zeta(3)(\frac{g_0}{g})^4 - \frac{27}{64}\zeta(4)(\frac{g_0}{g})^5 + O(g^{-6}) & \text{symplectic ensemble} \end{cases}$$

where all proportionality factors are incorporated in g_0, where $g = g_0/\tilde{t}$.

Fig. 23.1 Schematic plot of the function $\beta(g)$, for dimensions $d = 1, 2, 3$, for the symplectic (*upper blue dashed dotted line*), unitary (*middle red full line*) and orthogonal case (*lower black dashed line*)

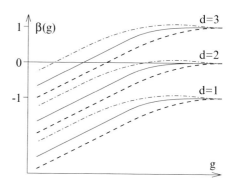

A renormalization for the conductance of the form (23.44) was predicted by Anderson et al.[1]. They gave the first correction to ϵ for the orthogonal case. $\beta(g)$ is schematically plotted in Fig. 23.1.

If, on a microscopic scale, $\beta(g)$ is positive, then g increases and finally β approaches ϵ with the behavior $g \propto L^\epsilon = L^{d-2}$, typical for metals. If, however, β is negative, then g decreases rapidly and the system insulates on a large scale. At the critical value g^*, where β vanishes, $\beta(g^*) = 0$, the system is at the mobility edge. Its conductance does not depend on the size L of the system. The critical values $\tilde{\imath}^*$ and g^* can be expanded in powers of ϵ,

$$\tilde{\imath}^* = \frac{g_0}{g^*} = \begin{cases} \epsilon - \frac{3}{4}\zeta(3)\epsilon^4 + O(\epsilon^5) & \text{orthogonal ensemble} \\ (2\epsilon)^{1/2} - \frac{3}{8}(2\epsilon)^{3/2} + O(\epsilon^{5/2}) & \text{unitary ensemble} \end{cases} \tag{23.45}$$

One obtains in the symplectic case already in two dimensions a $\tilde{\imath}^*$ of order one.

The localization exponent is obtained as

$$\nu = -1/\tilde{\beta}'(\tilde{\imath}^*) = \begin{cases} 1/\epsilon - \frac{9}{4}\zeta(3)\epsilon^2 + O(\epsilon^3) & \text{orthogonal ensemble} \\ 1/(2\epsilon) - \frac{3}{4} + O(\epsilon) & \text{unitary ensemble} \end{cases} \tag{23.46}$$

and thus, diverges as d approaches 2, whereas the exponent, $s = \epsilon \nu$, for the conductivity approaches finite values,

$$\xi \propto (g^* - g_{\text{micr}})^{-\nu}, \qquad \sigma \propto (g_{\text{micr}} - g^*)^s. \tag{23.47}$$

g_{micr} is the conductance on a microscopic scale.

One Dimensional Wires The β-function is negative for all values of g. Thus, the conductance of wires of a length large in comparison to their perpendicular extensions tends to zero. These considerations hold, however, only as long as inelastic scattering is negligible, which for long wires will become important.

Two-Dimensional Films In such films, β approaches 0 as g tends to infinity. If $g \gg g_0$, then β is small and the integration of (23.43) may be performed with the leading term in (23.44) only. Starting with g_{micr}, at a microscopic scale L_{micr}, one obtains, as long as $g, g_{\text{micr}} \gg g_0$,

$$
\begin{cases}
\frac{g}{g_0} = \frac{g_{\text{micr}}}{g_0} + \ln L_{\text{micr}} - \ln L & \text{orthogonal \quad ensemble} \\
\frac{g^2}{g_0^2} = \frac{g_{\text{micr}}^2}{g_0^2} + \ln L_{\text{micr}} - \ln L & \text{unitary \quad ensemble} \\
\frac{g}{g_0} = \frac{g_{\text{micr}}}{g_0} + \ln L - \ln L_{\text{micr}} & \text{symplectic \quad ensemble}
\end{cases}
\tag{23.48}
$$

Obviously the conductance decays slowly in the orthogonal and the unitary case. This regime is called the region of weak localization. However, the conductance increases in the symplectic case, which is called weak anti-localization. It appears when spin-orbit interactions become important. The effect of anti-localization in two dimensions was predicted by Hikami et al. [110]. Symmetries between the orthogonal and the symplectic ensemble and for the unitary ensemble in the replica limit were derived by Jüngling and Oppermann [126, 127]. However, in addition, one has to consider the effect of interactions and magnetic fields, which are responsible for dephasing. I refer the interested reader to the work by Bergmann [26–28].

The electron-electron interaction can be included in the disorder driven metal-insulator transition by means of replicas. Again a Hubbard-Stratonovich transformation can be performed and one obtains a nonlinear sigma-model with matrices Q depending, among other variables, on the Matsubara frequencies. See the work by Finkel'stein [81–83] and the reviews by Belitz and Kirkpatrick [22] and by Finkel'stein [84] for further details.

Three-Dimensional Samples Metal-insulator transitions are possible in all three-dimensional ensembles. Although one obtains an important insight from the $2 + \epsilon$-expansion, the exponents ν one can calculate from these expansions are not good estimates. It may be that the $2 + \epsilon$-expansion is an asymptotic one and, in several cases, an appropriate resummation has been tried [109]. On the basis of finite size scaling (See reviews by Kramer and Mac Kinnon [151], Kramer et al. [152]), the exponents in Table 23.1 were obtained. This table includes the exponent for the 2-dimensional metal-insulator transition in the symplectic ensemble.

Crossover Between Various Classes We have mainly considered special cases of ensembles. In reality, it may easily happen that some perturbations are admixed to these *pure* cases, which break their symmetry. The ensemble with the highest symmetry is the orthogonal one. Spin-orbit interactions shift it to the symplectic one,

Table 23.1 Localization exponent ν from finite size scaling

Class	d	ν	Ref.
Orthogonal	3	1.57 ± 0.02	[243]
Unitary	3	1.43 ± 0.04	[242]
Symplectic	3	1.375 ± 0.016	[12]
Symplectic	2	2.73 ± 0.02	[11]

magnetic impurities and magnetic fields to the unitary ensemble. These admixtures are governed by crossover exponents. Such exponents have been determined in $2 + \epsilon$-expansion by Khmelnitskii and Larkin [142], Oppermann [202], Wegner [275], and Pluhar et al.[211].

23.1.3   Density Fluctuations and Multifractality

The amplitudes $\psi(r)$ of the wave-functions are typically of order one in the localized regime and restricted to some small part of the system. In the metallic regime, they are of order $N^{-1/2}$. The fluctuations in the densities $|\psi(r)|^2$ become stronger as the mobility edge is approached.

Inverse Participation Ratio   We consider

$$P_k(E) = \overline{\sum_i |\psi_i(r)|^{2k} \delta(E - e_i)} / \rho(E) \tag{23.49}$$

and

$$1/p_k(E) = \lim_{\eta \to 0} \overline{[\sum_i |\psi_i(r)|^2 \delta_\eta (E - e_i)]^k} / \rho^k(E) \tag{23.50}$$

with the smeared δ-function

$$\delta_\eta(x) = \frac{\eta}{\pi(x^2 + \eta^2)} = \frac{1}{2\pi i} \left(\frac{1}{x - i\eta} - \frac{1}{x + i\eta} \right). \tag{23.51}$$

Here $\psi_i(r)$ is the amplitude of the eigenfunction with energy e_i at site r. P_k, and in particular P_2, are called inverse participation ratios, and p_k and p_2 the participation ratios. They are so called since they can be realized, if $1/P_2$ and Np_2 sites contribute to—that is participate in—the wave-function with $|\psi|^2 = P_2$ and $|\psi|^2 = 1/(Np_2)$, resp., and all other amplitudes vanish.

They can be obtained from

$$A_k(\eta, E) := \left(\frac{1}{2\pi i} \right)^k \overline{\left(\langle r| \frac{1}{E - i\eta - H} - \frac{1}{E + i\eta - H} |r\rangle \right)^k}, \tag{23.52}$$

which yields

$$\rho^k(E)/p_k(E) = \lim_{\eta \to 0} A_k(\eta, E) \tag{23.53}$$

and

$$P_k(E)\rho(E) = C_k^{-1} \lim_{\eta\to 0}(\eta^{k-1}A_k(\eta, E)), \tag{23.54}$$

$$C_k = \eta^{k-1}\int_{-\infty}^{+\infty}(\delta_\eta(x))^k dx = (2\pi)^{1-k}\frac{(2k-3)!!}{(k-1)!}.$$

A_k is evaluated as the expectation value of $(Q_{bb}^{RR} - Q_{bb}^{AA})^k$. Under a step of the renormalization group, which changes the length scale by a factor b, the operator A_k, distance $r - r'$, distance from criticality $E - E_c$, and the frequencies η and ω are multiplied by factors $b^{-\Delta_k}$, b^{-1}, $b^{1/\nu}$, and b^d, resp. Thus, if the size L of the system is large in comparison to the localization length ξ, then

$$P_k(E) \sim \xi^{-\tau_k} \sim |E - E_c|^{\tau_k\nu}, \quad p_k(E) \sim \xi^{\Delta_k} \sim |E_c - E|^{-\Delta_k\nu} \tag{23.55}$$

with

$$\tau_k = d(k-1) + \Delta_k. \tag{23.56}$$

Δ_k vanishes for $k = 0$ and $k = 1$,

$$\tau_0 = -d, \quad \Delta_0 = 0, \quad \tau_1 = 0, \quad \Delta_1 = 0. \tag{23.57}$$

If the localization length exceeds the size of the system L, then

$$P_k(E_c) \sim L^{-\tau_k}, \quad p_k(E_c) = L^{\Delta_k}. \tag{23.58}$$

Density Correlations The density correlations can be obtained from operators

$$O_k(r) = L^{dk}|\psi(r)|^{2k} \sim (Q_{bb}^{RR} - Q_{bb}^{AA})^k. \tag{23.59}$$

As mentioned above, under a change of length scale by a factor b, one obtains

$$\overline{O_k(r)O_{k'}(r')}_L = b^{-\Delta_k-\Delta_{k'}}\overline{O_k(r/b)O_{k'}(r'/b)}_{L/b}. \tag{23.60}$$

If we choose $b = |r - r'|/r_0$, where r_0 is the inverse of the ultraviolet cut-off, then the two operators are only a distance r_0 apart and one may use the operator product expansion

$$O_kO_{k'} = c_{k,k'}O_{k+k'}. \tag{23.61}$$

By this replacement, one obtains

$$L^{dk+dk'}\overline{|\psi(r)|^{2k}|\psi(r')|^{2k'}} \sim L^{-\Delta_{k+k'}}\left(\frac{|r-r'|}{r_0}\right)^{-\Delta_k-\Delta_{k'}+\Delta_{k+k'}}. \tag{23.62}$$

Correlation of two different eigenfunctions close in energy show the same scaling [46, 268]

$$\left. \begin{matrix} L^{2d}\overline{|\psi_i^2(r)\psi_j^2(r')|} \\ L^{2d}\overline{\psi_i^*(r)\psi_j(r)\psi_i(r')\psi_j^*(r')} \end{matrix} \right\} \sim \left(\frac{|r-r'|}{L_\omega} \right)^{\Delta_2} \tag{23.63}$$

with $\omega = E_i - E_j$, $L_\omega \sim (\rho\omega)^{-1/d}$, and $|r-r'| < L_\omega$.

Values and Symmetries for Δ Within the $2+\epsilon$-expansion Δ_k has been obtained [271, 276]

$$\Delta_k = \begin{cases} k(1-k)\epsilon + \frac{\zeta(3)}{4}k(k-1)(k^2-k+1)\epsilon^4 + O(\epsilon^5) & \text{orthogonal ens.} \\ k(1-k)(\epsilon/2)^{1/2} - \frac{3\zeta(3)}{8}k^2(k-1)^2\epsilon^2 + O(\epsilon^{5/2}) & \text{unitary ens.} \end{cases} \tag{23.64}$$

One observes that, up to the given order, Δ_k is a function of $k(1-k)$, thus,

$$\Delta_k = \Delta_{1-k}. \tag{23.65}$$

Mirlin et al. [187] have shown that (23.65) holds, since they derived

$$P_\rho(\tilde\rho) = (\tilde\rho)^{-3}P_\rho(\tilde\rho^{-1}), \tag{23.66}$$

for the probability of $\tilde\rho = |\psi|^2/\langle|\psi|^2\rangle$ for the conventional ensembles. More general symmetry relations were shown by Gruzberg et al. [102, 103]. For sufficiently negative exponents k, one has to average ψ over a small region. For further restrictions see the cited papers and the review by Evers and Mirlin [72].

Singularity Spectrum τ_q defines the increase of the moments of $|\psi^2|$ at the mobility edge. Its Legendre transform, $f(\alpha)$, yields the (envelope of the) eigenfunction density $p(|\psi|^2)$. From

$$p(|\psi|^2) \sim \frac{1}{|\psi^2|}L^{-d+f(\alpha)}, \tag{23.67}$$

one obtains, in the limit of large L,

$$P_k \sim \int d\alpha L^{-k\alpha+f(\alpha)} \tag{23.68}$$

with $\alpha = -\ln|\psi^2|/\ln L$. α and k are related by the Legendre transform

$$\tau_k + f(\alpha) = k\alpha, \quad k = f'(\alpha), \quad \alpha = \tau_k'. \tag{23.69}$$

The symmetry relation (23.65) yields

$$f(d - \alpha) = f(d + \alpha) - \alpha. \tag{23.70}$$

The Localized Regime Most explicit calculations have been performed in the regime of extended states, since they are directly obtained in the harmonic approximation of the nonlinear sigma-model. Apart from the considerations above, which directly relate exponents in the region of extended states with those at the mobility edge and in the region of localized states, the investigations by McKane and Stone [181] and by Zirnbauer [300] should be mentioned.

The supersymmetric sigma model approach was used to study various distribution functions characterizing wave function and energy level statistics in disordered systems. Reviews are given by Mirlin [184, 185].

23.2 Ten Symmetry Classes

Until now we have considered the three ensembles introduced by Dyson, which are called Wigner-Dyson classes. However, there are seven more ensembles. They differ from the Wigner-Dyson ensembles only at special energies. They were classified in a systematic way by Altland and Zirnbauer. These authors showed that all compact symmetric spaces correspond to a class of nonlinear sigma-models. These classes are classified according to their behaviour under time-reversal invariance, particle-hole symmetry, and sublattice symmetry. Depending on their topological properties they may be topological insulators and superconductors.

The ten classes were classified by Altland and Zirnbauer [8, 303]

The seven classes, besides the Wigner-Dyson classes, are known as *chiral* classes and *Bogolubov-de Gennes* classes.

An overview on the classes is also given in Sects. IV and VI of the review by Evers and Mirlin [72], in the article [304] by Zirnbauer, and in the paper by Chiu et al.[49].

23.2.1 Wigner-Dyson Classes

We start with the Wigner-Dyson classes. A system described by the Hamiltonian

$$H = c^\dagger \tilde{H} c, \tag{23.71}$$

without any constraint (apart from the hermiticity of $\tilde{H}$), is in the unitary symmetry class. If time-reversal invariance is imposed, then

$$\tilde{H} = \epsilon^\dagger \tilde{H} \epsilon^{-1} \tag{23.72}$$

is fulfilled, compare (22.162). If, moreover, one imposes SU(2) spin-rotation symmetry, then

$$\tilde{H} = {}^t\tilde{H} \tag{23.73}$$

holds.

23.2.2 Chiral Classes

The Hamiltonian of the chiral classes have the form

$$H = \frac{1}{2} \begin{pmatrix} c_a^\dagger & c_b^\dagger \end{pmatrix} \underbrace{\begin{pmatrix} 0 & h \\ h^\dagger & 0 \end{pmatrix}}_{\tilde{H}} \begin{pmatrix} c_a \\ c_b \end{pmatrix}. \tag{23.74}$$

c_a and c_b stand for annihilation operators of electrons on sites a of one sublattice and sites b of the other sublattice. Thus, electrons can hop from one sublattice to the other on a bipartite lattice, but not within the same sublattice. (Compare Sect. 18.2.1). The Hamiltonian is characterized by

$$\tilde{H} = -\tau_z \tilde{H} \tau_z, \quad \tau_z = \begin{pmatrix} 1 & 0 \\ 0 & -1 \end{pmatrix}. \tag{23.75}$$

Eigenenergies occur always in pairs, E and $-E$. If h is not a square matrix, but $h \in \mathcal{M}(N_1, N_2)$, then there are, at least, $|N_1 - N_2|$ vanishing eigenenergies.

As for the Wigner-Dyson classes the system may possess time-reversal and/or spin-rotation invariance with the conditions (23.72) and (23.73), resp. Correspondingly one obtains three chiral classes.

If time-reversal invariance is imposed, then (23.75), (23.72) yield

$$\tilde{H} = -(\tau_z \otimes \epsilon)\,{}^t\tilde{H}(\tau_z \otimes \epsilon)^{-1}, \quad (\tau_z \otimes \epsilon)^2 = -1. \tag{23.76}$$

Dyson [59] found, for the one-dimensional orthogonal case, a divergence of the density of states, Theodorou and Cohen [252] and Eggarter and Riedinger [69] a divergence of the averaged localization length,

$$\rho(E) \sim \frac{1}{|E(\ln|E|)^3}, \quad \xi(E) \sim |ln|E||. \tag{23.77}$$

The formulation in terms of a nonlinear σ-model was given by Gade and Wegner [93, 94] in the replica formulation. The model is mapped on a model of unitary matrices $Q \in U(n)$ if time-reversal invariance is broken. There appears a second term ($\mathrm{str}(Q^\dagger \nabla Q)^2$), called the Gade term, besides the conventional term $\mathrm{str}((\nabla Q)^2)$ [94]. Also, in two dimensions, the density of states and the localization length diverge at $E = 0$. This and related models have been investigated by Altland and Simons [6], Fukui [89], Fabrizio and Castellani [73], and Guruswami et al.[106].

23.2.3 Bogolubov-de Gennes Classes

The Bogolubov-de Gennes classes take the creation and annihilation of electron pairs into account. The Hamiltonian may be written as

$$H = \tfrac{1}{2} \begin{pmatrix} c^\dagger c \end{pmatrix} \underbrace{\begin{pmatrix} h & \Delta \\ -\Delta^* & -{}^t h \end{pmatrix}}_{\tilde{H}} \begin{pmatrix} c \\ c^\dagger \end{pmatrix}. \tag{23.78}$$

Hermiticity requires $h = h^\dagger$ and Fermi statistics $\Delta = -{}^t\Delta$. They can be combined in

$$\tilde{H} = -\tau_x \, {}^t\tilde{H}\tau_x, \quad \tau_x = \begin{pmatrix} 0 & 1 \\ 1 & 0 \end{pmatrix}. \tag{23.79}$$

Thus, the spectrum is particle-hole symmetric. With each eigenstate, $|\Phi\rangle$, with energy E exists $\tau_x|\Phi\rangle$ with eigenenergy $-E$.

Time-Reversal Invariance If the system is time-reversal invariant, then it obeys the additional condition (23.72). Combination with (23.79) yields

$$\tilde{H} = -(\tau_x \otimes \epsilon)\tilde{H}(\tau_x \otimes \epsilon)^{-1}, \quad (\tau_x \epsilon)^2 = -1. \tag{23.80}$$

Thus, the Hamiltonian is sublattice-symmetric in an appropriate basis.

Rotational Invariance Invariance under rotation around the z-axis in spin space allows the Hamiltonian to be written as

$$H = \begin{pmatrix} c_\uparrow^\dagger & c_\downarrow \end{pmatrix} \underbrace{\begin{pmatrix} a & b \\ b^\dagger & -{}^t a' \end{pmatrix}}_{\tilde{H}} \begin{pmatrix} c_\uparrow \\ c_\downarrow^\dagger \end{pmatrix} + \tfrac{1}{2}\,\mathrm{tr}\,(a' - a). \quad a^\dagger = a, \quad a'^\dagger = a'. \tag{23.81}$$

This Hamiltonian is, without further restriction, a member of the unitary class. If one requires invariance under SU(2), then $a' = a$ and $b = {}^t b$. This can be written as

$$\tau_y \, {}^t\tilde{H}\tau_y = -\tilde{H}. \tag{23.82}$$

The system belongs to class C.

Rotational Invariance and Time-Reversal Invariance Both conditions yield $a^* = a$ and $b^* = b$ in (23.81) which, besides (23.82), yields

$$\tilde{H} = {}^t\tilde{H}. \tag{23.83}$$

Combination of these two symmetries yields

$$\tilde{H} = -\tau_y \tilde{H} \tau_y \tag{23.84}$$

and thus a sublattice-symmetry.

23.2.4 Summary

The ten symmetry classes for the one-particle Hamiltonians, $\tilde{H}$, are listed in Table 23.2. The symmetries, TRS (time-reversal) and PHS (particle-hole), are determined by the symmetry relations

$$\tilde{H} = sC\,{}^t\tilde{H}C^{-1}, \quad C^t C = 1, \quad {}^t C = tC. \tag{23.85}$$

Time Reversal Invariance (TRS) Hamiltonians with time reversal invariance obey such a relation with $s = +1$ and $t = +1$ for the orthogonal case and $t = -1$ for the symplectic one [Eqs. (23.73) and (23.72)]. The column TRS gives t. If time-reversal invariance is violated, then we assign TRS $= 0$.

Particle Hole Symmetry (PHS) Particle-hole symmetry is obtained for $s = -1$ in (23.85). Then

$$\tilde{H}|\phi\rangle = E|\phi\rangle \succ {}^t\tilde{H}C^{-1}|\phi\rangle = -EC^{-1}|\phi\rangle. \tag{23.86}$$

Table 23.2 The ten symmetry classes and their symmetries

	Hamiltonian class	TRS	PHS	SLS
Wigner-Dyson	A (unitary)	0	0	0
	AI (orthogonal)	+1	0	0
	AII (symplectic)	−1	0	0
Chiral	AIII (chiral unitary)	0	0	1
	BDI (chiral orthogonal)	+1	+1	1
	CII (chiral symplectic)	−1	−1	1
Bogolubov-de Gennes	D	0	+1	0
	C	0	−1	0
	DIII	−1	+1	1
	CI	+1	−1	1

As explained in the text, TRS= ±1 indicates time-reversal symmetry, PHS=±1 indicates particle-hole symmetry and SLS= 1 indicates sublattice symmetry

Thus, to each eigenstate with energy E there is one with energy $-E$. We indicate $t = \pm 1$ for $s = -1$ in the column PHS. If the symmetry is missing, then 0 is indicated.

Sublattice Symmetry (SLS) If the system is both time-reversal invariant and particle-hole symmetric, then application of the two relations (23.85) denoted by C and C' yields the relation

$$\tilde{H} = -(CC')\tilde{H}(CC')^{-1}, \tag{23.87}$$

which implies a sublattice symmetry, that is, $\tilde{H}$ may be written in off-diagonal block form as in (23.74). These sublattice cases are indicated by 1 in SLS, otherwise 0 is given.

The ten classes differ by the values for TRS and PHS. Only A and AIII have the same ts. These two differ, however, since AIII obeys the sublattice symmetry SLS.

The Cartan symbols refer to the Hamilton class. For more details on the symmetric spaces see [8, 72, 232, 303].

23.2.5   Topological Insulators and Superconductors

A recent discovery is the physics of topological insulators and superconductors. It was already known for the quantum Hall effect, but now it turns out to be a more general phenomenon. Topological insulators are bulk insulators with delocalized, *i.e.* topologically protected, states on their surface. Protected means that small perturbations will not destroy the topological property.

The well known example is the quantized Hall conductivity, σ_{xy}. It arises from topologically protected edge currents. It can be interpreted as an integer Chern number [254]. We will not go through the complete classification of these systems, but refer the reader to the papers by Schnyder et al.[232, 233], by Kitaev [145], by Stone et al. [248], and by Chiu et al. [49].

The ten symmetry classes are divided in two groups: The two complex classes A and AIII and, the other eight real classes. These classes can be arranged in the Bott clock [35], Fig. 23.2. The classes are arranged according to the symmetries TRS and PHS. The two complex classes are located in the center. All classes are labelled by a number p: $p = 0, 1$ for the complex classes and $p = 0, .., 7$ for the real classes. The classes indicated in Fig. 23.2 are the Hamiltonian classes H_p. One also uses the symmetry classes $R_p = H_{p+1}$ of the classifying space introduced by Kitaev [145] and the symmetry class $S_p = R_{4-p}$ of the compact sector of the sigma-model manifold. Here the cyclic definition, modulo 2 for the complex and modulo 8 for the real classes, is assumed.

Fig. 23.2 The Bott clock.
The ten symmetry classes of
single-particle Hamiltonians
are arranged according to the
symmetries TRS and PHS

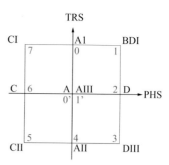

The homotopy groups π_d of the symmetry classes are essential for the classification of the topological insulators. The homotopy groups obey

$$\pi_d(R_p) = \pi_0(R_{p+d}). \tag{23.88}$$

Topological insulators of type $\mathbb{Z}$ and $\mathbb{Z}_2$ occur for

$$\pi_0(R_{p-d}) = \begin{cases} \mathbb{Z} & p = d, d+4 \bmod 8 & \text{real classes} \\ \mathbb{Z}_2 & p = d+1, d+2 \bmod 8 & \text{real classes} \\ \mathbb{Z} & p = d \bmod 2 & \text{complex classes} \end{cases} \tag{23.89}$$

The topological insulators of type $\mathbb{Z}$ are characterized by an integer (Chern number), which counts the edge states. The famous example is the integer quantum Hall effect.

Topological insulators of type $\mathbb{Z}_2$ occur in systems with strong spin-orbit coupling and time-reversal invariance, where an odd number of Kramers pairs guarantees metallic behavior.

23.3 More in Two Dimensions

Two dimensional systems (surfaces and interlayers) yield a large number of different phenomena. We mention some of them.

23.3.1 Integer Quantum Hall Effect

The description of the integer quantum Hall effect includes a topological term proportional to σ_{xy} in class A,

$$\mathscr{S} = \frac{1}{8} \int d^2r \, \mathrm{str}(-\sigma_{xx}(\nabla Q)^2 + 2\sigma_{xy}Q\nabla_x Q\nabla_y Q). \tag{23.90}$$

σ_{xx}, $\sigma_{xy} = \theta/(2\pi)$ are the conductivities in units of e^2/h. There are stable fixed points at integer σ_{xy} and $\sigma_{xx} = 0$. The quantum Hall transitions are described by the fixed points at half-integer σ_{xy} and some finite value of σ_{xx}. It is stable under variation of σ_{xx}, and unstable under variation of σ_{xy}. The exponent ν is estimated to be 2.35 ± 0.03 [114–116]. This model was first derived by Pruisken et al. in the replica formalism [166, 215, 217]. The supersymmetric generalization was given by Weidenmüller [278]. The flow diagram was proposed by Khmelnitskii [141] and Pruisken [216, 217]. See also [218].

23.3.2 Spin Quantum Hall Effect

The spin quantum Hall effect generates a spin current perpendicular to the gradient of the Zeeman field

$$j_x^s = -\sigma_{xy}^{sp} \frac{\partial B_z}{\partial y}. \tag{23.91}$$

It appears in superconductors with broken time-reversal invariance but preserved spin-rotation invariance (class C). The spin Hall conductivity σ_{xy}^{sp} appears in integer multiples of $\hbar/4\pi$. Here, I refer to the papers [9, 30, 131, 132, 222, 237, 239].

23.3.3 Quantum Spin Hall Effect

In topological insulators of class AII, a $\mathbb{Z}_2$ topological order allows for the quantum spin Hall effect. Typically there is a bulk gap, but a number, m, of gapless Kramers pairs at the edge. In the presence of disorder, the surface modes will get localized for even m. However, for odd m, one pair will be delocalized [131, 132]. A spin current flows perpendicular to the electric field proportional to it.

23.3.4 Spin Hall Effect

In these systems a spin current flows perpendicular to a charge current if topology allows. Otherwise the spins will polarize at the edges being opposite at opposite edges. The effect has been predicted by Dyakonov and Perel [57, 58] and independently by Hirsch [111]. It has been observed for example by Wunderlich et al.[291].

23.3.5 Thermal Quantum Hall Effect

Finally superconductors lacking time-reversal invariance and rotational symmetry (class D) may show the quantized thermal Hall effect, where thermal conductance divided by T, κ_{xy}/T is quantized in multiples of $\pi^2 k_B^2/6h$. See e.g. [238].

23.3.6 Wess-Zumino Term

Another term that may affect criticality is the Wess-Zumino term

$$i\mathscr{S}_{WZ} = \frac{ik}{24\pi} \int d^2r \int_0^1 ds \epsilon_{\mu\nu\lambda} \, \text{str}(Q^{-1}\partial_\mu Q Q^{-1}\partial_\nu Q Q^{-1}\partial_\lambda Q). \tag{23.92}$$

It may appear in classes AIII, CI, and DIII. with $Q(r,0) = 1$, $Q(r,1) = Q(r)$. The integer k is called the level of the Wess-Zumino-Witten-model.

23.3.7 Graphene

Graphene resides on a hexagonal lattice. A tight-binding model with nearest neighbor hopping yields a spectrum with two Dirac cones in the Brillouin zone. The three nearest neighbors of site (x_1, x_2), of one sublattice lie at $(x_1 + a, x_2)$, $(x_1 - \frac{1}{2}a, x_2 + \frac{1}{2}\sqrt{3}a)$, $(x_1 - \frac{1}{2}a, x_2 - \frac{1}{2}\sqrt{3}a)$. Then the Hamiltonian for waves with wave vector k has the form (23.74) with $h = t(e^{ik_1 a} + e^{i(-k_1+\sqrt{3}k_2)a/2} + e^{i(-k_1-\sqrt{3}k_2)a/2})$, t being the hopping matrix element. The two bands touch each other at $h = 0$, where the phases of the three exponentials differ by $\pm 2\pi/3$. Expansion of h around such a point in the Brillouin zone yields $|h| = \frac{3}{2}t|\delta k_1 \pm i\delta k_2|$ and thus the Dirac cone $E = \pm v_0|\delta k|$ with $v_0 = \frac{3}{2}t$.

Dirac Hamiltonians can realize all ten symmetry classes of disordered systems in the presence of randomness. A detailed classification has been given by Bernard and LeClair [29]. We refer the reader to the reviews by Mirlin et al.[72, 186]. An interesting feature is the experimentally found minimal conductivity at the Dirac point close to e^2/h per spin per valley [199, 295]. The superconducting gap in d-wave superconductors may even vanish at four points of the Brillouin zone [9, 197].

23.4 Superbosonization

Superbosonization allows the evaluation of integrals over functions of super-vectors, which are invariant under superunitary, and unitary-orthosymplectic transformations by replacing the integrals over the scalar products of the supervectors. Thus, superbosonization replaces the Hubbard-Stratonovich transformation in certain cases. We do not give a full derivation of superbosonization here, but some insight in the derivation and the results. It can be usefully applied to the n-orbital model.

Until now Gaussian integrals have been performed over vectors $S \in \mathcal{M}(n_1, m_1, 1, 0)$ yielding

$$\int [\mathrm{D}\, S] \mathrm{e}^{-S^\dagger A S} = (\mathrm{sdet} A)^{-1}, \quad A \in \mathcal{M}(n_1, m_1), \qquad (23.93)$$

where the bosonic sector of A is positive definite, compare Sect. 13.3. This may obviously be generalized by introducing n such vectors to $S \in \mathcal{M}(n_1, m_1, n, 0)$ as we have used for the n-orbital model,

$$I(A) = \int [\mathrm{D}\, S] \mathrm{e}^{-\mathrm{str}(S^\dagger A S)} = (\mathrm{sdet} A)^{-n}. \qquad (23.94)$$

Since $\mathrm{str}(S^\dagger A S) = \mathrm{str}(A S S^\dagger)$, one may integrate over the scalar products $W = S S^\dagger \in \mathcal{M}(n_1, m_1)$ and expect

$$I(A) = \int [\mathrm{D}\, W] J(W) \mathrm{e}^{-\mathrm{str}(AW)} = (\mathrm{sdet} A)^{-n}, \qquad (23.95)$$

where $[\mathrm{D}\, W]$ is the product of the differentials of all independent matrix elements of W. If they are complex, then integration over both real and imaginary part has to be performed. If the number of independent variables of S and W were equal, then $J(W)$ would be the Berezinian between W and $S, S^\dagger$, but this is generally not the case.

This step, from S to W is called superbosonization. Bosonization means usually, that one introduces a new entity for the product of two fermionic operators. This new entity has commutator relations close to bosons. The elements of W are bilinear in commuting and/or anticommuting components of S. Therefore the notion superbosonization is introduced.

Now $J(W)$ has to be determined. This was done rigorously by Littelmann, Sommers, and Zirnbauer [168]. Here some ideas are presented how to obtain $J(W)$ without being rigorous. In a first naïve step, the substitution

$$W = V W' V^\dagger, \quad V \in \mathcal{M}(n_1, m_1) \qquad (23.96)$$

is performed. It yields

$$I(A) = \int [\mathrm{D}\, W'] \frac{\partial W}{\partial W'} J(V W' V^\dagger) \mathrm{e}^{-\mathrm{str}(A V W' V^\dagger)} = (\mathrm{sdet} A)^{-n}. \qquad (23.97)$$

Using (10.2),

$$\frac{\partial W}{\partial W'} = (\text{sdet}(V^\dagger V))^{n_1 - m_1},$$

(23.98)

and comparing (23.97) with

$$I(V^\dagger AV) = \int [\text{D } W'] J(W') e^{-\text{str}(V^\dagger AVW')} = (\text{sdet}V^\dagger AV)^{-n} = I(A)(\text{sdet}(VV^\dagger))^n$$

(23.99)

we find that both equations agree, provided

$$\frac{J(VW'V^\dagger)}{J(W')} = (\text{sdet}(VV^\dagger))^{n - n_1 + m_1}.$$

(23.100)

Thus,

$$J(W) = (\text{sdet}W)^{n - n_1 + m_1}$$

(23.101)

fulfills, apart from a constant factor $\mathscr{N}_{n,n_1,m_1}$, Eq. (23.95) for all W and A. From Eq. (23.98), we see that

$$\text{d}\mu(W) = \frac{[\text{D } W]}{(\text{sdet}W)^{n_1 - m_1}}$$

(23.102)

is invariant under the transformation (23.96). Thus, $\text{d}\mu(W)$ is the invariant measure, or Haar measure, and

$$I(A) = \mathscr{N} \int \text{d}\mu(W)(\text{sdet}W)^n e^{-\text{str}(AW)}.$$

(23.103)

We have naïvely written down Eq. (23.97) without indicating the range over which W has to be integrated. The bosonic sector a of W is positive definite [as usual we use the representation (10.2)]. Thus, integration is performed over the positive definite hermitian matrices, a. This applies for $n \geq n_1$. For $n < n_1$, the matrix W has $n_1 - n$ zero eigenvalues. Then the present formulation does not work (see Problem 23.1). A way out has been given in [43].

Next, consider the fermionic sector b of W. Its matrix elements are bilinear in Grassmannians. A priori, it is not clear over which matrices, b, one should integrate. It turns out that integration can be done over the set of unitary matrices $b = U$. Consider the case $A \in \mathscr{M}(0, m_1)$,

$$I(A) = \mathscr{N} \int \text{d}\mu(U)(\det U)^{-n} e^{\text{tr}(AU)}.$$

(23.104)

Then $U = VU'$ yields

$$I(A) = \mathcal{N} \int d\mu(U')(\det VU')^{-n} e^{\operatorname{tr}(AVU')} = (\det V)^{-n} I(AV). \qquad (23.105)$$

From this, one concludes, with $V = A^{-1}$,

$$I(A) = (\det A)^n I(\mathbf{1}). \qquad (23.106)$$

This is valid for unitary V and thus, unitary A. Since, however, $\det A$ is linear in any component of A, it is also valid for non-unitary A. Thus, integration of the fermionic sector b of W runs over the circular unitary ensemble $\mathrm{CUE}(m_1)$ as introduced in Sect. 21.8.1.

Thus, for a matrix A, which contains only commuting components, we have

$$I\left(\begin{pmatrix} A_b & 0 \\ 0 & A_f \end{pmatrix}\right) \sim \int_{\mathrm{Herm}_+} d\mu(a) \int_{U_{m_1}} d\mu(b) \left(\frac{\det(a)}{\det(b)}\right)^n e^{-\operatorname{str}(AW)} \qquad (23.107)$$

with

$$d\mu(a) = \frac{[D\,a]}{(\det a)^{n_1}}, \quad d\mu(b) = \frac{[D\,b]}{(\det b)^{m_1}}. \qquad (23.108)$$

Comparing with (23.103), there are the following additional factors and integrations over the anticommuting variables

$$\int [D\,\alpha D\,\beta] \frac{(\operatorname{sdet} W)^{n-n_1+m_1} (\det b)^{n+m_1}}{(\det a)^{n-n_1}}$$

$$= \int [D\,\alpha D\,\beta] (\det a)^{m_1} (\det b)^{n_1} (\det(\mathbf{1} - a^{-1}\alpha b^{-1}\beta))^{n-n_1+m_1}. \qquad (23.109)$$

The substitution

$$\hat{\alpha} = a^{-1}\alpha, \quad \alpha = a\hat{\alpha}, \quad [D\,\alpha] = (\det a)^{-m_1}[D\,\hat{\alpha}],$$
$$\hat{\beta} = b^{-1}\beta, \quad \beta = b\hat{\beta}, \quad [D\,\beta] = (\det b)^{-n_1}[D\,\hat{\beta}]. \qquad (23.110)$$

reduces the integral (23.109) to

$$\int [D\,\hat{\alpha}] \int [D\,\hat{\beta}] (\det(\mathbf{1} - \hat{\alpha}\hat{\beta}))^{n-n_1+m_1}, \qquad (23.111)$$

which is a constant for given n, n_1, and m_1. All multiplicative constants have to be incorporated in the integral.

One can introduce non-commuting coefficients in A by means of a similarity transformation. $d\mu(W)$ is invariant under such a transformation. However, it is not apparent that the support of integration is invariant under such a transformation. This has been accomplished in [168]. For $n \geq n_1$, we obtain

$$I(A) = \mathcal{N} \int d\mu(W)(\text{sdet}(W))^n e^{-\text{str}(AW)}, \quad d\mu(W) = \frac{[D\,W]}{(\text{sdet}W)^{n_1-m_1}}.$$
(23.112)

If a function $f(S, S^\dagger) = F(W)$ can be written as a Laplace-integral,

$$f(S, S^\dagger) = \int [D\,A]\tilde{f}(A)e^{-\text{tr}(S^\dagger AS)},$$
(23.113)

then

$$\int [D\,S]f(S, S^\dagger) = \mathcal{N} \int d\mu(W)(\text{sdet}W)^n F(W),$$
(23.114)

which holds for functions decaying faster than any power in S. This equation, together with Eq. (23.102), is the superbosonization equation for functions invariant under superunitary transformations.

Orthogonal Case So far we have considered the unitary case. If we choose superreal vectors $S \in \mathcal{M}(n_1, 2r_1, n, 0)$, $S^\times = C^{-1}S$, (14.5), and $A \in \mathcal{M}(n_1, 2r_1)$—which obeys $A = C^{-1\,T}AC$, which implies that CA is supersymmetric $CA = {}^T(CA)\sigma$, (12.6)—with positive definite bosonic sector A_b, then

$$I = \int [D\,S]e^{-\frac{1}{2}\text{tr}(S^\ddagger AS)} = \int [D\,S]e^{-\frac{1}{2}\text{tr}({}^TSCAS)} = (\text{sdet}A)^{-n/2}$$

$$= \mathcal{N} \int d\mu(W)(\text{sdet}W)^{n/2}e^{-\frac{1}{2}\text{str}(AW)},$$
(23.115)

$$d\mu(W) = \frac{[D\,W]}{(\text{sdet}W)^{(n_1-2r_1+1)/2}}$$
(23.116)

with $W = SS^\ddagger$. Here, we have used Theorem 17.5.

Integration of the bosonic sector a of W runs over the positive definite hermitian real matrices. Integration of the fermionic sector b runs over the circular symplectic ensemble CSE$(2r)$, defined in Sect. 21.8.1.

Symplectic Case Finally we consider superreal vectors $S \in \mathcal{M}(n_1, 2r_1, 0, 2n)$, $S^\times = C^{-1}SC$, and $A \in \mathcal{M}(n_1, 2r_1)$, with the same symmetry properties as for the orthogonal case, but with positive-definite fermionic sector A_{f}. Then

$$I = \int [D\, S] e^{-\frac{1}{2}\,\mathrm{tr}\,(S^{\ddagger}AS)} = \int [D\, S] e^{-\frac{1}{2}\,\mathrm{tr}\,(C^{-1\,\mathrm{T}}SCAS)} = (\mathrm{sdet}A)^n$$

$$= \mathscr{N} \int d\,\mu(W)(\mathrm{sdet}W)^{-n} e^{-\frac{1}{2}\,\mathrm{str}(AW)}, \tag{23.117}$$

$$d\,\mu(W) = \frac{[D\,W]}{(\mathrm{sdet}W)^{(n_1 - 2r_1 - 1)/2}}, \tag{23.118}$$

where Theorem 17.5 has again been used. Integration of the bosonic sector, a, of W runs over the circular orthogonal ensemble COE(n), defined in Sect. 21.8.1. Integration of the fermionic sector b runs over the positive-definite hermitian quaternion matrices.

Application The technique of superbosonization was applied in [43, 168] to a unitary and orthogonal model, where n orbitals are located at each site. Particular cases were considered earlier [56, 68, 90]. Here we consider only the most simple application, the model of Sect. 4.4. Starting from (4.15), (4.16) and

$$W = \begin{pmatrix} a & \alpha \\ \beta & b \end{pmatrix} = \begin{pmatrix} x^{\dagger}x & x^{\dagger}\eta \\ x^{\dagger}\eta & \bar{\xi}\eta \end{pmatrix} \tag{23.119}$$

we obtain the integral

$$\int d\,a\, d\,b\, d\,\alpha\, d\,\beta\, \frac{\partial^2}{\partial\alpha\,\partial\beta} \exp\left(-sz\,\mathrm{str}W - \frac{1}{2gN}\,\mathrm{str}(W^2)\right) \mathrm{sdet}^N(W) \tag{23.120}$$

$$= \int d\,a\, d\,b\, a^N \exp(-sza - \frac{a^2}{2gN}) \underbrace{}\, b^{-N} \exp(szb + \frac{b^2}{2gN}) \underbrace{} (\frac{1}{gN} + \frac{N}{ab}).$$

The extrema of the underbraced functions are at

$$a_{\mathrm{ext}} = b_{\mathrm{ext}} = N\left(\frac{szg}{2} \pm \sqrt{g - \frac{g^2 z^2}{4}}\right). \tag{23.121}$$

The Green's function $G(z)$ is s/N times the maximum of a in agreement with (4.15). We realize that in Sect. 4.4 we have used the superbosonization idea in the bosonic sector, where it is evident, whereas in the fermionic sector we have used the Hubbard-Stratonovich transformation. It may appear that the extreme solution is the maximum for a and the minimum for b. This is not the case, since one integrates in different directions in the complex plane.

Problem

23.1 Suppose $n_1 = m_1$. Show that the integral (23.111) vanishes, if $n < n_1$. Hint: What is the highest power of $\hat{\alpha}\hat{\beta}$ in the expansion of the determinant?

References

[1] E. Abrahams, P.W. Anderson, D.C. Licciardello, T.V. Ramakrishnan, Scaling theory of localization: absence of quantum diffusion in two dimensions. Phys. Rev. Lett. **42**, 673 (1979)

[3] A. Aharony, Y. Imry, The mobility edge as a spin-glass problem. J. Phys. C **10**, L487 (1977)

[5] G. Akeman, J. Baik, P. di Francesco (eds.), *Handbook of Random Matrix Theory* (Oxford University Press, Oxford, 2011)

[6] A. Altland, B.D. Simons, Field theory of the random flux model. Nucl. Phys. B **562**, 445 (1999)

[8] A. Altland, M.R. Zirnbauer, Nonstandard symmetry classes in mesoscopic normal-superconducting hybrid structures. Phys. Rev. B **55**, 1142 (1997)

[9] A. Altland, B.D. Simons, M.R. Zirnbauer, Theories of low-energy quasi-particle states in disordered d-wave superconductors. Phys. Rep. **359**, 283 (2002)

[10] P.W. Anderson, Absence of diffusion in certain random lattices. Phys. Rev. **109**, (1958) 1492

[11] Y. Asada, K. Slevin, T. Ohtsuki, Anderson transition in two-dimensional systems with spin-orbit coupling. Phys. Rev. Lett. **89**, (2002) 256601

[12] Y. Asada, K. Slevin, T. Ohtsuki, Anderson transition in the three dimensional symplectic universality state. J. Phys. Soc. Jpn. **74**(Suppl), (2005) 238

[22] D. Belitz, T.R. Kirkpatrick, The Anderson-Mott transition. Rev. Mod. Phys. **66**, 261 (1994)

[26] G. Bergmann, Physical interpretation of weak localization: a time-of-flight experiment with conduction electrons. Phys. Rev. B **28**, 2914 (1983)

[27] G. Bergmann, Weak localization in thin films: a time-of-flight experiment with conduction electrons. Phys. Rep. **107**, 1 (1984)

[28] G. Bergmann, Weak localization and its applications as an experimental tool, in *50 Years of Anderson Localization*, ed. by E. Abrahams (World Scientific, Singapore, 2010), p. 231

[29] D. Bernard, A. LeClair, A classification of 2d random Dirac fermions. J. Phys. A **35**, 2555 (2002)

[30] B.A. Bernevig, S.C. Zhang, Quantum spin Hall effect. Phys. Rev. Lett. **96**, 106802 (2006)

[31] W. Bernreuther, F. Wegner, Four-loop order β-function for two dimensional non-linear σ models. Phys. Rev. Lett. **57**, 1383 (1986)

[35] R. Bott, The stable homotopy of the classical groups. Ann. Math. **70**, 313 (1959)

[43] J.E. Bunder, K.B. Efetov, V.E. Kravtsov, O.M. Yevtushenko, M.R. Zirnbauer, Superbosonization formula and its application to random matrix theory. J. Stat. Phys. **129**, (2007) 809

[46] J.T. Chalker, Scaling and eigenfunction correlations near a mobility edge. Physica A **167**, 253 (1990)

[49] C. Chiu, J.C.Y. Teo, A.P. Schnyder, S. Ryu, Classification of topological quantum matter with symmetries. arXiv 1505.03535 (2015)

[56] M. Disertori, H. Pinson, T. Spencer, Density of states of random band matrices. Commun. Math. Phys. **232**, 83 (2002)

[58] M.I. Dyakonov, V.I. Perel, Possibility of orienting electron spins with current. Pis'ma Zh. Eksp. Teor. Fiz. **13**, 657 (1971); Sov. Phys. JETP Lett. **13**, 467 (1971)

[57] M.I. Dyakonov, V.I. Perel, Current-induced spin orientation of electrons in semiconductors. Phys. Lett. A **25**, 459 (1971)

[59] F.J. Dyson, The dynamics of a disordered linear chain. Phys. Rev. **92**, 1331 (1953)

[63] S.F. Edwards, P.W. Anderson, Theory of spin glasses. J. Phys. F **5**, 965 (1975)

[64] K.B. Efetov, Supersymmetry method in localization theory. Zh. Eksp. Teor. Fiz. **82**, 872 (1982); Sov. Phys. JETP **55**, 514 (1982)

[65] K.B. Efetov, Supersymmetry and theory of disordered metals. Adv. Phys. **32**, 53 (1983)

[66] K.B. Efetov, *Supersymmetry in Disorder and Chaos* (Cambridge University Press, Cambridge, 1997)

[67] K.B. Efetov, A.I. Larkin, D.E. Khmel'nitskii, Interaction of diffusion modes in the theory of localization. Zh. Eksp. Teor. Fiz. **79**, 1120 (1980); Sov. Phys. JETP **52**, 568 (1980)

[68] K.B. Efetov, G. Schwiete, K. Takahashi, Bosonization for disordered and chaotic systems. Phys. Rev. Lett. **92**, 026807 (2004)

[69] T.P. Eggarter, R. Riedinger, Singular behavior of tight-binding chains with off-diagonal disorder. Phys. Rev. B **18**, 569 (1978)

[72] F. Evers, A.D. Mirlin, Anderson transitions. Rev. Mod. Phys. **80**, 1355 (2008)

[73] M. Fabrizio, C. Castellani, Anderson localization in bipartite lattices. Nucl. Phys. B **583**, 542 (2000)

[81] A.M. Finkel'stein, The influence of Coulomb on the properties of disordered metals. Zh. Eksp. Teor. Fiz. **84**, 168 (1983); Sov. Phys. JETP **57**, 97 (1983)

[82] A.M. Finkel'stein, Weak localization and Coulomb interactions in disordered systems. Z. Phys. B **56**, 189 (1984)

[83] A.M. Finkel'stein, Electron liquid in disordered conductors. Sov. Sci. Rev. A; Phys. Rev. **14**, 1 (1990)

[84] A.M. Finkel'stein, Disordered electron liquid with interactions, in *50 Years of Anderson Localization*, ed. by E. Abrahams (World Scientific, Singapore, 2010), p. 385

[89] T. Fukui, Critical behavior of two-dimensional random hopping fermions with π-flux. Nucl. Phys. B **562**, 477 (1999)

[90] Y.V. Fyodorov, Negative moments of characteristic polynomials of random matrices: Ingham-Siegel integral as an alternative to Hubbard-Stratonovich transformation. Nucl. Phys. B **621**, 643 (2002)

[93] R. Gade, Anderson localization for sublattice models. Nucl. Phys. B **398**, 499 (1993)

[94] R. Gade, F. Wegner, The $n = 0$ replica limit of U(n) and U(n)/SO(n) models. Nucl. Phys. B **360**, 213 (1991)

[97] L.P. Gorkov, A.I. Larkin, D.E. Khmelnitskii, Particle conductivity in a two-dimensional random potential, Pisma Zh. Eksp. Teor. Fiz. **30**, 248 (1979) ; JETP Lett. **30**, 228 (1979)

[102] I.A. Gruzberg, A.W.W. Ludwig, A.D. Mirlin, M.R. Zirnbauer, Symmetries of multifractal spectra and field theories of Anderson localization. Phys. Rev. Lett. **107**, 086403 (2011)

[103] I.A. Gruzberg, A.D. Mirlin, M.R. Zirnbauer, Classification and symmetry properties of scaling dimensions of Anderson transitions. Phys. Rev. B **87**, 125144 (2013)

[106] S. Guruswamy, A. LeClair, A.W.W. Ludwig, gl(N|N) Supercurrent algebras for disordered Dirac fermions in two dimensions. Nucl. Phys. **583**, 475 (2000)

[109] S. Hikami, Localization, nonlinear σ model and string theory. Prog. Theor. Phys. Suppl. **107**, 213 (1992)

[110] S. Hikami, A.I. Larkin, Y. Nagaoka, Spin-orbit interaction and magnetoresistance in the two dimensional random system. Prog. Theor. Phys. **63**, 707 (1980)

[111] J.E. Hirsch, Spin Hall effect. Phys. Rev. Lett. **83**, 1834 (1999)

[114] B. Huckestein, Scaling theory of the integer quantum Hall effect. Rev. Mod. Phys. **67**, 357 (1995)

[115] B. Huckestein, B. Kramer, One-parameter scaling in the lowest Landau band: Precise determination of the critical behavior of the localization length. Phys. Rev. Lett. **64**, 1437 (1990)

[116] B. Huckestein, B. Kramer, L. Schweitzer, Characterization of the electronic states near the centres of the Landau bands under quantum Hall conditions. Surf. Sci. **263**, 125 (1992)

[126] K. Jüngling, R. Oppermann, Random electronic models with spin-dependent hopping. Phys. Lett. A **76**, 449 (1980)

[127] K. Jüngling, R. Oppermann, Effects of spin-interactions in disordered electronic systems: loop expansions and exact relations among local gauge invariant models. Z. Phys. B **38**, 93 (1980)

[131] C.L. Kane, E.J. Mele, Z_2 topological order and the quantum spin Hall effect. Phys. Rev. Lett. **95**, 146802 (2005)

[132] C.L. Kane, E.J. Mele, Quantum spin Hall effect in graphene. Phys. Rev. Lett. **95**, 226801 (2005)

[141] D.E. Khmelnitskii, Quantization of Hall conductivity. JETP Lett. **38**, 552 (1984)

[142] D.E. Khmelnitskii, A.I. Larkin, Mobility edge shift in external magnetic field, Solid State Commun. **39**, 1069 (1981)

[145] A. Kitaev, Periodic table for topological insulators and superconductors. AIP Conf. Proc. **1134**, 22 (2009)

[151] B. Kramer, A. MacKinnon, Localization theory and experiment. Rep. Prog. Phys. **56**, 1469 (1993)

[152] B. Kramer, A. MacKinnon, T. Ohtsuki, K. Slevin, Finite size scaling analysis of the Anderson transition, in *50 Years of Anderson Localization*, ed. by E. Abrahams (World Scientific, Singapore, 2010), p. 347

[166] H. Levine, S.B. Libby, A.M.M. Pruisken, Electron delocalization by a magnetic field in two dimensions. Phys. Rev. Lett. **51**, 1915 (1983)

[168] P. Littelmann, H.-J. Sommers, M.R. Zirnbauer, Superbosonization of invariant random matrix ensembles. Commun. Math. Phys. **283**, 343 (2008)

[181] A.J. McKane, M. Stone, Localization as an alternative to Goldstone's theorem. Ann. Phys. **131**, 36 (1981)

[184] A.D. Mirlin, Statistics of energy levels and eigenfunctions in disordered and chaotic systems: supersymmetry approach, in *Proceedings of the International School of Physics "Enrico Fermi" on New Directions in Quantum Chaos*, Course CXLIII, ed. by G. Casati, I. Guarneri, U. Smilansky (IOS Press, Amsterdam, 2000), p. 223

[185] A.D. Mirlin, Statistics of energy levels and eigenfunctions in disordered systems. Phys. Rep. **326**, 259 (2000)

[186] A.D. Mirlin, F. Evers, I.V. Gornyi, P.M. Ostrovsky, *Anderson localization: Criticality, symmetries and topologies*, in *50 Years of Anderson Localization*, ed. by E. Abrahams (World Scientific, Singapore, 2010), p. 107

[187] A.D. Mirlin, Y.V. Fyodorov, A. Mildenberger, F. Evers, Exact relations between multifractal exponents at the Anderson transition. Phys. Rev. Lett. **97**, 046803 (2006)

[197] A.A. Nersesyan, A.M. Tsvelik, F. Wenger, Disorder effects in two-dimensional Fermi systems with conical spectrum: exact results for the density of states. Nucl. Phys. B **438**, 561 (1995)

[199] K.S. Novoselov, A.K. Geim, S.V. Morozov, D. Jiang, M.J. Katsnelson, I.V. Grigorieva, S.V. Dubonos, A.A. Firsov, Two-dimensional gas of massless Dirac fermions in graphene. Nature **438**, 197 (2005)

[202] R. Oppermann, Magnetic field induced crossover in weakly localized regimes and scaling of the conductivity. J. Phys. Lett. **45**, L-1161 (1984)

[203] R. Oppermann, F.J. Wegner, Disordered Systems with n orbitals per site: $1/n$ expansion. Z. Phys. B **34**, 327 (1979)

[211] Z. Pluhar, H.A. Weidenmüller, J.A. Zuk, C.H. Lewenkopf, F.J. Wegner, Crossover from orthogonal to unitary symmetry for ballistic electron transport in chaotic microstructures. Ann. Phys. **243**, 1 (1995)

[212] A.M. Polyakov, Interaction of Goldstone particles in two dimensions. Applications to ferromagnets and massive Yang-Mills fields. Phys. Lett. B **59**, 79 (1975)

[215] A.M.M. Pruisken, On localization in the theory of the quantized Hall effect: a two-dimensional realization of the θ-vacuum. Nucl. Phys. B **235**, 277 (1984)

[216] A.M.M. Pruisken, Dilute instanton gas as the precursor to the integral Hall quantum effect. Phys. Rev. B **32**, 2636 (1985)

[217] A.M.M. Pruisken, in *The Quantum Hall Effect*, ed. by R. Prange, S. Girvin (Springer, Berlin, 1987)

[218] A.M.M. Pruisken, *Topological principles in the theory of Anderson localization*, in *50 Years of Anderson Localization*, ed. by E. Abrahams (World Scientific, Singapore, 2010), p. 503

[222] N. Read, D. Green, Paired states of fermions in two dimensions with breaking of parity and time-reversal symmetries and the fractional quantum hall effect. Phys. Rev. B **61**, 10267 (2000)

[231] L. Schäfer, F. Wegner, Disordered System with n orbitals per site: Lagrange formulation, hyperbolic symmetry, and Goldstone modes. Z. Phys. B **38**, 113 (1980)

[232] A.P. Schnyder, S. Ryu, A. Furusaki, A.W.W. Ludwig, Classification of topological Insulators and superconductors in three dimensions. Phys. Rev. B **78**, 195125 (2008)

[233] A.P. Schnyder, S. Ryu, A. Furusaki, A.W.W. Ludwig, Classification of topological Insulators and superconductors. AIP Conf. Proc. **1134**, 10 (2009)

[237] T. Senthil, M.P.A. Fisher, Quasiparticle density of states in dirty high-T_c superconductors. Phys. Rev. B **60**, 6893 (1999)

[238] T. Senthil, M.P.A. Fisher, Quasiparticle localization in superconductors with spin-orbit scattering. Phys. Rev. B **61**, 9690 (2000)

[239] T. Senthil, M.P.A. Fisher, L. Balents, C. Nayak, Quasiparticle transport and localization in high-T_c superconductors. Phys. Rev. Lett. **81**, 4704 (1998)

[242] K. Slevin, T. Ohtsuki, The Anderson transition: Time reversal symmetry and universality. Phys. Rev. Lett. **78**, 4083 (1997)

[243] K. Slevin, T. Ohtsuki, Corrections to scaling at the Anderson transition. Phys. Rev. Lett. **82**, 382 (1999)

[248] M. Stone, C. Chiu, A. Roy, Symmetries, dimensions, and topological insulators: the mechanism behind the face of the Bott clock. J. Phys. A **44**, 045001 (2011)

[252] G. Theodorou, M.H. Cohen, Extended states in a one-dimensional system with off-diagonal disorder. Phys. Rev. B **13**, 4597 (1976)

[254] D.J. Thouless, M. Kohmoto, M.P. Nightingale, M. den Nijs, Quantized Hall conductance in a two-dimensional periodic potential. Phys. Rev. Lett. **49**, 405 (1982)

[268] F.J. Wegner, Electrons in disordered systems. Scaling near the mobility edge. Z. Phys. B **25**, 327 (1976)

[269] F. Wegner, Disordered systems with n orbitals per site: $n = \infty$ limit. Phys. Rev. B **19**, 783 (1979)

[270]   F. Wegner, *The mobility edge problem: continuous symmetry and a conjecture*. Z. Phys. B **35**, 207 (1979)

[271]   F. Wegner, Inverse participation ratio in $2 + \epsilon$ dimensions. Z. Phys. B **36**, 209 (1980)

[275]   F.J. Wegner, Crossover of the mobility edge behaviour. Nucl. Phys. B **270**[FS16], 1 (1986)

[276]   F. Wegner, Anomalous dimensions for the nonlinear sigma-model in $2 + \epsilon$ dimensions (I, II). Nucl. Phys. B **280**[FS18], 193/210 (1987)

[278]   H.A. Weidenmüller, Single electron in a random potential and a strong magnetic field. Nucl. Phys. B **290**, 87 (1987)

[291]   J. Wunderlich, B. Kaestner, J. Sinova, T. Jungwirth, Experimental observation of the spin-Hall effect in a two-dimensional spin-orbit coupled semiconductor system. Phys. Rev. Lett. **94**, 047204 (2004)

[295]   Y. Zhang, Y.-W. Tan, H.L. Stormer, P. Kim, Experimental observation of the quantum Hall effect and Berry's phase in graphene. Nature **438**, 201 (2005)

[300]   M.R. Zirnbauer, Anderson localization and non-linear sigma model with graded symmetry. Nucl. Phys. B **265**, 375 (1986)

[303]   M.R. Zirnbauer, Riemannian symmetric superspaces and their origin in random-matrix theory. J. Math. Phys. **37**, 4986 (1996)

[304]   M.R. Zirnbauer, Symmetry classes, in ed. by G. Akeman, J. Baik, P. di Francesco *Handbook of Random Matrix Theory* (Oxford University Press, Oxford, 2011), p. 43

Chapter 24
Summary and Additional Remarks

Abstract In this final chapter, I summarize the contents of this book and add a few remarks on further subjects and papers. They are related to the subjects considered in this book, but do not necessarily use Grassmann variables or supersymmetry.

Part I Grassmann Variables and Applications

In the first part, the mathematics of Grassmann variables, that is of anticommuting variables, was introduced. We started with the algebra and the analysis of Grassmann variables. Surprisingly, integration and differentiation of Grassmann variables is identical. Exterior algebra is briefly considered and used to formulate Maxwell's equations.

There are two operations of conjugation. These antilinear operations correspond to hermitian conjugation and to the operation of time-reversal.

Probably the most important application of Grassmann variables are fermionic path integrals. We gave a short introduction, but this field is very rich. It is used nearly everywhere, where quantum mechanics and quantum field theory has to be applied, unless only bosons are considered.

Other interesting applications are the solution of the two-dimensional dimer problem and of the two-dimensional Ising model, both without crossing bonds. We also considered a first application to the random matrix model, which we considered in more detail in Part III.

Part II Supermathematics

Supermathematics is the combination of commuting and anticommuting variables on an equal footing. Vectors and matrices, with even and odd components, were introduced and supervectors and supermatrices were formed.

Two types of transpositions of matrices, two types of adjoint, the definition of superreal and supersymmetric matrices were also introduced. The two types of superunitary groups related to the two types of adjoint were discussed.

© Springer-Verlag Berlin Heidelberg 2016
F. Wegner, *Supermathematics and its Applications in Statistical Physics*,
Lecture Notes in Physics 920, DOI 10.1007/978-3-662-49170-6_24

Integrals over functions invariant under superunitary transformations show interesting cancellation properties. An equal number of even and odd components of supervectors may be simply set to zero. Similarly the off-diagonal elements of pairs of columns and rows of supermatrices with different Z_2-degrees may be set to zero and the diagonal elements equal. Then the integral over the remaining components yields the same result.

Obviously more has to be done in supermathematics. Such a contribution is the Fourier analysis on hyperbolic supermanifolds by Zirnbauer [301].

Part III Supersymmetry in Statistical Physics

Supersymmetry in particle physics predicts bosons and fermions with equal masses, called supersymmetric partners. In this theory, two pairs of anticommuting spacetime components are added to the conventional four spacetime components. Although supersymmetric partners have not been found at present, the basic mathematical idea can be built into the theory of stochastic time-dependent equations, where the symmetry yields the fluctuation-dissipation theorem.

Supersymmetric quantum mechanics consists of pairs of Hamitonians with equal excitation spectrum. They can be obtained from the Hamiltonians $Q^\dagger Q$ and $QQ^\dagger$, which, with the exception of eigenstates at zero energy, have identical spectra.

The interaction of some disordered systems can be formulated in a supersymmetric way, which allows the reduction of d-dimensional disordered systems to $d-2$-dimensional pure systems. The odd components enter, since a Jacobian is expressed as an integral over Grassmann variables.

The next two chapters were devoted to the random matrix model and the diffusive model and their representation as nonlinear sigma-models. This allowed us to consider the level statistics, the diffusion (Diffusons) and Cooperons. Apart from the behaviour at the Anderson transition, which we considered in the following Sect. 23.1, there are more applications like quantum dots and persistent currents. The theory is also closely related to that of quantum chaos, compare [193]. Intensively considered systems are the stadium billiard and Rydberg atoms in strong magnetic fields. We also mention Gutzwillers trace formula for periodic orbits [107].

The electron-electron interaction can be included in the disorder driven metal-insulator transition by means of replicas. Finkel'stein obtained a nonlinear sigma-model with matrices Q depending, among other variables, on the Matsubara frequencies [81–83]. See also the reviews by Belitz and Kirkpatrick [22] and by Finkel'stein [84].

Aspects not considered here are the use of ratios of characteristic polynomials (see e.g. [38])

$$F_k(\lambda_1,\dots\lambda_k;\mu_1,\dots\mu_k) = \langle\prod_i \frac{\det(\lambda_i\mathbf{1}-f)}{\det(\mu_i\mathbf{1}-f)}\rangle, \qquad (24.1)$$

which allow the determination of Green's functions—the average runs over the matrices f—for example

$$\frac{\partial}{\partial\lambda}\langle\frac{\det(\lambda\mathbf{1}-f)}{\det(\mu\mathbf{1}-f)}\rangle_{\mu=\lambda} = \langle \operatorname{tr}\frac{1}{\lambda\mathbf{1}-f}\rangle, \tag{24.2}$$

and the use of Harish-Chandra-Itzykson-Zuber integrals [108, 124]

$$I(A, B) = \int \mathrm{d}\,g\exp(\operatorname{str}(Ag^{-1}Bg)), \tag{24.3}$$

where A and B are matrices and the integral runs over matrices g out of a group. These integrals serve for the determination of Green's functions [104], see also [105].

Random matrix theory is of importance for counting planar graphs [40] and critical phenomena on lattices with fluctuating geometry [150]. It is also related to knot theory [299].

There seems to be a surprising relation between the unitary random matrix model and the Riemann ζ-function. This function $\zeta(z)$ has zeros at negative even arguments z. The Riemann conjecture states that all other zeros are located at $z = \frac{1}{2} + i\gamma$, with real γ. Apparently the local distribution of the γs follows the distribution of the unitary random matrix ensemble [136, 192]. But even a bus system, whose schedule is not well observed or not existent, runs in time-intervals described by the unitary random matrix ensemble [14, 154, 155].

The use of the nonlinear sigma-model has now arrived in strict mathematics. A pioneering paper is that by Disertori, Pinson, and Spencer [56]. Shcherbina [240] shows the universality of spectral correlation functions of the n-orbital model, and Bao and Erdős [16] show delocalization of certain band matrices by means of the nonlinear sigma-model.

For a long time the theory of free probability of Voiculescu [263]—see also the paper by Speicher [245]—and the supersymmetry method ran parallel. Mandt and Zirnbauer [173] showed, inspired by a paper by Zinn-Justin [298], that they are related. See also [246].

Many results on random matrix models can be found in the review by Guhr, Müller-Groehling, and Weidenmüller [105], in the book by Bleher and Its [32], and in the handbook [5] edited by Akeman, Baik, and di Francesco.

In Chap. 23, we considered the Anderson transition, in particular, the scaling theory of conductivity and multifractality close to the mobility edge. Further we considered the ten classes of disordered systems: The three Wigner-Dyson classes, the three chiral classes, and the four Bogolubov-de Gennes classes. They are related to the ten symmetric spaces. Their relation to the Bott clock and the topological insulators and to metal-insulator and superconductor insulator transitions is a field of large present interest, see the review by Chiu et al.[49]. Chiral models are also of importance in quantum chromodynamics [259, 260].

338   24 Summary and Additional Remarks

Further, a short account of the physics of two-dimensional disordered systems, which is particularly rich, is given. The properties of these systems, in particular, those with graphene structure, are at present heavily investigated.

Finally the concept of superbosonization was considered, which can be well applied to the n-orbital model.

References

[5] G. Akeman, J. Baik, P. di Francesco, (eds.), *Handbook of Random Matrix Theory* (Oxford University Press, Oxford, 2011)

[14] J. Baik, A. Borodin, P. Deift, T. Suidan, A model for the bus system in Cuernavaca (Mexico). J. Phys. A Gen. **39**, 8965 (2006)

[16] Z. Bao, L. Erdős, Delocalization for a class of random block matrices. arXiv 1503.07510 (2015)

[22] D. Belitz, T.R. Kirkpatrick, The Anderson-Mott transition. Rev. Mod. Phys. **66**, 261 (1994)

[32] P.M. Bleher, A.R. Its (eds.), *Random Matrix Models and Their Applications.* Math Sciences Research Institute Publications (Cambridge University Press, Cambridge, 2001)

[38] E. Brézin, S. Hikami, Characteristic polynomials, in *Handbook of Random Matrix Theory*, ed. by G. Akeman, J. Baik, P. di Francesco (Oxford University Press, Oxford, 2011), p. 398

[40] E. Brézin, C. Itzykson, G. Parisi, J.B. Zuber, Planar diagrams. Commun. Math. Phys. **59**, 35 (1978)

[49] C. Chiu, J.C.Y. Teo, A.P. Schnyder, S. Ryu, Classification of topological quantum matter with symmetries. arXiv 1505.03535 (2015)

[56] M. Disertori, H. Pinson, T. Spencer, Density of states of random band matrices. Commun. Math. Phys. **232**, 83 (2002)

[81] A.M. Finkel'stein, The influence of Coulomb on the properties of disordered metals. Zh. Eksp. Teor. Fiz. **84**, 168 (1983); Sov. Phys. JETP **57**, 97 (1983)

[82] A.M. Finkel'stein, Weak localization and Coulomb interactions in disordered systems. Z. Phys. B **56**, 189 (1984)

[83] A.M. Finkel'stein, Electron liquid in disordered conductors. Sov. Sci. Rev. A; Phys. Rev. **14**, 1 (1990)

[84] A.M. Finkel'stein, Disordered electron liquid with interactions, in *50 years of Anderson Localization*, ed. by E. Abrahams (World Scientific, Singapore, 2010), p. 385

[104] T. Guhr, Dyson's correlation function and graded symmetry. J. Math. Phys. **32**, 336 (1991)

[105] T. Guhr, A. Müller-Groehling, H.A. Weidenmüller, Random-matrix theories in quantum physics: common concepts. Phys. Rep. **299**, 189 (1998)

[107] M.C. Gutzwiller, *Chaos in Classical and Quantum Mechanics* (Springer, New York, 1990)

[108] Harish-Chandra, Invariant differential operators on a semisimple Lie algebra. Proc. Natl. Acad. Sci. **42**, 252 (1956)

[124] C. Itzykson, J.-B. Zuber, The planar approximation. II. J. Math. Phys. **21**, 411 (1980)

[136] J.P. Keating, N.C. Snaith, Number theory, in *Handbook of Random Matrix Theory*, ed. by G. Akeman, J. Baik, P. di Francesco (Oxford University Press, Oxford, 2011), p. 491

[150] I. Kostov, Two-dimensional gravity, in *Handbook of Random Matrix Theory*, ed. by G. Akeman, J. Baik, P. di Francesco (Oxford University Press, Oxford, 2011), p. 619

[154] M. Krbalek, P. Seba, Statistical properties of the city transport in Cuernavaca (Mexico) and random matrix theory. J. Phys. A Gen. **33**, 229 (2000)

[155] M. Krbalek, P. Seba, Spectral rigidity of vehicular streams. J. Phys. A **42**, 345001 (2009)

[173] S. Mandt, M.R. Zirnbauer, Zooming in on local level statistics by supersymmetric extension of free probability. J. Phys. A **43**, 025201 (2010)

[192] H.L. Montgomery, The pair correlation of the zeta function. Proc. Symp. Pure Math. **24**, 181 (1973)

[193] S. Müller, M. Sieber, Quantum chaos and quantum graphs, in *Handbook of Random Matrix Theory*, ed. by G. Akeman, J. Baik, P. di Francesco (Oxford University Press, Oxford, 2011), p. 683

[240] T. Shcherbina, Univerality of the local regime for the block band matrices with a finite number of blocks. J. Stat. Phys. **155**, 466 (2014)

[245] R. Speicher, Multiplicative functions on the lattice of noncrossing partitions and free convolution. Math. Anal. **298**, 611 (1994)

[246] R. Speicher, Free probability theory, in *Handbook of Random Matrix Theory*, ed. by G. Akeman, J. Baik, P. di Francesco (Oxford University Press, Oxford, 2011), p. 452

[259] J.J.M. Verbaarschot, The spectrum of the QCD Dirac operator and chiral random matrix theory. Phys. Rev. Lett. **72**, 2531 (1994)

[260] J.J.M. Verbaarschot, Quantum chromodynamics, in *Handbook of Random Matrix Theory*, ed. by G. Akeman, J. Baik, P. di Francesco (Oxford University Press, Oxford, 2011), p. 661

[263] D. Voiculescu, Addition of certain non-commuting random variables. J. Funct. Anal. **66**, 323 (1986)

[298] P. Zinn-Justin, Adding and multiplying random matrices: generalization of Voiculescu's formulas. Phys. Rev. E **59**, 4884 (1999)

[299] P. Zinn-Justin, J.B. Zuber, Knot theory and matrix integrals, in *Handbook of Random Matrix Theory*, ed. by G. Akeman, J. Baik, P. di Francesco (Oxford University Press, Oxford, 2011), p. 557

[301] M.R. Zirnbauer, Fourier analysis on a hyperbolic supermanifold of constant curvature. Commun. Math. Phys. **141**, 503 (1991)

Solutions

Problems of Chap. 2

2.1 Result: $2(a_{12}a_{34} - a_{13}a_{24} + a_{14}a_{23})\xi_1\xi_2\xi_3\xi_4$.

2.2 Since ord $(a^2) = $ ord (x^2), one obtains ord $(x) = \pm a$. If we expand $x = \pm(a + a_1\zeta_1 + a_2\zeta_2 + a_{12}\zeta_1\zeta_2)$, then one finds $x^2 = a^2 + 2aa_1\zeta_1 + 2aa_2\zeta_2 + 2aa_{12}\zeta_1\zeta_2$. Thus one obtains

$$x = \pm(a + \frac{1}{a}\zeta_1\zeta_2), \quad a \neq 0.$$

Apparently there is no solution for $a = 0$.

2.3 Any $x = a_1\xi_1 + a_2\xi_2 + a_3\xi_3 + a_{12}\xi_1\xi_2 + a_{13}\xi_1\xi_3 + a_{23}\xi_2\xi_3 + a_{123}\xi_1\xi_2\xi_3$, which obeys $a_1a_{23} - a_2a_{13} + a_3a_{12} = \frac{1}{2}$ is solution.

Quite generally algebraic equations of order n have n solutions, if the ordinary parts differ. If they coincide, as in this problem and problem (2.2), then the situation can be quite different.

2.4 Substitution and expansion yields

$$g(z + \alpha\zeta_1 + \beta\zeta_2 + c\zeta_1\zeta_2, \zeta_1, \zeta_2) = g_\emptyset(z) + g'_\emptyset(z)(\alpha\zeta_1 + \beta\zeta_2 + c\zeta_1\zeta_2)$$
$$+ g''_\emptyset(z)(-\alpha\beta\zeta_1\zeta_2) + g_1(z)\gamma_1\zeta_1$$
$$+ g'_1(z)\beta\zeta_2\gamma_1\zeta_1 + g_2(z)\gamma_2\zeta_2 + g'_2(z)\alpha\zeta_1\gamma_2\zeta_2$$
$$+ g_{12}(z)\zeta_1\zeta_2.$$

© Springer-Verlag Berlin Heidelberg 2016
F. Wegner, *Supermathematics and its Applications in Statistical Physics*,
Lecture Notes in Physics 920, DOI 10.1007/978-3-662-49170-6

Thus one obtains

$$f_\emptyset(z) = g_\emptyset(z), \quad f_1(z) = g'_\emptyset(z)\alpha + g_1(z)\gamma_1, \quad f_2(z) = g'_\emptyset(z)\beta + g_2(z)\gamma_2,$$
$$f_{12}(z) = -g''_\emptyset(z) + g'_1(z)\beta\gamma_1 - g'_2(z)\alpha\gamma_2 + g_{12}(z).$$

2.5 $A = a_0 + \text{nil}\, a$ yields the Taylor expansion, which terminates

$$A^{-1} = a_0^{-1} - a_0^{-2}\text{nil}\, a + a_0^{-3}(\text{nil}\, a)^2$$
$$= a_0^{-1} - a_0^{-2}(\xi_1\eta_1 + \xi_2\eta_2 + a_2\xi_1\xi_2\eta_1\eta_2) - a_0^{-3}\xi_1\xi_2\eta_1\eta_2.$$

Problems of Chap. 3

3.1 From

$$f = f(0) + \zeta\sum_i \mathscr{P}(a_i)b_i = f(0) + \sum_i a_i\mathscr{P}(b_i)\zeta$$

one obtains

$$f_l = \sum_i \mathscr{P}(a_i)b_i = \sum_i (-)^{k_i}a_ib_i, \quad f_r = \sum_i a_i\mathscr{P}(b_i) = \sum_i a_i(-)^{l_i}b_i.$$

Since $\mathscr{P}(a_ib_i) = (-)^{l_i+k_i}a_ib_i$, the last Eq. (3.1) is fulfilled.

3.2 With $f(\xi) = f_0 + \xi f_1$ one obtains

$$(\xi - \eta)f(\xi) = \xi f_0 - \eta f_0 - \eta\xi f_1, \quad \int d\xi(\xi - \eta)f(\xi) = f_0 + \eta f_1 = f(\eta).$$

Thus $\xi - \eta$ is the delta function for Grassmann variables.

3.3

$$\int d\xi e^{i\alpha\xi} = -i\alpha.$$

In comparison to 3.2 the result is $-i$ times the delta function of α.

3.4 Use

$$f(\zeta) = f(0) + \zeta f_l, \quad g(\zeta) = g(0) + \zeta g_l$$

Then the l.h.s. of (3.6) reads

$$\frac{\partial}{\partial\zeta}(fg) = \frac{\partial}{\partial\zeta}(f(0)g(0) + \zeta f_1 g(0) + f(0)\zeta g_1) = f_1 g(0) + \mathscr{P}(f(0))g_1.$$

The r.h.s. reads

$$(\frac{\partial}{\partial\zeta}f)g + \mathscr{P}(f)\frac{\partial}{\partial\zeta}g = f_1 g(0) + \underbrace{f_1\zeta}_{\zeta\mathscr{P}(f_1)} g_1 + \mathscr{P}(f(0))g_1 + \underbrace{\mathscr{P}(\zeta f_1)}_{-\zeta\mathscr{P}f_1} g_1.$$

The two terms containing ζ cancel and both sides of (3.6) agree.

3.5 The expansion yields

$$\exp(\sum_{k,l=1}^{2,2} \xi_k a_{kl}\eta_l) = 1 + \xi_1 a_{11}\eta_1 + \xi_1 a_{12}\eta_2 + \xi_2 a_{21}\eta_1 + \xi_2 a_{22}\eta_2$$

$$+ \xi_1\eta_1\xi_2\eta_2(a_{11}a_{22} - a_{12}a_{21}).$$

Thus $I_- = a_{11}a_{22} - a_{12}a_{21}$.

3.6 Set $x = y + iz$, $x^* = y - iz$, then

$$(x^* - u^*)(x - v) = (y - \frac{v + u^*}{2})^2 + (z + \frac{i(v - u^*)}{2})^2.$$

Thus the integral does not depend on u and v.

3.7 $*1 = \omega$, $*\omega = g$.

Problem of Chap. 4

4.1 Lloyd model:

(i) Starting from (4.12) one obtains for the Green's function

$$G(r, r', z) = s \int \left\{ \begin{matrix} x_r x_{r'}^* \\ \eta_r \xi_{r'} \end{matrix} \right\} \exp\left(-\sum_{r,r'} s(x_r^*(z\delta_{r,r'} - t_{r,r'})x_{r'} + \xi_r(z\delta_{r,r'} - t_{r,r'})\eta_{r'}) \right)$$

$$\times \prod_r \exp(s\epsilon_r(x_r^* x_r + \xi_r\eta_r)) \prod_r (\frac{d\,\Re x_r\,d\,\Im x_r}{\pi} d\,\xi_r d\,\eta_r). \qquad (\text{S1})$$

The average over the second exponential yields

$$\int \frac{d\epsilon\, \Gamma}{\pi(\Gamma^2 + \epsilon^2)} \exp(s\epsilon(x^*x + \xi\eta)) = \exp(-\Gamma(x^*x + \xi\eta)). \tag{S2}$$

Since $x^*x \geq 0$, the contour of integration has to be shifted down (up) in the complex plane, if $\Im s = -\mathrm{sign}\Im z$ is negative (positive). Then the pole at $\epsilon = s\Gamma$ yields the integral. If one inserts this result in the first equation, then one realizes that z has to be replaced by $z - s\Gamma$ and ϵ_r by 0. Thus $\overline{G}(r, r', z) = G^{(0)}(r, r', z - s\Gamma)$.

(ii) For a product of Green functions $G(r_i, r_i', z_i)$ one has to introduce a product of integrals (S1). Then the average (S2) reads

$$\int \frac{d\epsilon\, \Gamma}{\pi(\Gamma^2 + \epsilon^2)} \exp(\sum_i s_i\epsilon(x_i^*x_i + \xi_i\eta_i)).$$

If all s_i are equal, then the same argument as before applies and $\prod_i G(r_i, r_i', z_i) = \prod_i G^{(0)}(r_i, r_i', z_i - s\Gamma)$. If the s_i differ, then $\Im(\sum_i s_i x_i^*x_i)$ may be positive or negative and one cannot obtain such a simple result.

Problems of Chap. 5

5.1 $-A = (-1)A$, where A and $\mathbf{1}$ are $n \times n$ matrices. Thus $\det(-A) = \det(-\mathbf{1}) \det A = (-)^n \det A$. An antisymmetric matrix obeys ${}^tA = -A$. Since $\det({}^tA) = \det A$, one obtains $\det A = (-)^n \det A$. This yields $\det A = 0$ for odd n.

5.2

$$
{}^t\!jaj = \begin{pmatrix} j_{11} & j_{21} \\ j_{12} & j_{22} \end{pmatrix} \begin{pmatrix} 0 & a_{12} \\ -a_{12} & 0 \end{pmatrix} \begin{pmatrix} j_{11} & j_{12} \\ j_{21} & j_{22} \end{pmatrix}
$$

$$
= \begin{pmatrix} 0 & a_{12}(j_{11}j_{22} - j_{12}j_{21}) \\ -a_{12}(j_{11}j_{22} - j_{12}j_{21}) & 0 \end{pmatrix}.
$$

Thus $\mathrm{pf}({}^t\!jaj) = a_{12}(j_{11}j_{22} - j_{12}j_{21}) = \mathrm{pf}(a)\det j$.

5.3 One obtains

$$\frac{\partial(\zeta_1, \zeta_2)}{\partial(\eta_1, \eta_2)} = D + D' - 2\beta\delta\eta_1\eta_2,$$

$$D = a_{11}a_{22} - a_{12}a_{21}, \qquad D' = a_{11}\delta\eta_1 + a_{12}\delta\eta_2 - a_{21}\beta\eta_1 - a_{22}\beta\eta_2.$$

The inverse yields $\hat{D} = D^{-1} - D'D^{-2}$. Since it does not contain an $\eta_1\eta_2$ term, the integral of a constant expressed in ηs vanishes, as it should. Multiplication by ζ_1 yields

$$\hat{D}\zeta_1 = D^{-1}(\alpha - \alpha D^{-1}D' + a_{11}\eta_1 + a_{12}\eta_2).$$

Thus also the integral of ζ_1 expressed in terms of the ηs vanishes. Similarly for ζ_2. Multiplication by $\zeta_1\zeta_2$ yields

$$\hat{D}\zeta_1\zeta_2 = \eta_1\eta_2 + D^{-1}(\alpha\gamma(1 - D^{-1}D') + (a_{21}\alpha - a_{11}\gamma)\eta_1 + (a_{22}\alpha - a_{12}\gamma)\eta_2)$$

and thus the correct result for the integral.

Problem of Chap. 6

6.1

$$(a^*b)^* = b^*a^{**} = b^*a, \quad (a^\times b)^\times = a^{\times\times}b^\times = (-)^{v_a}ab^\times = (-)^{v_a(1-v_b)}b^\times a.$$

If $a \in \mathscr{A}_0$ or $b \in \mathscr{A}_1$, then $(a^\times b)^\times = b^\times a$.

Problems of Chap. 7

7.1 $\langle c^*|C|c\rangle = \langle 0|e^{c^*b}e^{-kb^\dagger b}e^{cb^\dagger}|0\rangle$. One expands

$$e^{cb^\dagger} = \sum_n \frac{c^n}{n!}(b^\dagger)^n, \quad e^{c^*b} = \sum_n \frac{c^{*n}}{n!}b^n.$$

Only terms with equal number of creation and annihilation operators contribute. Thus one obtains

$$\langle c^*|C|c\rangle = \sum_n \langle 0| \sum_n \frac{c^{*n}}{n!}b^n e^{-kn}\frac{c^n}{n!}(b^\dagger)^n|0\rangle = \sum_n \frac{1}{n!}c^n e^{-kn}c^{*n} = \exp(cc^*e^{-k}).$$

Both sides of (7.9) yield $\frac{1}{1-e^{-k}}$.

7.2

$$\langle \gamma^*|C| - \gamma\rangle = ((\langle 0| + \gamma^*\langle 0|f)C(|0\rangle - f^\dagger|0\rangle\gamma)$$
$$= \langle 0|C|0\rangle + \gamma^*\langle 0|fC|0\rangle - \langle 0|Cf^\dagger|0\rangle\gamma - \gamma^*\langle 0|fCf^\dagger|0\rangle\gamma.$$

Thus

$$\int d\gamma^* d\gamma \langle \gamma^*|C| - \gamma\rangle e^{-\gamma^*\gamma} = \langle 0|C|0\rangle + \langle 0|fCf^\dagger|0\rangle = \text{tr}\,(C) = 2a + c.$$

7.3 Let us define the equal time one-particle Green's functions

$$G(r) = T\sum_\omega \int \frac{d^d p}{(2\pi)^d} \hat{G}_B(p,\omega) e^{ipr}.$$

Then the equal time expectation value reads

$$\langle b^\dagger(r_1)b^\dagger(r_2)b(r_3)b(r_4)\rangle = G(r_1 - r_3)G(r_2 - r_4) + G(r_1 - r_4)G(r_2 - r_3)$$
$$+ C(r_1, r_2, r_3, r_4).$$

where the cumulant C is given by

$$\frac{T^2}{\mathscr{V}^2} \sum_{p_1 p_2 q_1 q_2, \omega_1 \omega_2 \omega_1' \omega_2'} e^{i(p_1 r_1 + p_2 r_2 - q_1 r_3 - q_2 r_4)} \tilde{C}_B(p_1, p_2, q_1, q_2; \omega_1 \omega_2 \omega_1' \omega_2').$$

Due to translational invariance in space and time q_2 and ω_2' are restricted to $q_2 = p_1 + p_2 - q_1$ and $\omega_2' = \omega_1 + \omega_2 - \omega_1'$, and the cumulant C yields

$$\mathscr{V} \int \frac{d^d p_1 d^d p_2 d^d q_1}{(2\pi)^{3d}} e^{i(p_1(r_1-r_4)+p_2(r_2-r_4)-q_1(r_3-r_4))}$$
$$\times T^2 \sum_{\omega_1 \omega_2 \omega_1'} \hat{C}_B(p_1 p_2 q_1, p_1 + p_2 - q_1; \omega_1 \omega_2 \omega_1', \omega_1 + \omega_2 - \omega_1').$$

$\hat{C}_B$ is of order $1/\mathscr{V}$, but $C(r_1,\dots)$ has a finite limit in the thermodynamic limit $\mathscr{V} \to \infty$. It is not negligible.

Problems of Chap. 8

8.1

$$4g_h^2 \cos^2 q + 4g_v^2 \cos^2 p = (u(q) + v(q)\exp(2ip))(u(q) + v(q)\exp(-2ip))$$

with

$$\left.\begin{array}{c} u(q) \\ v(q) \end{array}\right\} = \sqrt{g_v^2 + g_h^2 \cos^2 q} \pm g_h \cos q.$$

Thus

$$\prod_{n=1}^{L_v} \left(4g_h^2 \cos^2 q + 4g_v^2 \cos^2\left(\frac{\pi n}{L_v + 1}\right) \right) = \left(\frac{u(q)^{L_v+1} + (-)^{L_v} v(q)^{L_v+1}}{u(q) + v(q)} \right)^2.$$

Since $n = 0$ does not contribute, we obtain the denominator $u + v$.

8.2 One obtains

$$\mathcal{N}(g_h, g_v) = (ag_h^2 + ag_v^2)(ag_h^2 + bg_v^2)(bg_h^2 + ag_v^2)(bg_h^2 + bg_v^2)$$
$$= ab(g_h^2 + g_v^2)^2(abg_h^4 + (a^2 + b^2)g_h^2 g_v^2 + abg_v^4)$$

with

$$a = 4\cos^2\left(\frac{\pi}{5}\right) = 4\left(\frac{\sqrt{5}+1}{4}\right)^2 = \frac{3 + \sqrt{5}}{2},$$

$$b = 4\cos^2\left(\frac{2\pi}{5}\right) = 4\left(\frac{\sqrt{5}-1}{4}\right)^2 = \frac{3 - \sqrt{5}}{2},$$

$$ab = 1, \quad a^2 + b^2 = 7,$$

which yields

$$\mathcal{N}(g_h, g_v) = (g_h^2 + g_v^2)^2(g_h^4 + 7g_h^2 g_v^2 + g_v^4).$$

8.3 With $\mathcal{N} = g_h^{32} \hat{\mathcal{N}}(g_v^2/g_h^2)$ one obtains for the checkerboard

$$\hat{\mathcal{N}}(x) = (1 + x)^4(1 + 66x + 1515x^2 + 15196x^3 + 73953x^4 + 187506x^5$$
$$+ 255327x^6 + 187506x^7 + 73953x^8 + 15196x^9 + 1515x^{10} + 66x^{11} + x^{12}).$$

The polynomial may be further factorized, since one obtains with $c_m = 4\cos^2(\pi m/9)$, $c_3 = 1$, $c_1 c_2 c_4 = 1$ and some further relations between the c_m

$$(1 + c_1 x)(1 + c_2 x)(1 + c_4 x) = 1 + 6x + 9x^2 + x^3,$$

$$(c_1 + x)(c_2 + x)(c_4 + x) = 1 + 9x + 6x^2 + x^3,$$

$$(c_1 + c_2 x)(c_2 + c_4 x)(c_4 + c_1 x) = 1 + 30x + 21x^2 + x^3,$$

$$(c_1 + c_4 x)(c_2 + c_1 x)(c_4 + c_2 x) = 1 + 21x + 30x^2 + x^3.$$

The total number for $g_h = g_v = 1$ is $\mathcal{N}(1, 1) = 12,988,816 = 2^4 \times 901^2$.

8.4 Denote the number of dimer configurations for a $2 \times n$ lattice by $\mathcal{N}_n$. Then $\mathcal{N}_1 = 1$ and $\mathcal{N}_2 = 2$. The configurations for larger n can be obtained by adding one dimer at the configurations for a $2 \times (n - 1)$ lattice ⬚ or two dimers at a

$2 \times (n-2)$ lattice ⬛. Thus $\mathcal{N}_n = \mathcal{N}_{n-1} + \mathcal{N}_{n-2}$, which is the recursion relation for Fibonacci numbers with correct initial conditions.

Problems of Chap. 9

9.1 From

$$0 = \mathrm{pf}(A(0,0))\mathrm{pf}(A(\pi,\pi)) = (1 - t_\mathrm{v} - t_\mathrm{h} - t_\mathrm{v}t_\mathrm{h})(1 + t_\mathrm{v} + t_\mathrm{h} - t_\mathrm{v}t_\mathrm{h})$$
$$= (1 - t_\mathrm{v}t_\mathrm{h})^2 - (t_\mathrm{v} + t_\mathrm{h})^2 = (1 - t_\mathrm{v}^2)(1 - t_\mathrm{h}^2) - 4t_\mathrm{v}t_\mathrm{h}$$

one obtains

$$\frac{2t_\mathrm{v}}{1 - t_\mathrm{v}^2}\frac{2t_\mathrm{h}}{1 - t_\mathrm{h}^2} = 1.$$

Since $2\tanh(\beta I)/(1 - \tanh^2(\beta I)) = \sinh(2\beta I)$, one obtains $\sinh(2\beta I_\mathrm{v})$ $\sinh(2\beta I_\mathrm{h}) = 1$. Starting with $\mathrm{pf}(A(0,\pi))\mathrm{pf}(A(\pi,0)) = 0$ one has to change the sign of t_v and finally obtains $\sinh(2\beta I_\mathrm{v})\sinh(2\beta I_\mathrm{h}) = -1$.

9.2 With

$$\left.\begin{array}{c}u(q) \\ v(q)\end{array}\right\} = \frac{1}{2}\sqrt{\kappa^2 + 2c_\mathrm{v}(1 - \cos q) + 4c_\mathrm{h}} \pm \frac{1}{2}\sqrt{\kappa^2 + 2c_\mathrm{v}(1 - \cos q)},$$

$$\left.\begin{array}{c}\hat{u}(p) \\ \hat{v}(p)\end{array}\right\} = \frac{1}{2}\sqrt{\kappa^2 + 2c_\mathrm{h}(1 - \cos p) + 4c_\mathrm{v}} \pm \frac{1}{2}\sqrt{\kappa^2 + 2c_\mathrm{h}(1 - \cos p)}$$

one obtains

$$Z_{s,s'} = \pm C\prod_q \left(u(q)^{L_\mathrm{h}} - sv(q)^{L_\mathrm{h}}\right)$$

$$= \pm C\prod_p \left(\hat{u}(p)^{L_\mathrm{v}} - s\hat{v}(p)^{L_\mathrm{v}}\right),$$

where q and p run over the allowed values given in (9.36). Thus one obtains

$$\frac{Z_{+,s'}}{Z_{-,s'}} = \prod_q \frac{1 - \left(\frac{v(q)}{u(q)}\right)^{L_\mathrm{h}}}{1 + \left(\frac{v(q)}{u(q)}\right)^{L_\mathrm{h}}} = \exp(-2\chi^{\mathrm{e},\mathrm{o}}), \tag{S3}$$

$$\frac{Z_{s,+}}{Z_{s,-}} = \prod_p \frac{1 - \left(\frac{\hat{v}(p)}{\hat{u}(p)}\right)^{L_\mathrm{v}}}{1 + \left(\frac{\hat{v}(p)}{\hat{u}(p)}\right)^{L_\mathrm{v}}} = \exp(-2\chi^{\mathrm{o},\mathrm{e}}). \tag{S4}$$

(S3, S4) yields in leading order

$$\chi^{e,o} = \sum_q \left(\frac{v(q)}{u(q)}\right)^{L_h}, \quad \chi^{o,e} = \sum_p \left(\frac{\hat{v}(p)}{\hat{u}(p)}\right)^{L_v}.$$

The maximum of $v(q)/u(q)$ lies for positive c_v at $q = q_0 = 0$, for negative c_v at $q = q_0 = \pi$. χ is given approximately by the maximum

$$\chi^{e,o} \approx \left(\frac{|pf(A(\pi,q_0))| - |pf(A(0,q_0))|}{|pf(A(\pi,q_0))| + |pf(A(0,q_0))|}\right)^{L_h}.$$

Similarly the maximum of $\hat{v}(p)/\hat{u}(p)$ lies for positive c_h at $p = p_0 = 0$, for negative c_h at $p = p_0 = \pi$, which yields

$$\chi^{o,e} \approx \left(\frac{|pf(A(p_0,\pi))| - |pf(A(p_0,0))|}{|pf(A(p_0,\pi))| + |pf(A(p_0,0))|}\right)^{L_v}.$$

9.3 Introduce the notation

$$R(1) = \zeta_u(1)\alpha(1) + B(1), \quad \alpha(1) = -\zeta_d(1) + \zeta_r(1) + \zeta_l(1),$$
$$B(1) = \zeta_l(1)\zeta_d(1) + \zeta_d(1)\zeta_r(1) + \zeta_r(1)\zeta_l(1)$$
$$R(2) = \zeta_d(2)\alpha(2) + B(2), \quad \alpha(2) = \zeta_u(2) - \zeta_l(2) + \zeta_r(2),$$
$$B(2) = \zeta_u(2)\zeta_r(2) + \zeta_u(2)\zeta_l(2) + \zeta_r(2)\zeta_l(2).$$

(i) Then one obtains for the triangular lattice

$$\int d\zeta_u(1)d\zeta_d(2)\exp(R(1)+R(2)-\zeta_u(1)\zeta_d(2)) = \exp(\alpha(1)\alpha(2)+B(1)+B(2)).$$

Thus each bilinear term of the remaining six ζs appears in the exponential.
(ii) One obtains for the honeycomb lattice

$$\int d\zeta_u(1)d\zeta_d(2)\exp(R(1)+R(2)) = \alpha(1)\alpha(2)\exp(B(1)+B(2)).$$

Since $\alpha(1)$ and $\alpha(2)$ no longer appear in the exponent, it is not a good idea to integrate over these variables in a first step.

9.4 Generalization for honeycomb and triangular lattice

$$G = t_1\zeta_r\zeta_l(1) + t_3\zeta_r(3)\zeta_l - t_2\zeta_u\zeta_d(2) - t_4\zeta_u(4)\zeta_d + R$$

Integration over the ζs of the central site in Fig. 9.6b yields

$$\int d\zeta_u d\zeta_d d\zeta_r d\zeta_l \exp(G) =$$

$$\exp\left(-t_1 t_3 \zeta_l(k+1, l+1)\zeta_r(k, l) + t_2 t_4 \zeta_d(k, l+1)\zeta_u(k+1, l)\right.$$

$$+ t_3 t_4 \zeta_r(k, l)\zeta_u(k+1, l) - t_1 t_2 \zeta_l(k+1, l+1)\zeta_d(k, l+1)$$

$$\left.+ t_1 t_4 \zeta_l(k+1, l+1)\zeta_u(k+1, l) + t_3 t_2 \zeta_r(k, l)\zeta_d(k, l+1)\right).$$

Fourier transformation of the exponent yields

$$\sum_{p,q}\left(-t_1 t_3 e^{i(p+q)}\xi_l(p, q)\xi_r(-p, -q) + t_2 t_4 e^{i(q-p)}\xi_d(p, q)\xi_u(-p, -q)\right.$$

$$+ t_3 t_4 e^{-ip}\xi_r(p, q)\xi_u(-p, -q) - t_1 t_2 e^{ip}\xi_l(p, q)\xi_d(-p, -q)$$

$$\left.+ t_1 t_4 e^{iq}\xi_l(p, q)\xi_u(-p, -q) + t_3 t_2 e^{-iq}\xi_r(p, q)\xi_d(-p, -q)\right)$$

and

$$A(p, q) = \begin{pmatrix} 0 & -1 - t_2 t_4 e^{i(p-q)} & 1 - t_3 t_4 e^{ip} & 1 - t_1 t_4 e^{-iq} \\ 1 + t_2 t_4 e^{i(q-p)} & 0 & 1 - t_2 t_3 e^{iq} & -1 + t_1 t_2 e^{-ip} \\ -1 + t_3 t_4 e^{-ip} & -1 + t_2 t_3 e^{-iq} & 0 & 1 + t_1 t_3 e^{-ie(p+q)} \\ -1 + t_1 t_4 e^{iq} & 1 - t_1 t_2 e^{ip} & -1 - t_1 t_3 e^{i(p+q)} & 0 \end{pmatrix}.$$

Thus

$$\det(A(p, q)) = 1 + t_1^2 t_2^2 + t_1^2 t_3^2 + t_1^2 t_4^2 + t_2^2 t_3^2 + t_2^2 t_4^2 + t_3^2 t_4^2 + 8t_1 t_2 t_3 t_4 + t_1^2 + t_2^2 + t_3^2 + t_4^2$$

$$- 2[t_1 t_2(1 - t_3^2)(1 - t_4^2) + t_3 t_4(1 - t_1^2)(1 - t_2^2)]\cos p$$

$$- 2[t_1 t_4(1 - t_2)^2(1 - t_3^2) + t_2 t_3(1 - t_1^2)(1 - t_4^2)]\cos q$$

$$- 2t_1 t_3(1 - t_2^2)(1 - t_4^2)\cos(p + q)$$

$$- 2t_2 t_4(1 - t_1^2)(1 - t_3^2)\cos(p - q)$$

and

$$\text{pf}(A(0, 0)) = 1 - t_1 t_2 - t_1 t_3 - t_1 t_4 - t_2 t_3 - t_2 t_4 - t_3 t_4 + t_1 t_2 t_3 t_4,$$

$$\text{pf}(A(0, \pi)) = 1 - t_1 t_2 + t_1 t_3 + t_1 t_4 + t_2 t_3 + t_2 t_4 - t_3 t_4 + t_1 t_2 t_3 t_4,$$

$$\text{pf}(A(0, 0)) = 1 + t_1 t_2 + t_1 t_3 - t_1 t_4 - t_2 t_3 + t_2 t_4 + t_3 t_4 + t_1 t_2 t_3 t_4,$$

$$\text{pf}(A(0, 0)) = 1 + t_1 t_2 - t_1 t_3 + t_1 t_4 + t_2 t_3 - t_2 t_4 + t_3 t_4 + t_1 t_2 t_3 t_4.$$

9.5

(i) Star-triangle transformation

$$S_1 + S_2 + S_3 = \pm 3 \; e^{3\beta I_h} + e^{-3\beta I_h} = c' e^{3\beta I_t},$$
$$S_1 + S_2 + S_3 = \pm 1 \; e^{\beta I_h} + e^{-\beta I_h} = c' e^{-\beta I_t}. \tag{S5}$$

From (S5) one obtains the relation

$$e^{2\beta I_h} - 1 + e^{-2\beta I_h} = e^{4\beta I_t}. \tag{S6}$$

(ii) The dual transformation yields

$$e^{4\beta I_t} = \left(\frac{1 + e^{-2\beta I_h}}{1 - e^{-2\beta I_h}} \right)^2. \tag{S7}$$

From Eqs. (S6, S7) one obtains for the critical temperature on the honeycomb lattice $e^{2\beta I_h} = 2 \pm \sqrt{3}$ and on the triangular lattice $e^{2\beta I_t} = \sqrt{3}$.

9.6 Duality:

(i)

$$t = \tanh(\beta I) = \frac{e^{\beta I} - e^{-\beta I}}{e^{\beta I} + e^{-\beta I}} = \frac{1 - \hat{t}}{1 + \hat{t}}$$

yields the relations between the ts and $\hat{t}$s

$$t_h + \hat{t}_v + t_h \hat{t}_v = 1, \quad t_v + \hat{t}_h + t_v \hat{t}_h = 1.$$

(ii) One obtains

$$\kappa = -\hat{q}\hat{\kappa}, \quad c_h = \hat{q}^2 \hat{c}_h, \quad c_v = \hat{q}^2 \hat{c}_v.$$

and

$$\mathrm{pf}(A(0,0)) = -\hat{q}\,\mathrm{pf}(\hat{A}(0,0)), \quad \mathrm{pf}(A(0,\pi)) = \hat{q}\,\mathrm{pf}(\hat{A}(0,\pi)),$$
$$\mathrm{pf}(A(\pi,0)) = \hat{q}\,\mathrm{pf}(\hat{A}(\pi,0)), \quad \mathrm{pf}(A(\pi,\pi)) = \hat{q}\,\mathrm{pf}(\hat{A}(\pi,\pi))$$

with

$$\hat{q} = \frac{2}{(1 + \hat{t}_v)(1 + \hat{t}_h)}, \quad q = \frac{2}{(1 + t_v)(1 + t_h)}, \quad q\hat{q} = 1.$$

(iii) Consequently

$$Z_{\pm,\pm} = \pm \hat{q}^N \hat{Z}_{\pm,\pm},$$

where the minus sign in front of the factor $\hat{q}^N$ applies for $Z_{+,+}$, otherwise the plus sign.

Problem of Chap. 10

10.1 Equations (10.19) and (10.23) yield

$$\frac{a - \alpha b^{-1}\beta}{b} = \frac{a}{b}(1 - a^{-1}\alpha b^{-1}\beta) = \frac{a}{b}(1 + b^{-1}\beta a^{-1}\alpha),$$

$$\frac{a}{b - \beta a^{-1}\alpha} = \frac{a}{b}(1 - b^{-1}\beta a^{-1}\alpha)^{-1} = \frac{a}{b}(1 + b^{-1}\beta a^{-1}\alpha).$$

Problems of Chap. 11

11.1

$$W^{-1} = \begin{pmatrix} a^{-1} + a^{-2}b^{-1}\alpha\beta & -a^{-1}b^{-1}\alpha \\ -a^{-1}b^{-1}\beta & b^{-1} + b^{-2}a^{-1}\beta\alpha \end{pmatrix}.$$

11.2

$$\begin{pmatrix} a & \alpha \\ 0 & 1 \end{pmatrix}^{-1} = \begin{pmatrix} a^{-1} & -a^{-1}\alpha \\ 0 & 1 \end{pmatrix}, \quad \begin{pmatrix} 1 & 0 \\ \beta & b \end{pmatrix}^{-1} = \begin{pmatrix} 1 & 0 \\ -b^{-1}\beta & b^{-1} \end{pmatrix}.$$

Thus

$$W^{-1} = \begin{pmatrix} a^{-1} & -a^{-1}\alpha \\ 0 & 1 \end{pmatrix} \begin{pmatrix} 1 & 0 \\ -b'^{-1}\beta a^{-1} & b'^{-1} \end{pmatrix}$$

$$= \begin{pmatrix} 1 & 0 \\ -b^{-1}\beta & b^{-1} \end{pmatrix} \begin{pmatrix} a'^{-1} & -a'^{-1}\alpha b^{-1} \\ 0 & 1 \end{pmatrix}.$$

11.3

$$W^2 = \begin{pmatrix} a^2 + \alpha\beta & 2a\alpha \\ 2a\beta & a^2 + \beta\alpha \end{pmatrix}, \quad f(a) = a^2, \quad f'(a) = 2a, \quad f''(a) = 2.$$

11.4

$$^\mathrm{T}W(^\mathrm{T}W)^{-1} = 1, \quad ^\mathrm{T}W(^{-1}W) = 1 \rightarrow \,^\mathrm{T}W\,^\mathrm{T}W^{-1} = 1.$$

Since both $(^\mathrm{T}W)^{-1}$ and $^\mathrm{T}(W^{-1})$ are the inverse of $^\mathrm{T}W$ they are equal.

Problems of Chap. 12

12.1 W is supersymmetric, $W = \,^\mathrm{T}W\sigma$. Thus $^\mathrm{T}(\mathrm{d}\,W) = \mathrm{d}\,W\sigma$, $^\mathrm{T}(W^{-1}) = \sigma W^{-1}$,

$$^\mathrm{T}(W^{-1}\mathrm{d}\,W) = \,^\mathrm{T}(\mathrm{d}\,W)\,^\mathrm{T}(W^{-1}) = (\mathrm{d}\,W\sigma)(\sigma W^{-1}) = \mathrm{d}\,WW^{-1}.$$

12.2 $^\mathrm{T}(^\mathrm{T}ACX) = \,^\mathrm{T}X\,^\mathrm{T}C\sigma A = \,^\mathrm{T}XCA$. Thus

$$\int [D\,X]\exp(-\tfrac{1}{2}\,^\mathrm{T}XWX + \,^\mathrm{T}ACX) = \mathrm{spf}(W)\exp(\tfrac{1}{2}\,^\mathrm{T}ACW^{-1}CA).$$

12.3 The orthosymplectic transformation obeys (12.25) $^\mathrm{T}UCU = C$, thus $(1 + \,^\mathrm{T}V)C(1 + V) = C$, which for infinitesimal V reads $^\mathrm{T}VC + CV = 0$. In components

$$\begin{pmatrix} {}^!a + a & {}^!\beta\epsilon + \alpha \\ -{}^!\alpha + \epsilon\beta & {}^!b\epsilon + \epsilon b \end{pmatrix} = 0,$$

which yields the relations between the matrices a, α, β, b.

12.4

$$^\mathrm{T}(^\mathrm{T}XWX)\sigma = \,^\mathrm{T}X\,^\mathrm{T}W^{\mathrm{TT}}X\sigma = \,^\mathrm{T}XW\sigma^{-1}(\sigma X\sigma)\sigma = \,^\mathrm{T}XWX.$$

12.5 1 and σ are not supersymmetric. C and C^{-1} are supersymmetric.

Problems of Chap. 13

13.1 One obtains

$$(^!\beta^*(^!b^*)^{-1}\,^!\alpha^*)_{il} = (^!\beta^*)_{ij}((^!b^*)^{-1})_{jk}(^!\alpha^*)_{kl} = \beta_{ji}^*(b^{-1*})_{kj}\alpha_{lk}^* \qquad \text{(S8)}$$

$$= (\alpha_{lk}b_{kj}^{-1}\beta_{ji})^* = ((\alpha b^{-1}\beta)^*)_{li},$$

where from the first to the second line Eq. (6.2) is used.

13.2 Substitute $X_i = y_i + iz_i$, $X_i^* = y_i - iz_i$, $\tilde{A}_i^* = k_i - il_i$, $\tilde{B}_i = k_i + il_i$, where k_i and l_i are complex. Then

$$(X - \tilde{A})^\dagger W(X - \tilde{B}) = \sum_{ij}(y_i - k_i - i(z_i - l_i))W_{ij}(y_j - k_j + i(z_j - l_j))$$

Under the integral y and z can be shifted by k and l, resp., which yields the second equation in problem 13.2.

Problems of Chap. 14

14.1 Equivalent to (i), (ii), (iii) are

$$W = {}^\mathsf{T}W^\times \quad (\mathrm{i}'), \qquad W^\times = \mathsf{C}^{-1}W\mathsf{C} \quad (\mathrm{ii}'), \qquad {}^\mathsf{T}W = \mathsf{C}W\mathsf{C}^{-1} \quad (\mathrm{iii}').$$

Given (ii) and (iii) one concludes ${}^\mathsf{T}W^\times = {}^\mathsf{T}(\mathsf{C}W\mathsf{C}) = \mathsf{C}^{-1}\,{}^\mathsf{T}W\mathsf{C} = W$, that is (i).
Given (i) and (iii) one concludes $W = {}^\mathsf{T}W^\times = (\mathsf{C}W\mathsf{C}^{-1})^\times = \mathsf{C}W^\times\mathsf{C}^{-1}$, that is (ii).
Given (i) and (ii) one concludes $\mathsf{C}W = \mathsf{C}\,{}^\mathsf{T}W^\times = \mathsf{C}\,{}^\mathsf{T}(\mathsf{C}^{-1}W\mathsf{C}) = {}^\mathsf{T}W\mathsf{C} = {}^\mathsf{T}(\mathsf{C}W)\sigma$, that is (iii).

14.2 The supertransposed of (14.4) yields ${}^\mathsf{T}A^\ddagger = \mathsf{C}A\mathsf{C}^{-1}$. This equation and (14.4) yield ${}^\mathsf{T}(BA^\ddagger) = {}^\mathsf{T}A^\ddagger\,{}^\mathsf{T}B = \mathsf{C}A\mathsf{C}^{-1}\mathsf{C}B^\ddagger\mathsf{C}^{-1}$ and thus (14.43).
Similarly ${}^\mathsf{T}(A^\ddagger B) = {}^\mathsf{T}B\,{}^\mathsf{T}A^\ddagger = \mathsf{C}B^\ddagger\mathsf{C}^{-1}\mathsf{C}A\mathsf{C}^{-1} = \mathsf{C}B^\ddagger A\mathsf{C}^{-1}$.

Problem of Chap. 15

15.1 The coefficient can be written $\partial_A G_A + \partial_B G_B$ with

$$G_A = -t\frac{AF'_{0A} + BF'_{0B}}{uA - tB} + c_A F'_{0B}, \qquad G_B = u\frac{AF'_{0A} + BF'_{0B}}{uA - tB} + c_B F'_{0A}$$

with constants $c_A + c_B = -1$. Then

$$\int_0 \mathrm{d}A \int_0 \mathrm{d}B(\partial_A G_A + \partial_B G_B)$$

$$= -\int \mathrm{d}B G_A(0, B) - \int \mathrm{d}A G_B(A, 0) = (2 + c_A + c_B)F_0(0, 0) = F(0, 0).$$

Problems of Chap. 18

18.1 $2V_{\pm}(x) = W^2 \pm h\delta(x - L) + \mp h\delta(-x - L)$. The wave-function decays exponentially for $|x| > L$

$$\psi(x) = \begin{cases} \psi(L)\exp(-\kappa(x - L)), & x \geq L, \\ \psi(-L)\exp(\kappa(x + L)), & x \leq -L, \end{cases} \qquad \kappa^2 = (W_0(\pm L) + h)^2 - 2E.$$

(i) Integration over an infinitesimal interval around $x = \pm L$ yields due to the δ-function

$$\psi'(L - 0) = \psi'(L + 0) \mp h\psi(L) = (-\kappa \mp h)\psi(L),$$
$$\psi'(-L + 0) = \psi'(-L - 0) \mp h\psi(-L) = (\kappa \mp h)\psi(-L).$$

This yields in the limit $h \to \infty$

$$\psi'_+(L - 0) = (-2h - W_0(L))\psi_+(L) \to \psi_+(L)=0, \ \psi'_-(L - 0) = -W_0(L)\psi_-(L),$$
$$\psi'_-(-L + 0) = (-2h + W_0(-L))\psi_-(-L) \to \psi_-(-L) = 0, \ \psi'_+(-L + 0) = W_0(-L)\psi_+(-L).$$

(ii) For constant W_0 one obtains

$$\psi_+(x) = \sin(k(x-L)), \quad \psi_-(x) = \sin(k(x+L)), \quad k = -W_0\tan(2kL), \quad 2E = W_0^2 + k^2.$$

Remark: Instead of $h \to +\infty$ one could use the different boundary condition $h \to -\infty$.

18.2

$$H = 2 + a(c_1^\dagger - c_2^\dagger)(c_1 - c_2), \quad a := v + v^* + vv^*.$$

Energies and eigenstates: $(2, |0\rangle, (c_1^\dagger + c_2^\dagger)|0\rangle/\sqrt{2})$, $(2(1 + a), c_1^\dagger c_2^\dagger|0\rangle, (c_1^\dagger - c_2^\dagger)|0\rangle/\sqrt{2})$.

18.3 One obtains

$$V = \tfrac{1}{2}(w^2\theta(x)\theta(a - x) + ws\delta(x) - ws\delta(x - a)).$$

The wave function is continuous at $x = 0$ and $x = a$, but shows cusps,

$$\psi'(+0) - \psi'(-0) = ws\psi(0), \quad \psi'(a + 0) - \psi'(a - 0) = -ws\psi(a).$$

In the interval $x = [0, a]$ $\psi(x)$ reads

$$\psi(x) = c_1 e^{\kappa x} + c_2 e^{-\kappa x}, \quad \kappa = \sqrt{w^2 - 2E}.$$

In the limit $w \to \infty$ we may write $\kappa = w$. Elimination of c_1 and c_2 yields

$$w(1 + s)\psi(a) + \psi'(a + 0) = e^{\omega}(w(1 + s)\psi(0) + \psi'(-0)),$$
$$w(1 - s)\psi(a) - \psi'(a + 0) = e^{-\omega}(w(1 - s)\psi(0) - \psi'(-0)).$$

If $|s| \neq 1$ then the limit $w \to \infty$ yields (for $a \to 0$) (18.21), $\psi(a) = \psi(0) = 0$. If $s = \pm 1$, then one of these equations yields the relation between $\psi(0)$ and $\psi(a)$, the other one the relation between $\psi'(-0)$ and $\psi'(a + 0)$, which in the limit $a \to 0$ yields (18.22).

Problems of Chap. 20

20.1 Differentiation yields

$$\triangle_{ss} f(r^2) = (2n - cc'r)f' + (4\sum_i x_i^2 + c^2c'\sum_i \theta_i^{\times}\theta_i)f''.$$

Thus $cc' = 4$ has to hold. The distribution to c and c' is a matter of definition.

20.2 The derivative of Eq. (20.78) with respect to a and a^* at $a = a^* = 0$ yields $-\partial f/\partial z = -z'^* f$, $-\partial f/\partial z'^* = -zf$. These differential equations yield (20.79).

Problems of Chap. 21

21.1 From (21.27) we have $z_i = 2\cos(\phi_i)/\sqrt{g} = (e^{s_i \phi_i} + e^{-s_i \phi_i})/\sqrt{g}$. Then with $G(z_i) = \sqrt{g}e^{s_i \phi_i}$ and (21.41) one obtains

$$\tilde{K}' = g\frac{e^{s_2 \phi_2} - e^{s_1 \phi_1}}{e^{s_1 \phi_1} + e^{-s_1 \phi_1} - e^{s_2 \phi_2} - e^{-s_2 \phi_2}}.$$

The denominator can be written $(e^{s_1 \phi_1} - e^{s_2 \phi_2})(1 - e^{-s_1 \phi_1 - s_2 \phi_2})$, which yields (21.43).

21.2 Complex square matrices $A \in \mathcal{M}(n, 0)$ whose matrix elements do not contain nilpotent contributions and which commute with their hermitian adjoint $[A^\dagger, A] = 0$, can be diagonalized by unitary matrices. The reason is that $A_+ = A^\dagger + A$ and $A_- = i(A^\dagger - A)$ are hermitian and commute and thus can be diagonalized simultaneously by unitary matrices. The unitary matrices U are of this type. Their eigenvalues are $e^{i\phi_i}$. Then a similar derivation as for hermitian matrices can be performed. The differences $|\lambda_i - \lambda_k|$ have to be replaced by

$$|e^{i\phi_i} - e^{i\phi_k}| = |e^{i(\phi_i - \phi_k)/2} - e^{i(\phi_k - \phi_i)/2}| = 2|\sin(\tfrac{1}{2}(\phi_i - \phi_k))|.$$

There are no exponential prefactors.

Problems of Chap. 22

22.1 $D_{q,kl}(E-, E+) = v^k v^l (e_{kl} + o_{kl})$, $e_{lk} = v^k v^l e_{kl}$, $o_{lk} = -v^k v^l o_{kl}$. $e_{kl} \in \mathbb{R}$, $o_{kl} \in \mathbb{I}$.

22.2 $e_{kl} = e_{kl}^{(0)} + e_{kl}^{(1)}(\eta - i\tfrac{1}{2}\omega)$.

Problem of Chap. 23

23.1 $\det(1 - \hat{\alpha}\hat{\beta})$ is a polynomial in $\hat{\alpha}\hat{\beta}$ of order n_1. The determinant to the power n is a polynomial of at most power nn_1. Since integration has to be performed over n_1^2 factors $\hat{\alpha}$, similarly for $\hat{\beta}$, the integrals vanish for $n_1^2 > nn_1$, that is for $n_1 > n$.

References

1. E. Abrahams, P.W. Anderson, D.C. Licciardello, T.V. Ramakrishnan, Scaling theory of localization: absence of quantum diffusion in two dimensions. Phys. Rev. Lett. **42**, 673 (1979)
2. A.A. Abrikosov, L.P. Gorkov, I.E. Dzyaloshinski, *Methods of Quantum Field Theory in Statistical Physics* (Prentice-Hall, Englewood Cliffs, NJ, 1963)
3. A. Aharony, Y. Imry, The mobility edge as a spin-glass problem. J. Phys. C **10**, L487 (1977)
4. A. Aharony, Y. Imry, S. Ma, Lowering of dimensionality in phase transitions with random fields. Phys. Rev. Lett. **37**,1364 (1976)
5. G. Akeman, J. Baik, P. di Francesco (eds.), *Handbook of Random Matrix Theory* (Oxford University Press, Oxford, 2011)
6. A. Altland, B.D. Simons, Field theory of the random flux model. Nucl. Phys. B **562**, 445 (1999)
7. A. Altland, B. Simons, *Condensed Matter Field Theory* (Cambridge University Press, Cambridge, 2010)
8. A. Altland, M.R. Zirnbauer, Nonstandard symmetry classes in mesoscopic normal-superconducting hybrid structures. Phys. Rev. B **55**, 1142 (1997)
9. A. Altland, B.D. Simons, M.R. Zirnbauer, Theories of low-energy quasi-particle states in disordered d-wave superconductors. Phys. Rep. **359**, 283 (2002)
10. P.W. Anderson, Absence of diffusion in certain random lattices. Phys. Rev. **109**, 1492 (1958)
11. Y. Asada, K. Slevin, T. Ohtsuki, Anderson transition in two-dimensional systems with spin-orbit coupling. Phys. Rev. Lett. **89**, 256601 (2002)
12. Y. Asada, K. Slevin, T. Ohtsuki, Anderson transition in the three dimensional symplectic universality state. J. Phys. Soc. Jpn. Suppl. **74**, 238 (2005)
13. J.E. Avron, H. van Beijeren, L.S. Schulman, R.K.P. Zia, Roughening transition, surface tension and equilibrium droplet shapes in a two-dimensional Ising system. J. Phys. A **15**, L81 (1982)
14. J. Baik, A. Borodin, P. Deift, T. Suidan, A model for the bus system in Cuernavaca (Mexico). J. Phys. A Gen. **39**, 8965 (2006)
15. R. Balian, G. Toulouse, Critical exponents for transitions with $n = -2$ components of the order parameter. Phys. Rev. Lett. **30**, 544 (1973)
16. Z. Bao, L. Erdös, Delocalization for a class of random block matrices. arXiv:1503.07510 (2015)
17. R. Bauerschmidt, H. Duminil-Copin, J. Goodman, G. Slade, Lectures on self-avoiding walks. Clay Math. Proc. **15**, 395 (2012). arXiv:1206.2092
18. G. Baym, L.P. Kadanoff, Conservation laws and correlation functions. Phys. Rev. **124**, 287 (1961)

© Springer-Verlag Berlin Heidelberg 2016

F. Wegner, *Supermathematics and its Applications in Statistical Physics*, Lecture Notes in Physics 920, DOI 10.1007/978-3-662-49170-6

359

19. G. Baym, N.D. Mermin, Determination of thermodynamic Green's functions. J. Math. Phys. **2**, 232 (1961)

20. C. Becchi, A. Rouet, R. Stora, The Abelian Higgs Kibble model, unitarity of the S-operator. Phys. Lett. B **52**, 344 (1974)

21. C. Becchi, A. Rouet, R. Stora, Renormalization of the abelian Higgs-Kibble model. Commun. Math. Phys. **42**, 127 (1975)

22. D. Belitz, T.R. Kirkpatrick, The Anderson-Mott transition. Rev. Mod. Phys. **66**, 261 (1994)

23. F.A. Berezin, Canonical transformations in the representation of second quantization. Dok. Akad. Nauk SSSR **137**, 311 (1961)

24. F.A. Berezin, *The Method of Second Quantization* (Academic, New York, 1966)

25. F.A. Berezin, *Introduction to Superanalysis* (Springer, Reidel, Dordrecht, 1987)

26. G. Bergmann, Physical interpretation of weak localization: a time-of-flight experiment with conduction electrons. Phys. Rev. B **28**, 2914 (1983)

27. G. Bergmann, Weak localization in thin films: a time-of-flight experiment with conduction electrons. Phys. Rep. **107**, 1 (1984)

28. G. Bergmann, Weak localization and its applications as an experimental tool, in *50 Years of Anderson Localization*, ed. by E. Abrahams (World Scientific, Singapore, 2010), p. 231

29. D. Bernard, A. LeClair, A classification of 2D random Dirac fermions. J. Phys. A **35**, 2555 (2002)

30. B.A. Bernevig, S.C. Zhang, Quantum spin Hall effect. Phys. Rev. Lett. **96**, 106802 (2006)

31. W. Bernreuther, F. Wegner, Four-loop order β-function for two dimensional non-linear σ models. Phys. Rev. Lett. **57**, 1383 (1986)

32. P.M. Bleher, A.R. Its (eds.), *Random Matrix Models and Their Applications* (Math Sciences Research Institute Publications, Cambridge University Press, Cambridge, 2001)

33. O. Bohigas, M.J. Gianoni, C. Schmitt, Characterization of chaotic quantum spectra and universality of level fluctuation laws. Phys. Rev. Lett. **52**, 1 (1984)

34. O. Bohigas, H.-A. Weidenmüller, History - an overview, in *Handbook of Random Matrix Theory*, ed. by G. Akeman, J. Baik, P. di Francesco (Oxford University Press, Oxford, 2011), p. 15

35. R. Bott, The stable homotopy of the classical groups. Ann. Math. **70**, 313 (1959)

36. E. Brézin, C. de Dominicis, New phenomena in the random field Ising model. Europhys. Lett. **44**, 13 (1998)

37. E. Brézin, C. de Dominicis, Interactions of several replicas in the random field Ising model. Eur. Phys. J. B **19**, 467 (2001)

38. E. Brézin, S. Hikami, Characteristic polynomials, in *Handbook of Random Matrix Theory*, ed. by G. Akeman, J. Baik, P. di Francesco (Oxford University Press, Oxford, 2011), p. 398

39. E. Brézin, D.J. Gross, C. Itzykson, Density of states in the presence of a strong magnetic field and random impurities. Nucl. Phys. B **235**, 24 (1984)

40. E. Brézin, C. Itzykson, G. Parisi, J.B. Zuber, Planar diagrams. Commun. Math. Phys. **59**, 35 (1978)

41. J. Bricmont, A. Kupiainen, Lower critical dimension for the random-field Ising model. Phys. Rev. Lett. **59**, 1829 (1987)

42. D.C. Brydges, J.Z. Imbrie, Branched Polymers and dimensional reduction. Ann. Math. **158**, 1019 (2003)

43. J.E. Bunder, K.B. Efetov, V.E. Kravtsov, O.M. Yevtushenko, M.R. Zirnbauer, Superbosonization formula and its application to random matrix theory. J. Stat. Phys. **129**, 809 (2007)

44. D.J. Candlin, On sums over trajectories for systems with Fermi statistics. Nuovo Cimento **4**, 231 (1956)

45. J. Cardy, Nonperturbative effects in a scalar supersymmetric theory. Phys. Lett. B **125**, 470 (1983)

46. J.T. Chalker, Scaling and eigenfunction correlations near a mobility edge. Physica A **167**, 253 (1990)

47. S. Chaturvedi, A.K. Kapoor, V. Srinivasan, Ward Takahashi identities and fluctuation-dissipation theorem in a superspace formulation of the Langevin equation. Z. Phys. B **57**, 249 (1984)

48. P. Chauve, P. Le Doussal, K.J. Wiese, Renormalization of pinned elastic systems: how does it work beyond one loop? Phys. Rev. Lett. **86**, 1785 (2001)

49. C. Chiu, J.C.Y. Teo, A.P. Schnyder, S. Ryu, Classification of topological quantum matter with symmetries. arXiv:1505.03535 (2015)

50. A. Comtet, C. Texier, Y. Tourigny, Product of random matrices and generalized quantum point scatterers. J. Stat. Phys. **140**, 427 (2010)

51. F. Cooper, A. Khare, U. Sukhatme, Supersymmetry and quantum mechanics. Phys. Rep. **251**, 267 (1995)

52. F. Constantinescu, H.F. de Groote, The integral theorem for supersymmetric invariants. J. Math. Phys. **30**, 981 (1989)

53. P.G. de Gennes, Exponents for the excluded volume problem as derived by the Wilson method. Phys. Lett. A **38**, 339 (1972)

54. R. Delbourgo, Superfield perturbation theory and renormalization. Nuovo Cimento A **25**, 646 (1975)

55. B. DeWitt, *Supermanifolds* (Cambridge University Press, Cambridge,1984)

56. M. Disertori, H. Pinson, T. Spencer, Density of states of random band matrices. Commun. Math. Phys. **232**, 83 (2002)

57. M.I. Dyakonov, V.I. Perel, Current-induced spin orientation of electrons in semiconductors. Phys. Lett. A **25**, 459 (1971)

58. M.I. Dyakonov, V.I. Perel, Possibility of orienting electron spins with current. Pis'ma Zh. Eksp. Teor. Fiz. **13**, 657 (1971) ; Sov. Phys. JETP Lett. **13**, 467 (1971)

59. F.J. Dyson, The dynamics of a disordered linear chain. Phys. Rev. **92**, 1331 (1953)

60. F.J. Dyson, Statistical theory of energy levels of complex systems. I, II, III. J. Math. Phys. **3**, 140, 157, 166 (1962)

61. F.J. Dyson, The threefold way. Algebraic structure of symmetry groups and ensembles in quantum mechanics. J. Math. Phys. **3**,1199 (1962)

62. F.J. Dyson, M.L. Mehta, Statistical theory of energy levels of complex systems. IV, V. J. Math. Phys. **4**, 701, 713 (1963)

63. S.F. Edwards, P.W. Anderson, Theory of spin glasses. J. Phys. F **5**, 965 (1975)

64. K.B. Efetov, Supersymmetry method in localization theory. Zh. Eksp. Teor. Fiz. **82**, 872 (1982) ; Sov. Phys. JETP **55**, 514 (1982)

65. K.B. Efetov, Supersymmetry and theory of disordered metals. Adv. Phys. **32**, 53 (1983)

66. K.B. Efetov, *Supersymmetry in Disorder and Chaos* (Cambridge University Press, Cambridge, 1997)

67. K.B. Efetov, A.I. Larkin, D.E. Khmel'nitskii, Interaction of diffusion modes in the theory of localization. Zh. Eksp. Teor. Fiz. **79**, 1120 (1980) ; Sov. Phys. JETP **52**, 568 (1980)

68. K.B. Efetov, G. Schwiete, K. Takahashi, Bosonization for disordered and chaotic systems. Phys. Rev. Lett. **92**, 026807 (2004)

69. T.P. Eggarter, R. Riedinger, Singular behavior of tight-binding chains with off-diagonal disorder. Phys. Rev. B **18**, 569 (1978)

70. E. Egorian, S. Kalitzin, A superfield formulation of stochastic quantization with fictitious time. Phys. Lett. B **129**, 320 (1983)

71. L. Erdös, Universality of Wigner random matrices: a survey of recent results. arXiv:1004.0861 [math-ph]; Russ. Math. Surv. **66**, 507 (2011)

72. F. Evers, A.D. Mirlin, Anderson transitions. Rev. Mod. Phys. **80**, 1355 (2008)

73. M. Fabrizio, C. Castellani, Anderson localization in bipartite lattices. Nucl. Phys. B **583**, 542 (2000)

74. P. Fayet, S. Ferrara, Supersymmetry. Phys. Rep. **32**, 249 (1977)

75. M.V. Feigel'man, A.M. Tsvelik, Hidden supersymmetry of stochastic dissipative dynamics. Sov. Phys. JETP **56**, 823 (1982) ; Zh. Eksp. Teor. Fiz. **83**, 1430 (1982)

76. P. Fendley, K. Schoutens, Exact results for strongly-correlated fermions in 2+1 dimensions. Phys. Rev. Lett. **95**, 046403 (2005)

77. P. Fendley, K. Schoutens, J. de Boer, Lattice models with $N = 2$ supersymmetry. Phys. Rev. Lett. **90**, 120402 (2003)

78. S. Ferrara, J. Wess, B. Zumino, Supergauge multiplets and superfields. Phys. Lett. B **51**, 239 (1974)

79. A.L. Fetter, J.D. Walecka, *Quantum Theory of Many-Particle Systems* (McGraw Hill, New York, 1971)

80. R.P. Feynman, Space-time approach to quantum electrodynamics. Phys. Rev. **76**, 769 (1949)

81. A.M. Finkel'stein, The influence of Coulomb on the properties of disordered metals. Zh. Eksp. Teor. Fiz. **84**, 168 (1983); Sov. Phys. JETP **57**, 97 (1983)

82. A.M. Finkel'stein, Weak localization and Coulomb interactions in disordered systems. Z. Phys. B **56**, 189 (1984)

83. A.M. Finkel'stein, Electron liquid in disordered conductors. Sov. Sci. Rev./Sect. A: Phys. Rev. **14**, 1 (1990)

84. A.M. Finkel'stein, Disordered electron liquid with interactions, in *50 Years of Anderson Localization*, ed. by E. Abrahams (World Scientific, Singapore, 2010), p. 385

85. M.E. Fisher, Statistical mechanics of dimers on a plane lattice. Phys. Rev. **124**, 1664 (1961)

86. M.E. Fisher, On the dimer solution of planar Ising models. J. Math. Phys. **7**, 1776 (1966)

87. M.E. Fisher, Yang-Lee edge singularity and ϕ^3 field theory. Phys. Rev. Lett. **40**, 1610 (1978)

88. D.S. Fisher, Random fields, random anisotropies, nonlinear σ models, and dimensional reduction. Phys. Rev. B **31**, 7233 (1985)

89. T. Fukui, Critical behavior of two-dimensional random hopping fermions with π-flux. Nucl. Phys. B **562**, 477 (1999)

90. Y.V. Fyodorov, Negative moments of characteristic polynomials of random matrices: Ingham-Siegel integral as an alternative to Hubbard-Stratonovich transformation. Nucl. Phys. B **621**, 643 (2002)

91. Y.V. Fyodorov, On Hubbard-Stratonovich transformations over hyperbolic domains. J. Phys. Condens. Matter **17**, S1915 (2005)

92. Y.V. Fyodorov, Y. Wei, M.R. Zirnbauer, Hyperbolic Hubbard-Stratonovich transformations made rigorous. J. Math. Phys. **49**, 053507 (2008)

93. R. Gade, Anderson localization for sublattice models. Nucl. Phys. B **398**, 499 (1993)

94. R. Gade, F. Wegner, The $n = 0$ replica limit of U(n) and U(n)/SO(n) models. Nucl. Phys. B **360**, 213 (1991)

95. J.L. Gervais, B. Sakita, Field theory interpretation of supergauges in dual models. Nucl. Phys. B **34**, 632 (1971)

96. Y.A. Golfand, E.P. Likhtman, Extension of the algebra of Poincaré group operators and violation of P-invariance. ZhETF Pis. Red. **12**, 452 (1971); JETP Lett. **13**, 323 (1971)

97. L.P. Gorkov, A.I. Larkin, D.E. Khmelnitskii, Particle conductivity in a two-dimensional random potential. Pisma Zh. Eksp. Teor. Fiz. **30**, 248 (1979) ; JETP Lett. **30**, 228 (1979)

98. E. Gozzi, Dimensional reduction in parabolic stochastic equations. Phys. Lett. B **143**, 183 (1984)

99. H. Grassmann, *Lineare Ausdehnungslehre* (Wigand, Leipzig, 1844)

100. D.A. Greenwood, The Boltzmann equation in the theory of electrical conduction in metals. Proc. Phys. Soc. Lond. **71**, 585 (1958)

101. G. Grinstein, Ferromagnetic phase transitions in random fields: the breakdown of scaling laws. Phys. Rev. Lett. **37**, 944 (1976)

102. I.A. Gruzberg, A.W.W. Ludwig, A.D. Mirlin, M.R. Zirnbauer, Symmetries of multifractal spectra and field theories of Anderson localization. Phys. Rev. Lett. **107**, 086403 (2011)

103. I.A. Gruzberg, A.D. Mirlin, M.R. Zirnbauer, Classification and symmetry properties of scaling dimensions of Anderson transitions. Phys. Rev. B **87**, 125144 (2013)

104. T. Guhr, Dyson's correlation function and graded symmetry. J. Math. Phys. **32** (1991) 336

105. T. Guhr, A. Müller-Groehling, H.A. Weidenmüller, Random-matrix theories in quantum physics: common concepts. Phys. Rep. **299**, 189 (1998)

106. S. Guruswamy, A. LeClair, A.W.W. Ludwig, gl(N|N) Supercurrent algebras for disordered Dirac fermions in two dimensions. Nucl. Phys. B **583**, 475 (2000)
107. M.C. Gutzwiller, *Chaos in Classical and Quantum Mechanics* (Springer, Berlin, 1990)
108. Harish-Chandra, Invariant differential operators on a semisimple Lie algebra. Proc. Natl. Acad. Sci. **42**, 252 (1956)
109. S. Hikami, Localization, nonlinear σ model and string theory. Prog. Theor. Phys. Suppl. **107**, 213 (1992)
110. S. Hikami, A.I. Larkin, Y. Nagaoka, Spin-orbit interaction and magnetoresistance in the two dimensional random system. Prog. Theor. Phys. **63**, 707 (1980)
111. J.E. Hirsch, Spin Hall effect. Phys. Rev. Lett. **83**, 1834 (1999)
112. A. Houghton, A. Jevicki, R.D. Kenway, A.M.M. Pruisken, Noncompact σ models and the existence of a mobility edge in disordered electronic systems near two dimensions. Phys. Rev. Lett. **45**, 394 (1980)
113. H. Hsu, W. Nadler, P. Grassberger, Statistics of lattice animals. Comp. Phys. Commun. **169**, 114 (2005)
114. B. Huckestein, Scaling theory of the integer quantum Hall effect. Rev. Mod. Phys. **67**, 357 (1995)
115. B. Huckestein, B. Kramer, One-parameter scaling in the lowest Landau band: precise determination of the critical behavior of the localization length. Phys. Rev. Lett. **64**, 1437 (1990)
116. B. Huckestein, B. Kramer, L. Schweitzer, Characterization of the electronic states near the centres of the Landau bands under quantum Hall conditions. Surf. Sci. **263**, 125 (1992)
117. L. Hujse, N. Moran, J. Vala, K. Schoutens, Exact ground state of a staggered supersymmetric model for lattice fermions. Phys. Rev. B **84**, 115124 (2011)
118. J.Z. Imbrie, Lower critical dimension of the random-field Ising model. Phys. Rev. Lett. **53**, 1747 (1984)
119. J.Z. Imbrie, The ground state of the three-dimensional random-field Ising model. Commun. Math. Phys. **98**, 145 (1985)
120. Y. Imry, S.K. Ma, Random-field instability of the ordered state of continuous symmetry. Phys. Rev. Lett. **35**, 1399 (1975)
121. C. Itzykson, Ising fermions (II). Three dimensions. Nucl. Phys. B **210**, 477 (1982)
122. C. Itzykson, J.-M. Drouffe, *Statistical Field Theory*, vols. 2 (Cambridge University Press, Cambridge, 1989)
123. C. Itzykson, J.-B. Zuber, *Quantum Field Theory* (Mc-Graw Hill, New York, 1980)
124. C. Itzykson, J.-B. Zuber, The planar approximation. II. J. Math. Phys. **21**, 411 (1980)
125. W. Jokusch, Perfect matchings and perfect squares. J. Combin. Theory A **67**, 100 (1994)
126. K. Jüngling, R. Oppermann, Random electronic models with spin-dependent hopping. Phys. Lett. A **76**, 449 (1980)
127. K. Jüngling, R. Oppermann, Effects of spin-interactions in disordered electronic systems: loop expansions and exact relations among local gauge invariant models. Z. Phys. B **38**, 93 (1980)
128. L.P. Kadanoff, G. Baym, *Quantum Statistical Mechanics* (Benjamin, New York, 1962)
129. L.P. Kadanoff, H. Ceva, Determination of an operator algebra for the two-dimensional Ising model. Phys. Rev. B **3**, 3918 (1971)
130. A. Kamenev, *Field Theory of Non-equilibrium Systems* (Cambridge University Press, Cambridge, 2011)
131. C.L. Kane, E.J. Mele, Z_2 topological order and the quantum spin Hall effect. Phys. Rev. Lett. **95**, 146802 (2005)
132. C.L. Kane, E.J. Mele, Quantum spin Hall effect in graphene. Phys. Rev. Lett. **95**, 226801 (2005)
133. P.W. Kasteleyn, The statistics of dimers on a lattice, the number of dimer arrangements on a quadratic lattice. Physica **27**, 1209 (1961)
134. P.W. Kasteleyn, Dimer statistics and phase transitions. J. Math. Phys. **4**, 287 (1963)

135. B. Kaufman, Crystal statistics. II. Partition function evaluated by spinor analysis. Phys. Rev. **76**, 1232 (1949)

136. J.P. Keating, N.C. Snaith, Number theory, in *Handbook of Random Matrix Theory*, ed. by G. Akeman, J. Baik, P. di Francesco (Oxford University Press, Oxford, 2011), p. 491

137. L.V. Keldysh, Diagram technique for nonequilibrium processes. Zh. Eksp. Teor. Fiz. **47**, 1515 (1964); Sovj. Phys. JETP **20**, 1018 (1965)

138. R. Kenyon, *Dimer Problems*, in Encyclopedia of Mathematical Physics, ed. J.-P. Françoise, G.L. Naber, T.S. Tsun, (Academic Press, Amsterdam, 2006)

139. R. Kenyon, Lectures on dimers. arXiv:0910.3129v1 (2009)

140. R. Kenyon, A. Okounkov, What is a dimer? Not. AMS **52**, 342 (2005)

141. D.E. Khmelnitskii, Quantization of Hall conductivity. JETP Lett. **38**, 552 (1984)

142. D.E. Khmelnitskii, A.I. Larkin, Mobility edge shift in external magnetic field. Sol. St. Comm. **39**, 1069 (1981)

143. M. Kieburg, H. Kohler, T. Guhr, Integration of Grassmann variables over invariant functions in flat superspaces. J. Math. Phys. **50**, 013528 (2009)

144. R. Kirschner, Quantization by stochastic relaxation processes and supersymmetry. Phys. Lett. B **139**, 180 (1984)

145. A. Kitaev, Periodic table for topological insulators and superconductors. AIP Conf. Proc. **1134**, 22 (2009)

146. D. Klarner, J. Pollack, Domino tilings of rectangles with fixed width. Discrete Math. **32**, 44 (1980)

147. A. Klein, J.F. Perez, Supersymmetry and dimensional reduction: a non-perturbative proof. Phys. Lett. B **125**, 473 (1983)

148. H. Kleinert, *Path Integrals in Quantum Mechanics, Statistics and Polymer Physics* (World Scientific, Singapore, 1990); *Pfadintegrale in der Quantenmechanik, Statistik und Polymerphysik* (BI Wissenschaftsverlag, Mannheim, 1993)

149. P.J. Kortmann, R.B. Griffiths, Density of zeroes on the Lee-Yang circle for two ising ferromagnets. Phys. Rev. Lett. **27**, 1439 (1971)

150. I. Kostov, Two-dimensional gravity, in *Handbook of Random Matrix Theory*, ed. by G. Akeman, J. Baik, P. di Francesco (Oxford University Press, Oxford, 2011), p. 619

151. B. Kramer, A. MacKinnon, Localization theory and experiment. Rep. Prog. Phys. **56**, 1469 (1993)

152. B. Kramer, A. MacKinnon, T. Ohtsuki, K. Slevin, Finite size scaling analysis of the Anderson transition, in *50 Years of Anderson Localization*, ed. by E. Abrahams (World Scientific, Singapore, 2010), p. 347

153. H.A. Kramers, G.H. Wannier, Statistics of the two-dimensional ferromagnet. Phys. Rev. **60**, 252–262 (1941)

154. M. Krbalek, P. Seba, Statistical properties of the city transport in Cuernavaca (Mexico) and random matrix theory. J. Phys. A Gen. **33**, 229 (2000)

155. M. Krbalek, P. Seba, Spectral rigidity of vehicular streams. J. Phys. A **42**, 345001 (2009)

156. R. Kubo, A general expression for the conductivity tensor. Can. J. Phys. **34**, 1274 (1956)

157. D.A. Kurtze, M.E. Fisher, Yang-Lee edge singularities at high temperatures. Phys. Rev. B **20**, 2785 (1979)

158. S. Lai, M.E. Fisher, The universal repulsive-core singularity and Yang-Lee edge criticality. J. Chem. Phys. **103**, 8144 (1995)

159. I.D. Lawrie, S. Sarbach, Theory of tricritical points, in *Phase Transitions and Critical Phenomena*, vol. 9, ed. by C. Domb, J.L. Lebowitz (Academic, London, 1984), p. 1

160. P. Le Doussal, K.J. Wiese, Functional renormalization group at large N for random manifolds. Phys. Rev. E **67**, 016121 (2003)

161. P. Le Doussal, K.J. Wiese, Random field spin models beyond one loop: a mechanism for decreasing the lower critical dimension. Phys. Rev. Lett. **96**, 197202 (2006)

162. P. Le Doussal, K.J. Wiese, Functional renormalization for disordered systems: basic recipes and gourmet dishes. Markov Process. Relat. Fields **13**, 777 (2007)

163. P. Le Doussal, K.J. Wiese, P. Chauve, 2-Loop-renormalization group theory of the depinning transition. Phys. Rev. B **66**, 174201 (2002)

164. T.D. Lee, C.N. Yang, Statistical theory of equation of state and phase transitions. II. Lattice gas and Ising model. Phys. Rev. **87**, 410 (1952)

165. J.M.H. Levelt-Sengers, From van der Waals' equation to the scaling laws. Physica **73**, 73 (1974)

166. H. Levine, S.B. Libby, A.M.M. Pruisken, Electron delocalization by a magnetic field in two dimensions. Phys. Rev. Lett. **51**, 1915 (1983)

167. A.L. Lewis, F.W. Adams, Tricritical behavior in two dimensions. II. Universal quantities from the ϵ expansion. Phys. Rev. B **18**, 5099 (1978)

168. P. Littelmann, H.-J. Sommers, M.R. Zirnbauer, Superbosonization of invariant random matrix ensembles. Commun. Math. Phys. **283**, 343 (2008)

169. P. Lloyd, Exactly solvable model of electronic states in a three-dimensional Hamiltonian: non-existence of localized states. J. Phys. C **2**, 1717 (1969)

170. T.C. Lubensky, J. Isaacson, Field theory of statistics of branched polymers, gelation, and vulcanization. Phys. Rev. Lett. **41**, 829 (1978); Erratum Phys. Rev. Lett. **42**, 410 (1979)

171. T.C. Lubensky, J. Isaacson, Statistics of lattice animals and branched polymers. Phys. Rev. A **20**, 2130 (1979)

172. S. Luther, S. Mertens, Counting lattice animals in high dimensions. J. Stat. Mech. **2011**, P09026 (2011). arXiv:1106.1078

173. S. Mandt, M.R. Zirnbauer, Zooming in on local level statistics by supersymmetric extension of free probability. J. Phys. A **43**, 025201 (2010)

174. J.L. Martin, General classical dynamics, and the 'classical analogue' of a Fermi Oscillator. Proc. Roy. Soc. A **251**, 536 (1959)

175. J.L. Martin, The Feynman principle for a Fermi system. Proc. Roy. Soc. A **251**, 543 (1959)

176. S.P. Martin, A supersymmetry primer. arXiv:hep-ph/9709356 (1997)

177. T. Matsubara, A new approach to quantum-statistical mechanics. Prog. Theor. Phys. **14**, 351 (1955)

178. B. McClain, A. Niemi, C. Taylor, L.C.R. Wijewardhana, Super space, dimensional reduction, and stochastic quantization. Nucl. Phys. B **217**, 430 (1983)

179. B. McCoy, T.T. Wu, *The Two-Dimensional Ising Model* (Harvard, Cambridge, 1973)

180. A.J. McKane, Reformulation of $n \to 0$ models using anticommuting scalar fields. Phys. Lett. A **76**, 22 (1980)

181. A.J. McKane, M. Stone, Localization as an alternative to Goldstone's theorem. Ann. Phys. **131**, 36 (1981)

182. M.L. Mehta, *Random Matrices and the Statistical Theory of Energy Levels* (Academic, New York, 1967)

183. M.L. Mehta, *Random Matrices* (Academic, Boston, 1991)

184. A.D. Mirlin, Statistics of energy levels and eigenfunctions in disordered and chaotic systems: supersymmetry approach, in *Proceedings of the International School of Physics "Enrico Fermi" on New Directions in Quantum Chaos, Course CXLIII*, ed. by G. Casati, I. Guarneri, U. Smilansky (IOS Press, Amsterdam, 2000), p. 223

185. A.D. Mirlin, Statistics of energy levels and eigenfunctions in disordered systems. Phys. Rep. **326**, 259 (2000)

186. A.D. Mirlin, F. Evers, I.V. Gornyi, P.M. Ostrovsky, Anderson localization: criticality, symmetries and topologies, in *50 Years of Anderson Localization*, ed. by E. Abrahams (World Scientific, Singapore, 2010), p. 107

187. A.D. Mirlin, Y.V. Fyodorov, A. Mildenberger, F. Evers, Exact relations between multifractal exponents at the Anderson transition. Phys. Rev. Lett. **97**, 046803 (2006)

188. C.W. Misner, K.S. Thorne, J.A. Wheeler, *Gravitation* (Freeman, NY, 2008)

189. G.E. Mitchell, A. Richter, H.A. Weidenmüller, Random matrices and chaos in nuclear physics: nuclear reactions. Rev. Mod. Phys. **82**, 2845 (2010)

190. H. Miyazawa, Baryon number changing currents. Progr. Theor. Phys. **36**, 1266 (1966)

191. H. Miyazawa, Spinor currents and symmetries of Baryons and Mesons. Phys. Rev. **170**, 1586 (1968)

192. H.L. Montgomery, The pair correlation of the zeta function. Proc. Symp. Pure Math. **24**, 181 (1973)

193. S. Müller, M. Sieber, Quantum chaos and quantum graphs, in *Handbook of Random Matrix Theory*, ed. by G. Akeman, J. Baik, P. di Francesco (Oxford University Press, Oxford, 2011), p. 683

194. J. Müller-Hill, M.R. Zirnbauer, Equivalence of domains for hyperbolic Hubbard-Stratonovich transformations. J. Math. **22** (2011). arXiv:1011.1389 053506

195. H. Nakazato, M. Nakimi, I. Okba, K. Okano, Equivalence of stochastic quantization method to conventional field theories through supertransformation invariance. Prog. Theor. Phys. **70**, 298 (1983)

196. J.W. Negele, H. Orland, *Quantum Many-Particle Systems*, 5th edn. (Westview Press, Reading, 1998)

197. A.A. Nersesyan, A.M. Tsvelik, F. Wenger, Disorder effects in two-dimensional Fermi systems with conical spectrum: exact results for the density of states. Nucl. Phys. B **438**, 561 (1995)

198. A. Neveu, J.H. Schwarz, Factorizable dual model of pions. Nucl. Phys. B **31**, 86 (1971)

199. K.S. Novoselov, A.K. Geim, S.V. Morozov, D. Jiang, M.J. Katsnelson, I.V. Grigorieva, S.V. Dubonos, A.A. Firsov, Two-dimensional gas of massless Dirac fermions in graphene. Nature (London) **438**, 197 (2005)

200. L. Onsager, Crystal statistics. I. A two-dimensional model with an order-disorder transition. Phys. Rev. **65**, 117 (1944)

201. L. Onsager, Discussion remark on p. 261 in G.S. Rushbrooke, On the theory of regular solutions. Nuovo Cimento (Series 9) **6** (Suppl.), 251 (1949)

202. R. Oppermann, Magnetic field induced crossover in weakly localized regimes and scaling of the conductivity. J. Phys. Lett. **45**, L-1161 (1984)

203. R. Oppermann, F.J. Wegner, Disordered systems with n orbitals per site: $1/n$ expansion. Z. Phys. B **34**, 327 (1979)

204. G. Parisi, N. Sourlas, Random magnetic fields, supersymmetry, and negative dimensions. Phys. Rev. Lett. **43**, 744 (1979)

205. G. Parisi, N. Sourlas, Selfavoiding walk and supersymmetry. J. Phys. Lett. **41**, L403 (1980)

206. G. Parisi, N. Sourlas, Critical behavior of branched polymers and the Lee-Yang edge singularity. Phys. Rev. Lett. **46**, 871 (1981)

207. G. Parisi, Y. Wu, Perturbation theory without gauge fixing. Sci. Sin. **24**, 483 (1981)

208. Y. Park, M.E. Fisher, Identity of the universal repulsive-core singularity with Yang-Lee edge criticality. Phys. Rev. E **60**, 6323 (1999) [condmat/9907429]

209. H.-J. Petsche. *Graßmann* (German). Vita Mathematica, vol. 13 (Springer, Birkhäusser, Basel, 2006)

210. H.-J. Petsche, M. Minnes, L. Kannenberg, *Hermann Grassmann: Biography* (English) (Birkhäusser, Basel, 2009)

211. Z. Pluhar, H.A. Weidenmüller, J.A. Zuk, C.H. Lewenkopf, F.J. Wegner, Crossover from orthogonal to unitary symmetry for ballistic electron transport in chaotic microstructures. Ann. Phys. (NY) **243**, 1 (1995)

212. A.M. Polyakov, Interaction of Goldstone particles in two dimensions. Applications to ferromagnets and massive Yang-Mills fields. Phys. Lett. B **59**, 79 (1975)

213. V.N. Popov, *Functional Integrals in Quantum Field Theory and Statistical Physics* (Reidel, Dordrecht, 1983)

214. C.E. Porter, *Statistical Theories of Spectra* (Academic, London, 1965)

215. A.M.M. Pruisken, On localization in the theory of the quantized Hall effect: a two-dimensional realization of the θ-vacuum. Nucl. Phys. B **235**, 277 (1984)

216. A.M.M. Pruisken, Dilute instanton gas as the precursor to the integral Hall quantum effect. Phys. Rev. B **32**, 2636 (1985)

217. A.M.M. Pruisken, in *The Quantum Hall Effect*, ed. by R. Prange, S. Girvin (Springer, Berlin, 1987)

218. A.M.M. Pruisken, Topological principles in the theory of Anderson localization, in *50 Years of Anderson Localization*, ed. by E. Abrahams (World Scientific, Singapore, 2010), p. 503

219. A.M.M. Pruisken, L. Schäfer, Field theory and the Anderson model for disordered electronic systems. Phys. Rev. Lett. **46**, 490 (1981)

220. A.M.M. Pruisken, L. Schäfer, The Anderson model for electron localisation non-linear σ model, asymptotic gauge invariance. Nucl. Phys. B **200** [FS4], 20 (1982)

221. P. Ramond, Dual theory for fermions. Phys. Rev. D **3**, 2415 (1971)

222. N. Read, D. Green, Paired states of fermions in two dimensions with breaking of parity and time-reversal symmetries and the fractional quantum Hall effect. Phys. Rev. B **61**, 10267 (2000)

223. K. Reich, Über die Ehrenpromotion Hermann Grassmanns an der Universität Tübingen im Jahre 1876, in P. Schreiber (ed.) *Hermann Grassmanns Werk und Wirkung*, (Ernst-Moritz-Arndt-Universität Greifswald, Fachrichtungen Mathematik/Informatik, Greifswald, 1995), S. 59

224. V. Rittenberg, M. Scheunert, Elementary construction of graded Lie groups. J. Math. Phys. **19**, 709 (1978)

225. M.J. Rothstein, Integration on noncompact supermanifolds. Trans. Am. Math. Soc. **299**, 387 (1987)

226. A. Salam, J. Strathdee, Super-gauge transformations. Nucl. Phys. B **76**, 477 (1974)

227. M. Salmhofer, *Renormalization – An Introduction*. Texts and Monographs in Physics (Springer, Berlin, Heidelberg, 1998)

228. S. Samuel, The use of anticommuting variable integrals in statistical mechanics. I. The computation of partition functions. J. Math. Phys. **21**, 2806 (1980)

229. S. Samuel, The use of anticommuting variable integrals in statistical mechanics. II. The computation of correlation functions. J. Math. Phys. **21**, 2815 (1980)

230. L. Schäfer, *Excluded Volume Effects in Polymer Solutions as Explained by the Renormalization Group* (Springer, Berlin, 1999)

231. L. Schäfer, F. Wegner, Disordered system with n orbitals per site: Lagrange formulation, hyperbolic symmetry, and Goldstone modes. Z. Phys. B **38**, 113 (1980)

232. A.P. Schnyder, S. Ryu, A. Furusaki, A.W.W. Ludwig, Classification of topological Insulators and superconductors in three dimensions. Phys. Rev. B **78**, 195125 (2008)

233. A.P. Schnyder, S. Ryu, A. Furusaki, A.W.W. Ludwig, Classification of topological Insulators and superconductors. AIP Conf. Proc. **1134**, 10 (2009)

234. E. Schrödinger, A method of determining quantum-mechanical eigenvalues and eigenfunctions. Proc. R. Ir. Acad. A **46**, 9 (1940)

235. E. Schrödinger, Further studies on solving eigenvalue problems by factorization. Proc. R. Ir. Acad. A **46**, 183 (1940)

236. F. Schwabl, *Quantenmechanik*, 2nd ed. (Springer, Berlin, Heidelberg, 1990)

237. T. Senthil, M.P.A. Fisher, Quasiparticle density of states in dirty high-T_c superconductors. Phys. Rev. B **60**, 6893 (1999)

238. T. Senthil, M.P.A. Fisher, Quasiparticle localization in superconductors with spin-orbit scattering. Phys. Rev. B **61**, 9690 (2000)

239. T. Senthil, M.P.A. Fisher, L. Balents, C. Nayak, Quasiparticle transport and localization in high-T_c superconductors. Phys. Rev. Lett. **81**, 4704 (1998)

240. T. Shcherbina, Universality of the local regime for the block band matrices with a finite number of blocks. J. Stat. Phys. **155**, 466 (2014)

241. A.A. Slavnov, Ward identities in gauge theories. Theor. Math. Phys. **19**, 99 (1972)

242. K. Slevin, T. Ohtsuki, The Anderson transition: time reversal symmetry and universality. Phys. Rev. Lett. **78**, 4083 (1997)

243. K. Slevin, T. Ohtsuki, Corrections to scaling at the Anderson transition. Phys. Rev. Lett. **82**, 382 (1999)

244. N. Sourlas, Introduction to supersymmetry in condensed matter physics. Physica D **15**, 115 (1985)

245. R. Speicher, Multiplicative functions on the lattice of noncrossing partitions and free convolution. Math. Anal. **298**, 611 (1994)

246. R. Speicher, Free probability theory, in *Handbook of Random Matrix Theory*, ed. by G. Akeman, J. Baik, P. di Francesco (Oxford University Press, Oxford, 2011), p. 452

247. M.J. Stephen, J.L. McCauley, Feynman graph expansion for tricritical exponents. Phys. Lett. A **44**, 89 (1973)

248. M. Stone, C. Chiu, A. Roy, Symmetries, dimensions, and topological insulators: the mechanism behind the face of the Bott clock. J. Phys. A **44**, 045001 (2011)

249. M. Suzuki, A theory of the second order phase transition in spin systems. II. Complex magnetic field. Prog. Theor. Phys. **38**, 1225 (1967)

250. J.C. Taylor, Ward identities and charge renormalization of the Yang-Mills field. Nucl. Phys. B **33**, 436 (1971)

251. H.N.V. Temperley, M.E. Fisher, Dimer problem in statistical mechanics - an exact result. Phil. Mag. **6**, 1061 (1961)

252. G. Theodorou, M.H. Cohen, Extended states in a one-dimensional system with off-diagonal disorder. Phys. Rev. B **13**, 4597 (1976)

253. W. Thirring, *A Course in Mathematical Physics. 2. Classical Field Theory* (Springer, New York, 1979,1986); *Lehrbuch der mathematischen Physik. 2. Klassische Feldtheorie* (Springer, Wien, 1978,1990)

254. D.J. Thouless, M. Kohmoto, M.P. Nightingale, M. den Nijs, Quantized Hall conductance in a two-dimensional periodic potential. Phys. Rev. Lett. **49**, 405 (1982)

255. M. Tissier, G. Tarjus, Nonperturbative function renormalization group for random field models and related disordered systems. IV. Phys. Rev. B **85**, 104203 (2012)

256. G.F. Tuthill, J.F. Nicoll, H.E. Stanley, Renormalization-group calculation of the critical-point exponent η for a critical point of arbitrary order. Phys. Rev. B **11**, 4579 (1975)

257. R. van Leeuwen, N.E. Dahlen, G. Stefanucci, C.-O. Almbladh, U. von Barth, Introduction to the Keldysh formalism, in *Time-Dependent Density Functional Theory*, ed. by M.A.L. Marques et al. Lecture Notes in Physics, vol. 706 (Springer, Berlin, 2006), pp. 33–59

258. B. Velicky, Theory of electronic transport in disordered binary alloys: coherent-potential approximation. Phys. Rev. **184**, 614 (1969)

259. J.J.M. Verbaarschot, The spectrum of the QCD Dirac operator and chiral random matrix theory. Phys. Rev. Lett. **72**, 2531 (1994)

260. J.J.M. Verbaarschot, Quantum chromodynamics, in *Handbook of Random Matrix Theory*, ed. by G. Akeman, J. Baik, P. di Francesco (Oxford University Press, Oxford, 2011), p. 661

261. J.J.M. Verbaarschot, H.A. Weidenmüller, M.R. Zirnbauer, Grassmann integration in stochastic quantum physics: the case of compound-nucleus scattering. Phys. Rep. **129**, 367 (1985)

262. J.J.M. Verbaarschot, M.R. Zirnbauer, Critique of the replica trick. J. Phys. A **17**, 1093 (1985)

263. D. Voiculescu, Addition of certain non-commuting random variables. J. Funct. Anal. **66**, 323 (1986)

264. D.V. Volkov, V.P. Akulov, Possible universal neutrino interaction. ZhETF Pis. Red. **16**, 621 (1972); JETP Lett. **16**, 438 (1972)

265. D.V. Volkov, V.P. Akulov, Is the neutrino a Goldstone particle? Phys. Lett. B **46**, 109 (1973)

266. F.J. Wegner, Exponents for critical points of higher order. Phys. Lett. A **54**, 1 (1975)

267. F.J. Wegner, The critical state, general aspects, in *Phase Transitions and Critical Phenomena*, vol. 6, ed. by C. Domb, M.S. Green (1976), p. 7

268. F.J. Wegner, Electrons in disordered systems. Scaling near the mobility edge. Z. Phys. B **25**, 327 (1976)

269. F. Wegner, Disordered systems with n orbitals per site: $n = \infty$ limit. Phys. Rev. B **19**, 783 (1979)

270. F. Wegner, The mobility edge problem: continuous symmetry and a conjecture. Z. Phys. B **35**, 207 (1979)

271. F. Wegner, Inverse participation ratio in $2 + \epsilon$ dimensions. Z. Phys. B **36**, 209 (1980)

272. F. Wegner, Algebraic derivation of symmetry relations for disordered electronic systems. Z. Phys. B **49**, 297 (1983)

273. F. Wegner, Exact density of states for lowest landau level in white noise potential. superfield representation for interacting systems. Z. Phys. B **51**, 279 (1983)

274. F. Wegner, unpublished notes (1983/84), compare acknowledgment in [52], ref. [5] in [143], ref. [17] in [261]

275. F.J. Wegner, Crossover of the mobility edge behaviour. Nucl. Phys. B **270** [FS16], 1 (1986)

276. F. Wegner, Anomalous dimensions for the nonlinear sigma-model in $2 + \epsilon$ dimensions (I, II). Nucl. Phys. B **280** [FS18], 193, 210 (1987)

277. Y. Wei, Y.V. Fyodoroy, A conjecture on Hubbard-Stratonovich transformations for the Pruisken-Schäfer parameterizations of real hyperbolic domains. J. Phys. A **40**, 13587 (2007)

278. H.A. Weidenmüller, Single electron in a random potential and a strong magnetic field. Nucl. Phys. B **290**, 87 (1987)

279. H.A. Weidenmüller, G.E. Mitchell, Random matrices and chaos in nuclear physics: nuclear structure. Rev. Mod. Phys. **81**, 539 (2009)

280. J. Wess, Fermi-Bose-supersymmetry, in *Trends in Elementary Particle Systems*, edited by H. Rollnik. Lecture Notes in Physics, vol. 37 (Springer, Berlin, 1975), p. 352

281. J. Wess, J. Bagger, *Supersymmetry and Supergravity*. Princeton Series in Physics (Princeton University Press, Princeton, 1983)

282. J. Wess, B. Zumino, A Lagrangian model invariant under supergauge transformations. Phys. Lett. B **49**, 52 (1974)

283. K.J. Wiese, Disordered systems and the functional renormalization group: a pedagogical introduction. Acta Phys. Slov. **52**, 341 (2002)

284. E.P. Wigner, On a class of analytic functions from the quantum theory of collisions. Ann. Math. **53**, 36 (1951)

285. E.P. Wigner, Characteristic vectors of bordered matrices with infinite dimensions. Ann. Math. **62**, 548 (1955)

286. E.P. Wigner, On the distribution of the roots of certain symmetric matrices. Ann. Math. **67**, 325 (1958)

287. E.P. Wigner, Results and theory of resonance absorption, in *Gatlinburg Conf. on Neutron Physics*, Oak Ridge Natl. Lab. Rept. No. ORNL-2309 (1957) 59; reprint in C.E. Porter, *Statistical Theories of Spectra* (Academic, London, 1965)

288. E.P. Wigner, Random matrices in physics. SIAM Rev. **9**, 1 (1967)

289. E. Witten, Dynamical breaking of supersymmetry. Nucl. Phys. B **188**, 513 (1981)

290. E. Witten, Constraints on supersymmetry breaking, Nucl. Phys. B **202**, 253 (1982)

291. J. Wunderlich, B. Kaestner, J. Sinova, T. Jungwirth, Experimental observation of the spin-Hall effect in a two-dimensional spin-orbit coupled semiconductor system. Phys. Rev. Lett. **94**, 047204 (2004)

292. C.N. Yang, The spontaneous magnetization of a two-dimensional Ising model. Phys. Rev. **85**, 808 (1952)

293. A.P. Young, On the lowering of dimensionality in phase transitions with random fields. J. Phys. C **10**, L257 (1977)

294. A.P. Young, M. Nauenberg, Quasicritical behavior and first-order transition in the $d = 3$ random field Ising model. Phys. Rev. Lett. **54**, 2429 (1985)

295. Y. Zhang, Y.-W. Tan, H.L. Stormer, P. Kim, Experimental observation of the quantum Hall effect and Berry's phase in graphene. Nature (London) **438**, 201 (2005)

296. J. Zinn-Justin, Renormalization and stochastic quantization. Nucl. Phys. B **275**, 135 (1986)

297. J. Zinn-Justin, *Quantum Field Theory and Critical Phenomena* (Clarendon Press, Oxford, 1993)

298. P. Zinn-Justin, Adding and multiplying random matrices: generalization of Voiculescu's formulas. Phys. Rev. E **59**, 4884 (1999)

299. P. Zinn-Justin, J.B. Zuber, Knot theory and matrix integrals, in *Handbook of Random Matrix Theory*, ed. by G. Akeman, J. Baik, P. di Francesco (Oxford University Press, Oxford, 2011), p. 557

300. M.R. Zirnbauer, Anderson localization and non-linear sigma model with graded symmetry. Nucl. Phys. B **265**, 375 (1986)

301. M.R. Zirnbauer, Fourier analysis on a hyperbolic supermanifold of constant curvature, Commun. Math. Phys. **141**, 503 (1991)

302. M.R. Zirnbauer, Supersymmetry for systems with unitary disorder: circular ensembles. J. Phys. A **29**, 7113 (1996)
303. M.R. Zirnbauer, Riemannian symmetric superspaces and their origin in random-matrix theory. J. Math. Phys. **37**, 4986 (1996)
304. M.R. Zirnbauer, Symmetry classes in *Handbook of Random Matrix Theory*, ed. by G. Akeman, J. Baik, P. di Francesco (Oxford University Press, Oxford, 2011), p. 43
305. D. Zwanziger, Covariant quantization of gauge fields without Gribov ambiguity. Nucl. Phys. B **192**, 259 (1981)

Index

© Springer-Verlag Berlin Heidelberg 2016
F. Wegner, *Supermathematics and its Applications in Statistical Physics*,
Lecture Notes in Physics 920, DOI 10.1007/978-3-662-49170-6